Basic Concepts of Orbital Theory in Organic Chemistry

Basic Concepts of Orbital Theory in Organic Chemistry

Eusebio Juaristi

C. Gabriela Ávila-Ortiz

Alberto Vega-Peñaloza

WILEY

Contents

Preface

During a postdoctoral stay at the University of California in Berkeley some years ago, one of us (EJ) had the opportunity to attend the graduate course on molecular orbital theory that was taught by Professor Andrew Streitwieser. The clarity and sequential order with which even the more difficult concepts of the course were presented motivated EJ to teach the subject at the graduate chemistry program offered by CINVESTAV (National Polytechnic Institute) following his return to Mexico. In this regard, the notes that EJ made in Berkeley were complemented and reorganised to result in a text that was further enriched by Claudia Gabriela Ávila-Ortiz and Alberto Vega-Peñaloza, who are co-authors of 'Basic Concepts of Orbital Theory in Organic Chemistry'.

The material covered in this textbook is arranged in a convenient sequence: atomic orbitals → hybrid atomic orbitals → bond formation → molecular orbitals → spectroscopic observations → applications in synthetic organic chemistry → Woodward and Hoffmann rules based on the conservation of molecular orbital symmetry.

Although several textbooks cover some of the topics that are presented here, neither one of them presents the subject in a way that all the different aspects are treated in a gradual and encompassing manner. We hope that this characteristic will render this textbook attractive to those students and professionals interested in this important area of organic chemistry.

We would like to thank several persons who helped us in various ways to make possible the preparation of this book: Professor Andrew Streitwieser as mentioned earlier, Professors Gabriel Gójon-Zorrilla, Bárbara Gordillo Román and Eugene Bratoeff for useful comments, Andrea Hernández Ávila and Alejandra Vega Peñaloza for technical assistance, as well as Sarah Higginbotham and Priyadarshini Natarajan, Wiley editors, for their confidence and continuous encouragement in this project.

Eusebio Juaristi
Claudia Gabriela Ávila-Ortiz
Alberto Vega-Peñaloza

Ciudad de México and Barcelona, Spain
15 May 2025

CHAPTER 1

Introduction and History of the Molecular Orbital Theory

INTRODUCTION

Chemistry is the science that studies the composition, structure and transformations of matter. As soon as it was established that virtually all the matter that surrounds us and builds us up is the result of combinations of atoms forming molecules, it became necessary to understand how the bonds between these atoms are formed and broken. To do this, it became essential to understand the models that describe the arrangement of electrons in atoms and molecules (chemical bonding theories).

A fundamental breakthrough was made in the early twentieth century by G.N. Lewis, who observed the special stability of inert gases. He proposed that the constituent atoms of molecules tend to present the same electronic configuration of such inert gases, which is eight electrons in their valence shell (Figure 1.1). Indeed, Lewis' bonding theory was the only general description of bond formation in chemical compounds until the mid-1920s. Lewis' bonding theory made it possible to interpret and predict important properties of molecules in a simple manner. In fact, this model still provides a good qualitative description of the nature of bonds in organic compounds and is therefore still in use.

However, Lewis' bonding theory has serious limitations and, for example, cannot accurately describe or predict the geometry and reactivity of certain types of compounds. Some of these limitations were alleviated by extending Lewis' theory to include resonance theory. However, this model was still an intuitive and imprecise approximation. Quantum mechanics, a new branch of physics that revolutionised science by allowing a better understanding of various physical, chemical and biological processes, was developed in the 1920s and 1930s. The application of quantum mechanics to distinct issues in chemistry has been in the form of valence bond theory and molecular orbital (MO) theory. The basic idea of valence bond theory is that covalent bonds can be formed by the

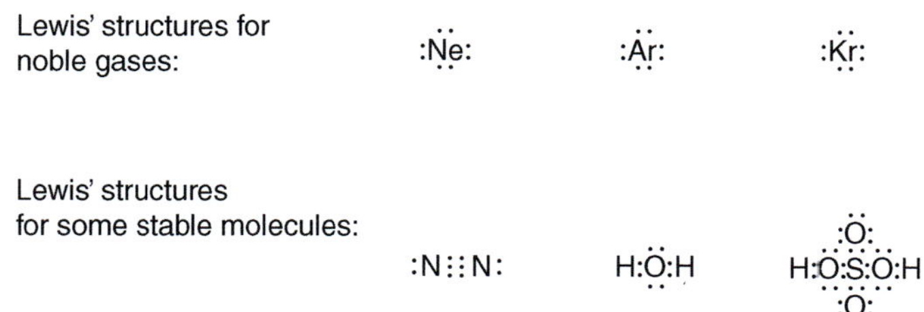

FIGURE 1.1   Some examples of Lewis' molecular structures.

overlapping of localised half-occupied atomic orbitals. This model provides a simple rationalisation of some molecular aspects, such as the geometries of molecules. However, valence bond theory cannot explain other important aspects of bonding and reactivity, such as those originating from the excited states of molecules. In other words, not all chemical bonding phenomena and not all molecular properties can be satisfactorily explained by localised bonding models.

Some properties of chemical bonds are best explained by a more complex quantum model, MO theory. The MO theory allows description of the electronic arrangement of molecules in terms of MOs, whose role is analogous to that of atomic orbitals in atoms. The fundamental characteristic of MOs is that they are distributed over all the atoms in a molecule; that is, the MOs comprise the whole molecule, rather than being associated with just one atom or being localised in a particular region between a pair of atoms.

In order to explain MO theory (and valence bond theory, discussed later in Chapters 3 and 4), it is necessary to understand the basic concepts of quantum mechanics. These concepts and their historical development are discussed in this chapter. In particular, the study of light absorbed or emitted by chemical species, such as atoms and molecules, has provided much of the basic knowledge of quantum mechanics. Therefore, we will begin by studying the properties of light. We will then go on to analyse concepts such as quantised energy, wave-particle duality, the uncertainty principle and others.

NATURE OF ELECTROMAGNETIC RADIATION

As mentioned earlier, different approaches to understanding the nature of electromagnetic radiation and its interaction with matter led to the development of quantum mechanics. To understand current atomic theory, it is necessary to comprehend the properties of electromagnetic radiation. Visible light, X-rays, radio waves and microwaves are some of the types of electromagnetic radiation. They all consist of energy propagated by electric and magnetic fields that increase and decrease in intensity as they move through space. The nature of light was debated as early as the seventeenth century. For the Dutch astronomer and physicist Christiaan Huygens (1678), light had to be a wave, such as the waves in water. By contrast, Englishman Isaac Newton (1704) suggested that light had to consist of small luminous 'corpuscles' (particles) travelling in a straight line at a finite speed and with an associated momentum. Scientific debate on this subject continued for many years later and culminated in the realisation of the dual behaviour of electromagnetic radiation. Before entering a more detailed discussion, let us first describe the properties of electromagnetic radiation in its wave-like behaviour.

THE WAVE NATURE OF LIGHT

Light is a form of electromagnetic radiation, consisting of electric and magnetic fields oscillating as a wave and travelling through empty space at about 3×10^8 m s^{-1} (Figure 1.2). The wave properties of electromagnetic radiation are described by three variables and a constant.

Frequency (ν, Greek nu). The frequency of electromagnetic radiation, expressed in the unit 1/second (s^{-1}; also called Hertz [Hz]), is the number of complete waves or cycles passing through a given point per second. For example, a 960 kHz AM radio station transmits waves with a frequency of 960 000 cycles per second.

Wavelength (λ, Greek lambda). The wavelength is the distance between any one point of a wave and the corresponding point of the next oscillation (Figure 1.3). For example, the distance from one peak to another, or valley-to-valley. In other words, it is the distance the wave travels in one cycle.

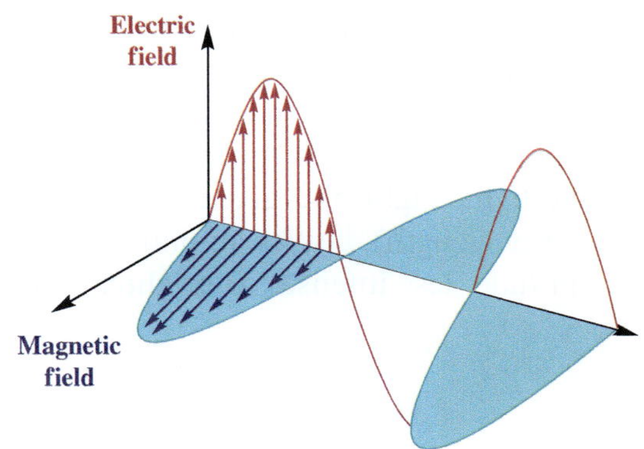

FIGURE 1.2   Light as both electric and magnetic fields.

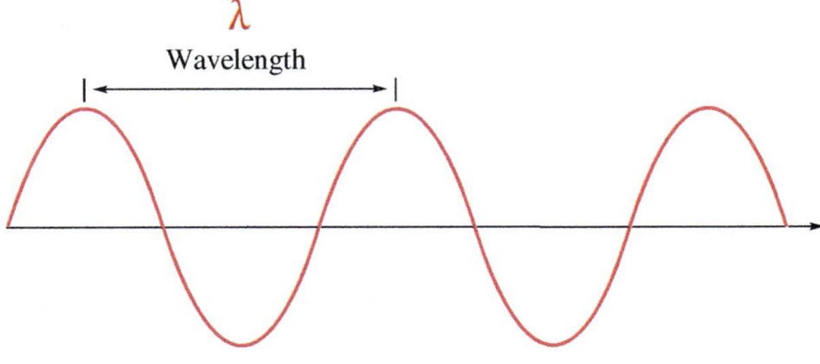

FIGURE 1.3   Wavelength concept.

The wavelength can have units of length as large as metres or, for very short wavelengths, nanometres (nm, 10^{-9} m) or picometres (pm, 10^{-12} m). An alternative unit is angstroms (Å, 10^{-10} m).

Velocity (u). The speed of a wave is the distance travelled per unit of time, which is the product of its frequency in cycles second^{-1} and its wavelength in meters cycle^{-1}:

$$u = \frac{\text{cycles}}{\text{s}} \times \frac{m}{\text{cycles}} = \frac{m}{\text{s}} \tag{1.1}$$

In vacuum, the speed of electromagnetic radiation is 2.99792458×10^8 m s^{-1}, which can be rounded to 3.0×10^8 m s^{-1} with three significant figures (i.e. 300 000 km s^{-1}). This is a physical constant called the speed of light (c) expressed in the formula

$$c = \nu \times \lambda \tag{1.2}$$

Since the product of ν and λ is a constant, ν and λ have an inverse relationship. Take the waves shown in Figure 1.4, which travel at the speed of light (c). If the wavelength of the light is very short, a large number of complete oscillations will pass through a given point in one second. If, on the other hand, the wavelength is long, a smaller number of complete oscillations will pass through the point in one second. Therefore, a short wavelength corresponds to high-frequency (high energy) radiation and a long wavelength corresponds to low-frequency (low energy) radiation.

The amplitude is the wave height above the centreline (Figure 1.5). For an electromagnetic wave, amplitude is related to radiation intensity, or brightness in the case of visible light. Light of a given colour presents a specific frequency (and therefore wavelength), but its amplitude can vary. The light can be softer (lower amplitude, less intense) or brighter (higher amplitude, more intense).

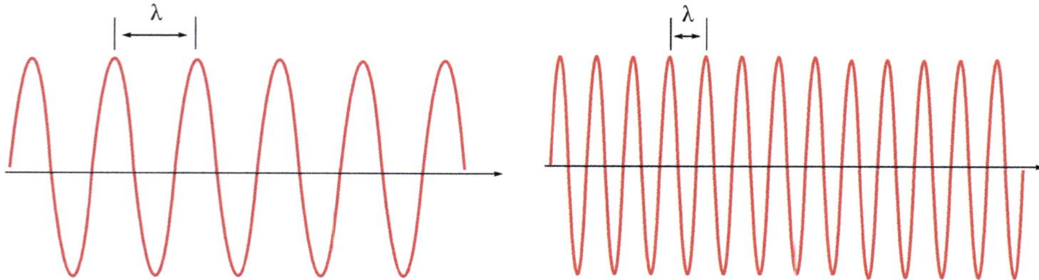

FIGURE 1.4   The relationship between frequency (ν) and wavelength (λ): Longer λ corresponds to a reduced frequency (lower energy, left side) and smaller λ corresponds to an increased frequency (higher energy, right side).

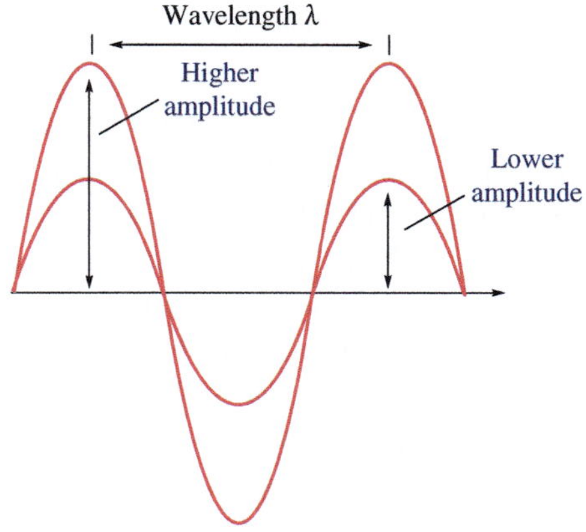

FIGURE 1.5 The concept of amplitude of a wave.

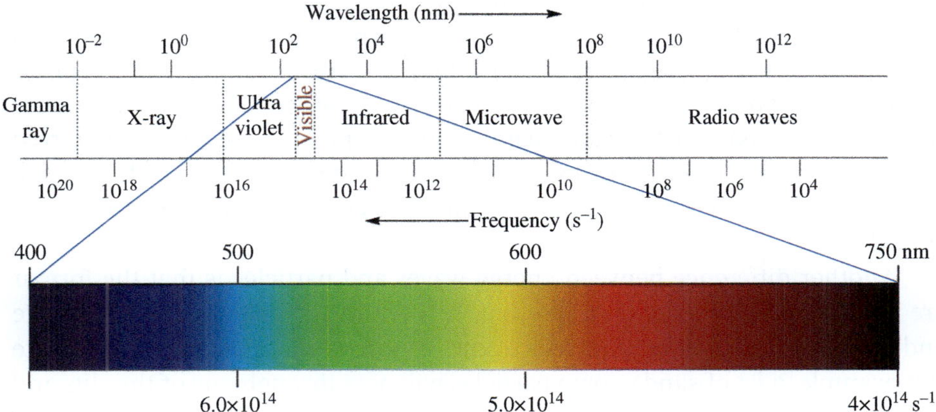

FIGURE 1.6 The electromagnetic spectrum.

ELECTROMAGNETIC SPECTRUM

In a vacuum, all electromagnetic waves travel at the same speed and differ in frequency and therefore wavelength. The electromagnetic spectrum (Figure 1.6) is the classification of electromagnetic waves by frequency or wavelength. The types of radiation vary in the order of increasing wavelength (decrease in frequency). Visible light represents only a tiny region of the spectrum, between about 750 and 400 nm, corresponding to red and violet light, respectively. This represents less than one-millionth of one percent of the measured electromagnetic spectrum. At frequencies lower than red, we have infrared radiation, microwaves and radio waves (in the kilohertz and megahertz range). At frequencies above violet, we have ultraviolet radiation (these higher frequency waves cause sunburn), X-rays and finally gamma rays. These present frequencies above 10 exahertz (10^{19} Hz).

THE DISTINCTION BETWEEN ENERGY AND MATTER

Understanding the difference between the properties of energy waves and particles (matter) is essential to grasping the dual behaviour of light. In a world where we are used to observing objects moving, it is more complex to understand the nature of radiant energy, which does not contain matter and travels in diffuse waves. The differences between the behaviour of particles and that of energy waves will be the subject of discussion in the following.

Light travels at different speeds in different transparent media. We know that it travels at $300\,000\,\text{km s}^{-1}$ in a vacuum, slightly slower in air $(299\,708\,\text{km s}^{-1})$ and about three-quarters of that speed in water. When a light wave changes direction and speed as it passes obliquely from one medium to another, it undergoes a phenomenon known as refraction. A schematic representation of refraction is shown in Figure 1.7. The angle of refraction depends on the two media and the wavelength of the light. Light of different frequencies (such as white light) travels at different speeds in transparent materials and is refracted to different degrees. The separation of white into frequency-ordered colours is called dispersion and is a phenomenon observed in rainbows and diamonds (refracting objects). On the other hand, none of these phenomena is observed in the case of particles. If, for example, a stone is thrown through the air and falls into a body of water, its speed will simply decrease gradually as it follows a curved path.

Another difference between energy waves and particles is that the former are subject to the process of diffraction. If two slits are opened in a solid plate and a screen is placed behind them, and we throw particles against the plate (for example, a jet of sand), some particles will pass through one of the slits and others will pass through the other, thus forming two piles, one in front of each slit. In the case of energetic waves, the situation is completely different.

When a wave (of any kind) hits the edge of an object, it bends in a phenomenon called diffraction (Figure 1.8). If the wave passes through a slit that is wide compared to its wavelength, the waves continue through the slit, bending only at the corners. If the wave passes through a slit that is similar in

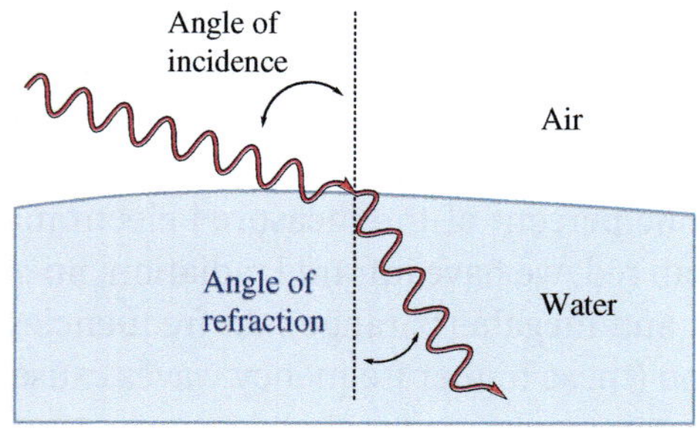

FIGURE 1.7   Representation of the refraction phenomenon.

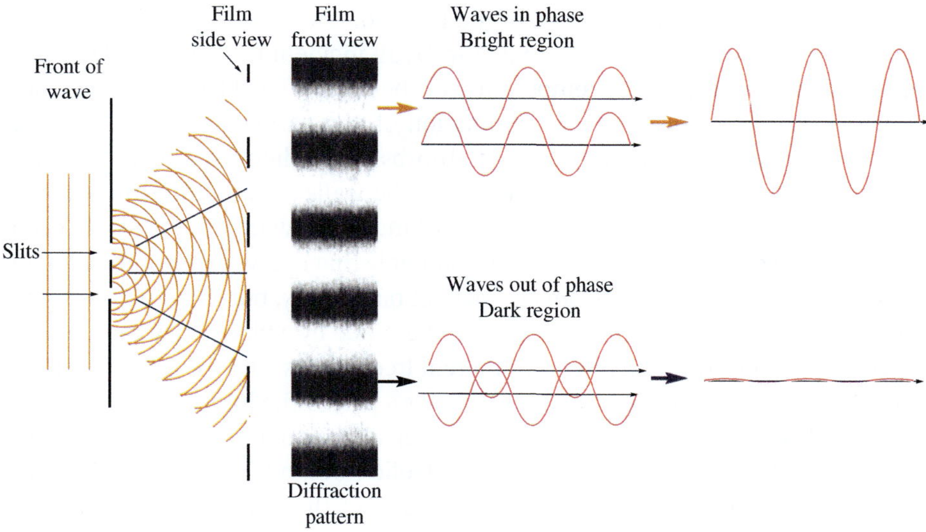

FIGURE 1.8   Constructive and destructive interference of waves.

width to its wavelength, it will bend completely around both edges of the slit, forming a semi-circular wave on the other side of the slit. Now suppose a wave hits the plate with two slits. As it arrives, the wave splits into two, each wave exiting through one of the slits. Nearby circular waves interact by the process of interference.

Figure 1.8 shows constructive and destructive interference. We can see that the superposition in phase of a pair of identical waves produces a wave of the same frequency but twice the amplitude, the bright spots in the diffraction pattern. When the waves are exactly out of phase, that is, when a peak coincides with a trough, their superposition cancels out completely. This cancellation produces the dark areas of the pattern. If they are out of phase by other amounts, partial cancellation occurs.

THE PARTICLE NATURE OF LIGHT

Towards the end of the nineteenth century, scientists were puzzled as they gathered more and more data on electromagnetic radiation that could not be explained by classical mechanics. The observation of diverse phenomena such as blackbody radiation and the photoelectric effect led to a radically new view of energy.

Every object emits electromagnetic radiation, and the hotter it is, the more intense and higher the frequency of the radiation. For example, at high temperatures, an iron bar begins to glow, emitting red light, and as it heats up further, the emitted light becomes brighter and whiter (this phenomenon is called incandescence). These changes in the intensity and wavelength of the light emitted as an object heats up are characteristic of the radiation emitted by a hot black body. A black body is an idealised object that absorbs all incident radiation, and its light emission depends only on the temperature of the object. The black body does not favour the absorption or emission of radiation of any particular wavelength. We could imagine a tightly closed oven that absorbs all the light that reaches it and emits nothing to the outside. Inside the oven, the walls are at a uniform temperature and are constantly emitting and absorbing light, which bounces off each other. In practice, it is possible to make a small hole that allows some of the light to escape from the inside; it is no longer a perfect black body, but if the hole is small enough, the disturbance is minimal. The light emitted by a black body can then be studied by passing it through filters of different frequencies, and the intensity of the filtered light can be measured. Figure 1.9 shows a plot of the intensity of the radiation emitted (spectral radiance) by a heated black body as a function of wavelength and at various temperatures. Experimentally (solid lines in Figure 1.9), as the temperature increases (for example, from 3000 to 5000 K), the peak of the emitted radiation appears at shorter wavelengths. This observation could not be explained by classical physics since the emitted radiation was predicted to increase indefinitely at higher frequencies into the ultraviolet range (dashed line in Figure 1.9). This hypothetical phenomenon is known as 'ultraviolet catastrophe'.

In 1900, the German physicist Max Planck developed a formula that fitted the results obtained from the black body emission. He proposed that the exchange of energy between matter and radiation occurs in quanta, or packets,

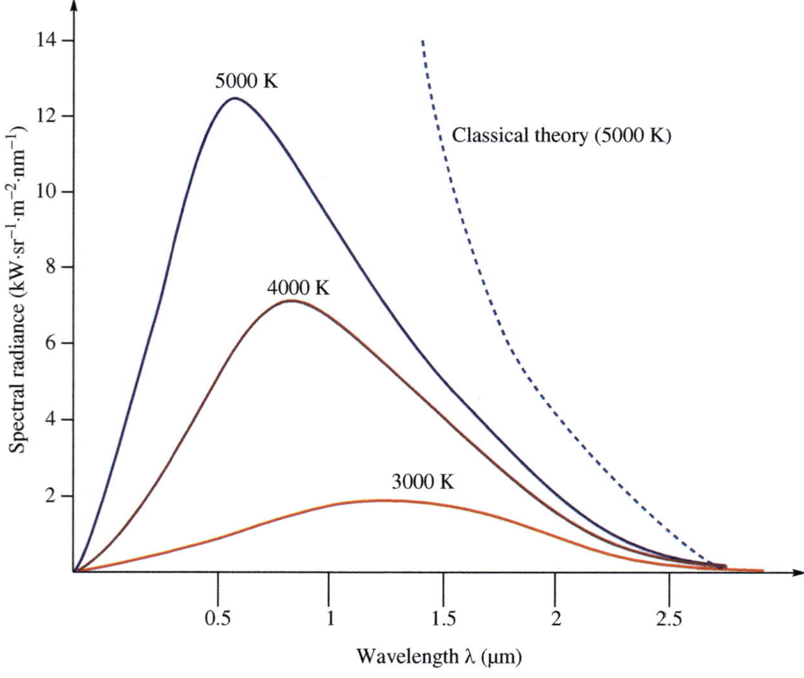

FIGURE 1.9   Spectral radiance emitted by a heated black body.

of energy. The central idea was that within matter there are microscopic oscillators (atoms, molecules) vibrating at certain frequencies that absorb and emit energy in the form of electromagnetic waves. Each oscillator can only absorb or emit energy that is an integer multiple of its fundamental energy, which is directly proportional to its frequency of vibration:

$$E = h\nu \tag{1.3}$$

where h is Planck's constant and has a value of 6.626×10^{34} J · s. If the oscillating atom releases an energy packet of magnitude E into the surroundings, then radiation of frequency $\nu = E/h$ will be detected. From Eq. (1.2) $c = \nu \times \lambda$ therefore,

$$\frac{c}{\lambda} = \nu \tag{1.4}$$

and substituting c/λ for frequency, ν, in Eq. (1.3), we find the relationships which indicate that energy is directly proportional to frequency and inversely proportional to wavelength.

$$E = h\nu = \frac{hc}{\lambda} \tag{1.5}$$

The Planck hypothesis states that radiation of a given frequency is produced only when an oscillator has acquired the minimum energy necessary to start oscillating at that frequency and then emits it as a packet of electromagnetic radiation of energy $h\nu$. At low temperatures, there is not enough energy to drive oscillations at very high frequencies as ultraviolet radiation. In contrast, classical physics assumed that an oscillator could oscillate at any energy, and therefore even at low temperatures, high-frequency oscillators could contribute to the emitted radiation causing the 'ultraviolet catastrophe'.

However, Planck was convinced that his premise of 'energy packets' was just a mathematical trick to arrive at the corrected result. A trick that worked for a while but had to be replaced by a more rigorous postulate. The postulate that consolidated Planck's theory came from the explanation of the photoelectric effect. The photoelectric effect is the phenomenon that occurs when monochromatic light of sufficient frequency strikes a metal plate and promotes the flow of an electric current, i.e. electrons are released and set in motion in the metal.

The experimental observations of the photoelectric effect were as follows: the current is generated only if the incident light has a frequency above a certain threshold, which depends on the metal; otherwise, the effect does not take place, no matter how intense the light. Another observation was that electrons were ejected immediately, regardless of how low the intensity of the radiation was. However, wave theory predicted that with low-intensity light there would be a time lag before the current flowed because the electrons had to absorb enough energy prior to their release (Figure 1.10).

Albert Einstein (1905) found an explanation for these observations that profoundly changed our understanding of the electromagnetic field. He proposed that electromagnetic radiation is made up of particles, later called photons. In this theory, light is also quantised, and each photon is a packet of energy related to the frequency of the radiation by the equation $E = h\nu$ (Eq. 1.3). For example, a beam of red light can be thought of as a group of photons of equal energy, while a beam of blue light can be thought of as a group of photons of equal energy but higher than that of red light. It is important to note that the intensity of the radiation is an indication of the number of photons present, whereas $E = h\nu$ is a measure of the energy of each individual photon.

Observations of the photoelectric effect have been explained as follows. If the incident radiation is of frequency ν, it consists of photons of energy $h\nu$. When the photons collide with the electrons of the metal, they absorb part of their energy. The energy required to extract an electron from a metal is called the work function of the metal and is denoted ϕ (phi, Figure 1.10). If the energy of an incident photon is less than the energy required to extract an electron from the metal, then no electron will be ejected no matter how many energy packets the metal receives (Figure 1.10a). On the other hand, if the photon energy is greater than ϕ, the electrons are immediately released from the metal (Figure 1.10b). An electron cannot be released by accumulating energy from several photons of lower energy than required and then expelled with a time delay. A photon of sufficient energy will rapidly eject an electron. Nevertheless, the electric current can be weak under radiation with light of sufficient energy but low intensity. This is because fewer photons

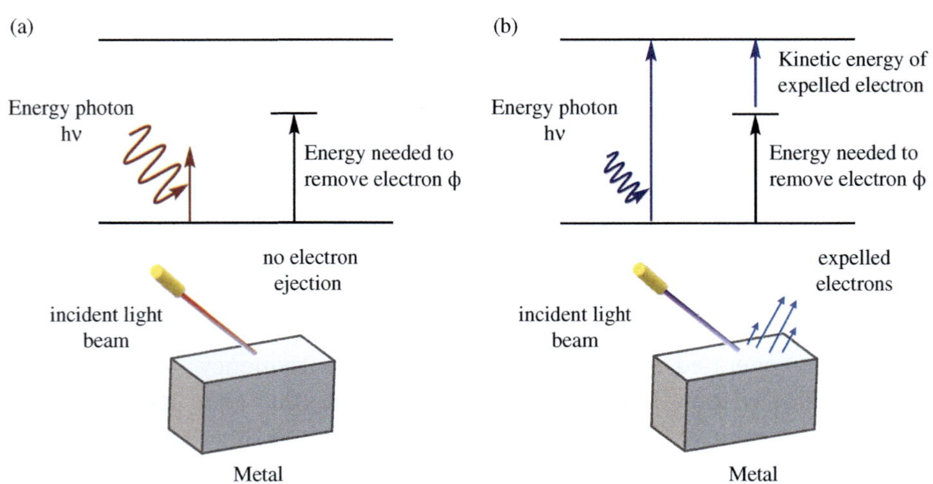

FIGURE 1.10   The photoelectric effect.

release fewer electrons per unit of time, although the current (electron flow) is still generated.

Planck's hypothesis of the quantisation of electromagnetic radiation and Einstein's proposal that light waves also exhibit particle-like behaviour led to the development of quantum mechanics.

MASS AND MOMENTUM ASSOCIATED WITH A LIGHT QUANTUM

Until 1900, the principles of classical mechanics dictated the analysis of the motion of bodies and its consequences. Everything from the interaction of two grains of sand to the orbits of the planets was studied according to the rules derived from classical mechanics. In contrast, electrical and magnetic phenomena, such as light, belonged to a separate field governed by ideas derived from electromagnetic theory. This separation between mechanical and electromagnetic phenomena came to an end when, in 1905, A. Einstein developed his equation,

$$E = mc^2 \tag{1.6}$$

Equation (1.6) implies, of course, that mass and energy are just different manifestations of the same thing. That is, every transfer of energy is accompanied by matter, and every mass (at rest or in motion) represents energy. As discussed earlier, the energy of these 'packets' is related to frequency (ν) or wavelength (λ) by Eqs. (1.3) and (1.5), $E = h\nu$ and $E = hc/\lambda$, respectively. Both equations indicate that Planck's constant h limits the size of light quanta in the same way that Avogadro's number limits the size of atoms. For instance, it can be verified that one mole of water can be heated from $0°$ to $100\,°C$ with 2/3 moles of quanta of infrared light with $\nu = 3 \times 10^3$ cycles s^{-1}. That is, the quanta are small.

In this context, the photon possesses what is known in physics as a quantity of motion or momentum. In classical physics, the momentum of a body is a very important concept and is defined as the product of its mass times its velocity

$$p = mu \tag{1.7}$$

If it is true that, according to Einstein's equation, light has the properties typically ascribed to matter, then, from the combination of Eqs. (1.6) and (1.5), we obtain that

$$m = \frac{h\nu}{c^2} \tag{1.8}$$

and as the momentum is $p = mc$ (Eq. 1.7, considering velocity u as the velocity of light c), then

$$p = \frac{h\nu}{c} = \frac{h}{\lambda}$$ (1.9)

An experimental verification of this reasoning was achieved by observing that X-ray (light) radiation behaves physically as expected for particles with mass. Let us first consider the behaviour of two billiard balls colliding with each other. As shown in Figure 1.11, the moving red ball

$$p_1 = m_1 u_1$$ (1.10)

hits the stationary black ball and sets it in motion

$$p_2 = m_2 u_2$$ (1.11)

When the ball m_1 collides with m_2 through a point that is not its centre, the direction they take in their subsequent displacement is expressed by angles θ and ϕ, relative to the collinear axis of approach (Figure 1.11). These trajectories possess, of course, projections on the y-axis, perpendicular to the axis of contact; vectorially, the motions in these projections on the y-axis are

$$p_1^y = m_1 u_1' \sin\phi$$ (1.12)

$$p_2^y = m_2 u_2 \sin\theta$$ (1.13)

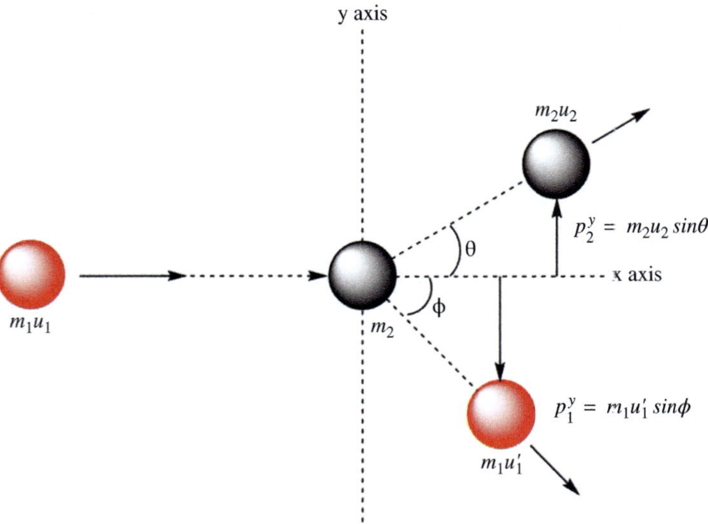

FIGURE 1.11   Collision of two billiard balls or two particles.

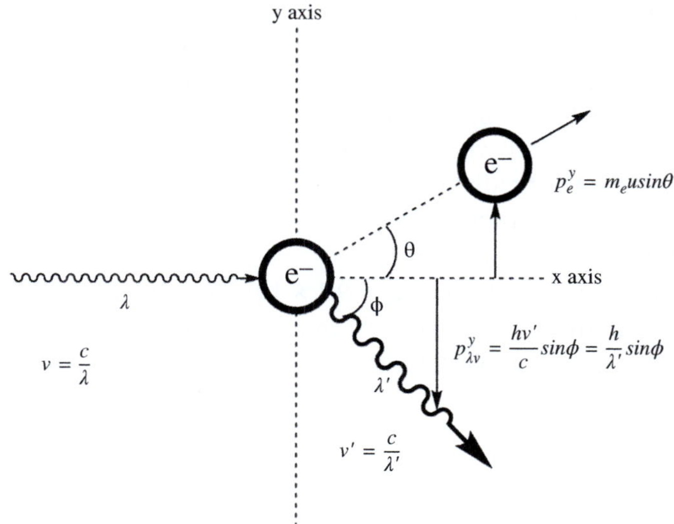

FIGURE 1.12   The Compton effect.

$$p_{hv}^y = \frac{hv'}{c}\sin\phi \tag{1.14}$$

$$p_e^y = m_e u \sin\theta \tag{1.15}$$

A similar behaviour of electromagnetic light waves was demonstrated by A. H. Compton in 1922, who observed that when X-ray radiation $\left(v = \frac{c}{\lambda}(\text{Eq. }1.4)\right)$ collides with an electron, it continues its path at an angle ϕ that varies from the original one, with a lower momentum than the initial, and with a projection on the y-axis that is exactly as predicted by Einstein's equation $p_{hv}^y = \frac{hv'}{c}\sin\phi$ (Eq. 1.14). The resulting effect is an increase in the wavelength of the photon after the collision, an indication that it has lost energy. Compton found that the energy loss of the photon is equal to the energy gain by the electron. Similarly, following collision, the electron acquires a momentum whose projection on the y-axis is equal to $p_e^y = m_e u \sin\theta$ (Eq. 1.15 and Figure 1.12). Just as billiard balls transfer momentum when colliding, photons transfer momentum to electrons. In this experiment, photons behaved like particles.

WAVE-PARTICLE DUALITY

In 1924, Louis de Broglie reasoned that if radiation could exhibit the properties of both particles and waves, then electrons could as well (indeed, any moving particle, whether a planet, a baseball, or an electron). This phenomenon is known as wave-particle duality. The de Broglie relation $\lambda = h/mu$ or $\lambda = h/p$ is derived from the combination of the mass–energy equivalence relation ($E = mc^2$, Eq. 1.6) and the energy of a photon ($E = hc/\lambda$, Eq. 1.5). It dictates that a particle of momentum (mu) has an associated wave of wavelength λ (or that a particular wavelength λ can be associated with a particle). In other words, the wavelength associated with matter is inversely proportional to the mass of the particle, m and its velocity, u. It is known that the phenomenon of light diffraction in crystalline lattices is possible when the spacings in the crystal are comparable to the wavelength of the light radiation. This observation led Davisson and Germer in 1927 to search for the appropriate crystal spacing for a particle to give rise to the phenomenon of scattering.

Specifically, an electron of mass $m_e = 9.1 \times 10^{-28}$ g, moving at 0.01 times the speed of light $u = 0.01c = 3 \times 10^8$ cm s^{-1} (an electron acquires this speed when a potential of $E = 26$ V is applied to it), has a momentum $p_e = m_e u_e = 27.3 \times 10^{-20}$ g cm s^{-1}. Consequently,

$$\lambda_e = \frac{h}{p_e} = \frac{6.6 \times 10^{-27}\,\text{erg s}}{27.3 \times 10^{-20}\,\text{g cm s}^{-1}} \tag{1.16}$$

$$\lambda_e = 0.24 \times 10^{-7}\,\text{cm} = 2.4\ \text{Å}$$

Indeed, Davisson and Germer found that a crystal with a lattice spacing = 2.4 Å caused diffraction patterns of the studied electrons, just as electromagnetic waves do!

The discovery of the wave-particle duality not only changed our understanding of electromagnetic radiation and matter, but also revolutionised physics. In classical physics, a particle in motion has a definite position at any given moment. In contrast, we cannot simultaneously determine the position and linear momentum of a wavelike object. This phenomenon is known as the uncertainty principle, formulated around 1927 by the German physicist Werner Heisenberg. It states that the product of the uncertainty in each quantity is less than $h/2\pi$, and that the position and momentum of a particle cannot be known simultaneously with such precision.

Under this premise, we must think of the electron from a different perspective and consider the *probability* of finding the electron in a given volume of space rather than attempting to define its exact position and momentum.

APPLICATION OF QUANTUM MECHANICS TO ATOMIC STRUCTURE

The simplest atom we can study is the hydrogen atom, which has only one proton and one electron. A proper atomic theory should explain, for instance, the reason for the characteristic reactivity of each element. In the case of hydrogen, one might ask, for example, why the reaction 2 H· $\rightarrow$ H$_2$ is exothermic $(\Delta H^\circ = -104.2\,\text{kcal mol}^{-1})$, but why 3H· $\nrightarrow$ H$_3$.

The determination of the visible spectrum of the hydrogen atom required quantum theory. As shown in Figure 1.13, photolysis of the hydrogen molecule gives rise to hydrogen atoms in excited states, H·*; a deexcitation (relaxation) process then occurs in which the emission of energy brings the hydrogen atoms back to their basal state.

The recording of the emitted light (the spectrum of the hydrogen atom) showed a surprising peculiarity: the emissions corresponded only to certain energy values (spectral lines); that is, they were quantised. This result is not consistent with Rutherford's incipient planetary model (1911) for atoms, in which the Coulombic attraction of the nucleus/electron is compensated by the centrifugal force of the electron in its orbitals because in this model any amount of energy would be absorbed and emitted, varying only the distance between the nucleus and the electron. Niels Bohr proposed a model for the hydrogen atom that predicted the existence of line spectra. Bohr used Planck's and Einstein's ideas about quantised energy and proposed the following postulates: (1) The H atom has only certain energy levels, called stationary states, which are associated with a fixed circular orbit of the electron around the nucleus (Figure 1.14). The higher the energy level, the further away from the nucleus the electron orbit is. The atom does not change its energy while the electron is orbiting. In other words, it does not release energy while it is in one of its stationary states. The atom only changes to another stationary state (the electron moves to another orbit) by absorbing or emitting a photon. When an electron 'jumps' from an orbit of higher energy to one of lower energy, it emits a quantum of light with exactly the energy difference between the two orbits (Figure 1.14a). This quantum has an associated frequency and wavelength λ, given by Planck's formula $E = h\nu$. Similarly, if a quantum of light with just the right energy comes along, it will be absorbed by an electron by 'hopping' to a higher-energy orbital (Figure 1.14b). In the first case, emission lines are produced; in the second, absorption lines are observed. This is how the origin of spectral lines was explained by Bohr's atomic model.

Thus, the spectrum of the hydrogen atom gave rise to the quantum model shown in Figure 1.15 (the y-axis represents the energy of the stationary states [orbitals]). By means of this model, it is understood how the recorded emissions result from energetic jumps between different stationary states (E_1, E_2, E_3, etc.), which are related by a simple formula involving the quantum number, $n = 1, 2, 3$, etc. Thus, certain emissions between nearby energy levels are of low energy (visible region) whereas decay to the basal state E_1 are of higher energy (235–313 kcal mol^{-1}) and occur in the ultraviolet region.

$$H_2 \xrightarrow{\ h\nu\ } 2\,H\overset{*}{\cdot}$$

$$\downarrow {-}h\nu' \quad \begin{cases} 43.6 & \text{kcal/mol} \\ 235.2 & " \\ 278.8 & " \\ \text{etc.} \end{cases}$$

$$2\,H\cdot$$

FIGURE 1.13 Photolysis of a hydrogen molecule.

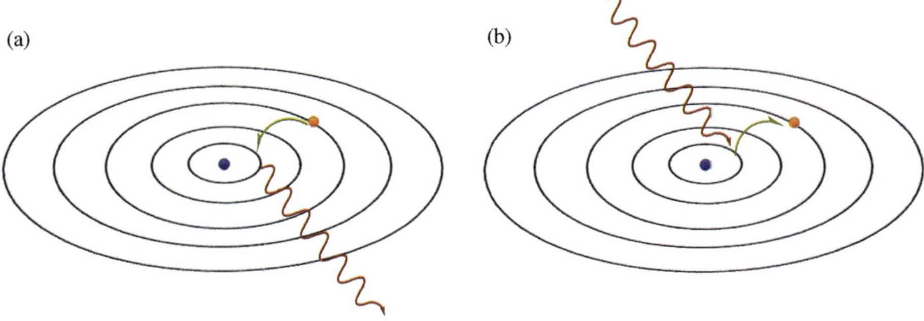

FIGURE 1.14   (a) Quantum of light emission when an electron 'jumps' from an orbit of higher energy to another of lower energy. (b) A quantum of light is absorbed by an electron when it goes to a higher-energy orbital.

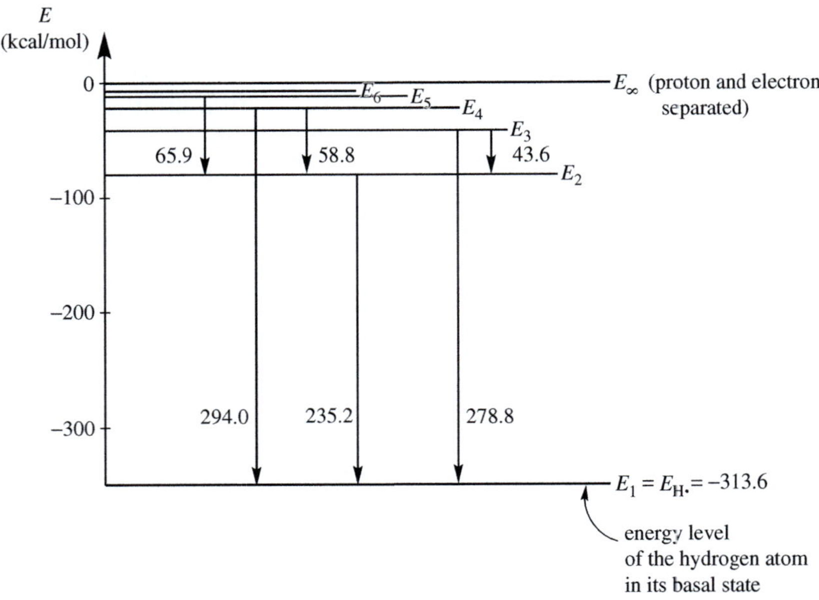

FIGURE 1.15   Quantum model of the hydrogen spectrum.

Despite its success in predicting the spectral lines of the hydrogen atom, Bohr's model could not predict the spectrum of any other atom with more than one electron. This is because it did not consider electron–electron repulsions and nucleus-electron attractions that give rise to rather complex electrostatic interactions. Furthermore, since any spinning electric charge emits electromagnetic radiation, an electron spinning around the nucleus should lose energy. Since all energy in the atom is conserved, any radiation must take place at the expense of the motional energy of the charged particle. The result would be that, following rotations, the electron would eventually 'fall' back into the nucleus. The most fundamental deficiency of Bohr's model is that electrons do not move in fixed and defined orbits (as we will see later). Moreover, this model does not explain why the energy levels of atoms should be quantised. Nevertheless, this model gave rise to the concept of quantum number, which is so important in more recent theories, and we retain the central idea that the energy of an atom is produced at discrete levels and that it changes when the atom absorbs or emits a photon of a particular energy. Indeed, the term orbital is still used in wave mechanics.

The definitive step was then taken by Erwin Schrödinger in 1926, who understood the connection between the existence of the stationary states of hydrogen and the wave properties of the electron. For this, some systems that absorb or release energy in special quantities have been known since then. For example, the guitar string and the drum produce sounds (vibrations) of characteristic energy, which contrasts with other systems that can accept or release energy in any magnitude: a baseball, the weights of a balance, etc. Unlike these systems that can be perfectly studied by classical mechanics, a guitar string of a certain length is only allowed certain vibrations; hence its characteristic tones (Figure 1.16).

As can be seen in Figure 1.16, the length of the string determines the vibrations that can be produced; that is, the vibrations are quantised. The displacement of the string at its peaks and troughs corresponds to the amplitude of the vibration, which also has a characteristic number of nodes: the points where the peaks become troughs and vice versa, i.e. where the phase of the vibration changes sign. It is clear from Figure 1.16 that the higher the number of nodes, the higher the vibrational energy. It can also be seen that the wavelength of the vibration (and therefore its energy E) is determined by the length of the string

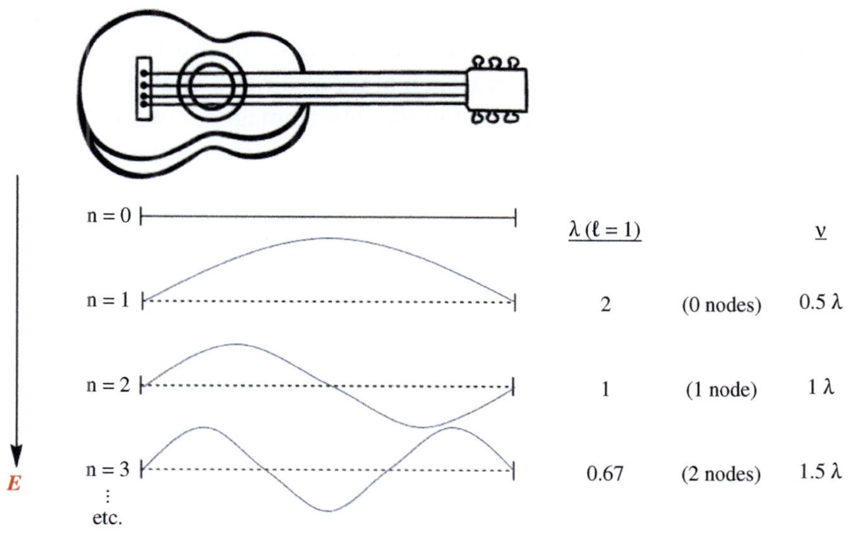

FIGURE 1.16   Vibrations allowed by the string of a guitar.

and the quantum number n (Eq. 1.17). The variable n is an integer number that indicates whether it is the fundamental ($n = 1$) or an overtone or harmonic ($n = 2, 3, 4, \ldots$). This number is strictly analogous to atomic quantum numbers.

$$\lambda = \left(\frac{2}{n}\right)\ell \tag{1.17}$$

It should be noted that the points at which the string is attached to the guitar, which determine its length, also limit the number of possible vibrations by preventing certain unsymmetrical or illogical vibrations from being allowed. Figure 1.17 shows two examples of prohibited vibrations.

Thus, the vibrations of a guitar string are quantised and defined by an integer number n, a node pattern (vibration frequency), and the frequency of the vibration. In the same way that the vibrations of a guitar string and the motion of an electron in a hydrogen atom produce characteristic spectra, both can be described mathematically using a *wave equation*.

The wave equation for a standing wave in two dimensions is more complex than the wave equation for a vibrating string because the vibration takes place in more dimensions. The best analogy for a standing wave in two dimensions is probably a vibrating drumhead. Like a guitar string, the vibrations of the leather on a drum have characteristic and defined tones. The circumference of the drum is fixed and represents the boundary condition that limits how the leather can vibrate; that is, it limits the number of specific frequencies that are possible. As shown in Figure 1.18, instead of nodal points (guitar string), we now have nodal lines associated with the different two-dimensional vibrations. In the Figure, the positive wave-like phases and negative wave-like phases are moving in opposite directions with respect to the viewer.

Like the guitar string, the drum produces line spectra corresponding to quantised vibrations of specific energy. These musical phenomena can be described mathematically by a wave equation. Therefore, when the Austrian physicist Erwin Schrödinger learnt that the hydrogen atom also has a characteristic line spectrum and that electrons are diffracted like electromagnetic waves, he concluded that the electron's motion must be governed by a wave equation.

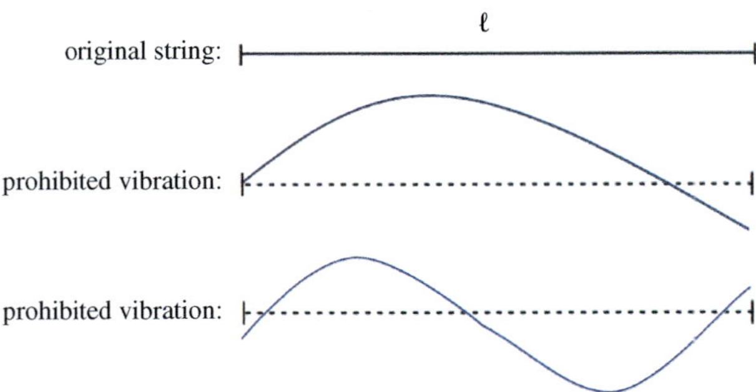

FIGURE 1.17   Examples of forbidden vibrations.

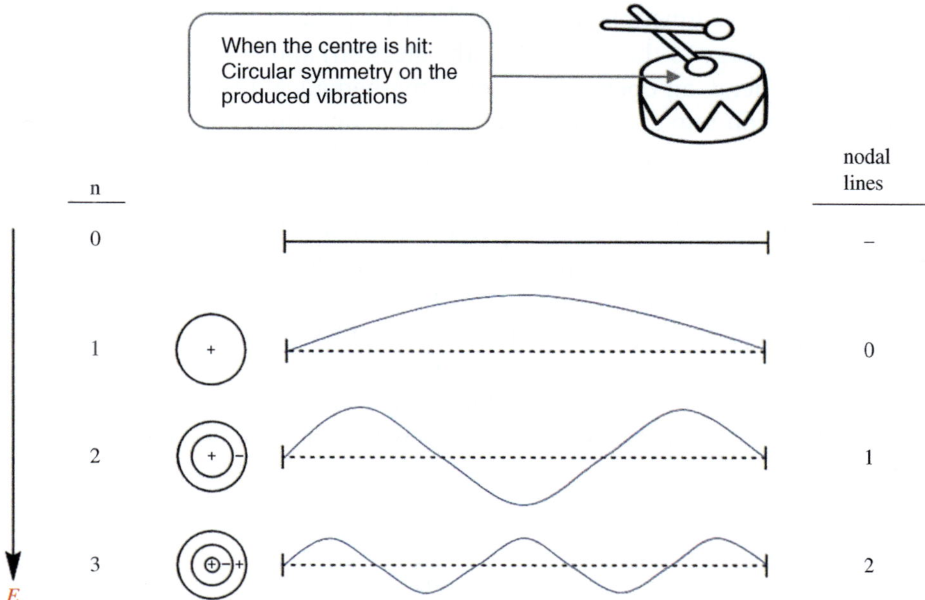

FIGURE 1.18 Two-dimensional vibrations allowed by a drum.

Central to quantum mechanics is the premise that electrons in an atom or molecule can be reasonably treated as standing waves in three dimensions. In three dimensions, the picture becomes much more complex, although the same rules apply. The nodes now become surfaces rather than points or lines of the guitar string and the drum, respectively. In a 3D standing wave, there are two types of nodal surfaces: spherical surface nodes and flat surface nodes. The Schrödinger equation for the energy of an electron in an atom is a second-order differential equation for a standing wave in three dimensions and is the mathematical cornerstone of the quantum mechanical model. Being a standing wave equation means that only certain solutions can exist. This is the first explanation of the origin of the quantisation of atomic energy levels. The Schrödinger equation, in its basic form, is described in the following.

SCHRÖDINGER'S EQUATION

The determination of the spectrum of the hydrogen atom with its line pattern, which is similar to that of a guitar string, and the demonstration of the wave properties of the electron led Schrödinger to develop a wave equation suitable for describing the motion of the electron in the hydrogen atom. In its condensed form, the Schrödinger equation can be reduced to the following:

$$H\psi = E\psi \tag{1.18}$$

where ψ is the wave function, a function describing the wave nature of the electron. E is the energy of the system (i.e. the electron) and H is the Hamiltonian operator. In general, an operator is a set of mathematical instructions to transform one function into another. In this case, the operator H acts on a function (ψ) and returns the same function (ψ) multiplied by a scalar quantity (E). In quantum mechanics, each observable (e.g. position, momentum and kinetic energy) is represented by an independent operator. This operator is used to obtain physical information in quantum mechanical formalism. For example, in classical mechanics, the energy of the system can be expressed as the sum of the kinetic and potential energy. For quantum mechanics, the kinetic and potential energies are transformed into their corresponding quantum mechanical operators, which in turn correspond to the Hamiltonian (H) operator. For a particle of mass m moving in one dimension in a region where the potential energy is $V(x)$, the term H is:

$$H = -\frac{h^2}{8\pi^2 m}\frac{\partial^2 \psi(x)}{\partial x^2} + V(x)\psi$$

$$\underbrace{}_{\substack{\text{Hamiltonian} \\ \textit{operator}}} \qquad \underbrace{\phantom{-\frac{h^2}{8\pi^2 m}\frac{\partial^2 \psi(x)}{\partial x^2}}}_{\substack{\text{Operator} \\ \text{associated with} \\ \text{kinetic energy}}} \quad \underbrace{}_{\substack{\text{Term associated} \\ \text{with potential} \\ \text{energy}}} \qquad (1.19)$$

The Hamiltonian (H) of a system is an operator corresponding to the total energy of a given system; including both kinetic and potential energy (H for hydrogen will be different from that for helium because of the extra electron involved in the latter atom and different for the other atoms). The kinetic energy is related to the momentum (expressed as its corresponding operator) and mass (m) of the particle, whereas the potential energy is the energy associated with electrostatic interactions. Therefore, the potential energy depends only on the charge and position of the particles as dictated by Coulomb's law (that is, the nucleus-electron attraction, nucleus-nucleus repulsion, and electron-electron repulsive interactions). Replacing H from Eq. (1.19) in Eq. (1.18) gives Schrödinger equation in the form:

$$-\frac{h^2}{8\pi^2 m}\frac{\partial^2 \psi(x)}{\partial x^2} + V(x)\psi = E\psi(x) \qquad (1.20)$$

The wave function (ψ) and its associated energies are solutions of the Schrödinger equation. The German physicist Max Born proposed a physical interpretation of ψ. He stated that the probability density (probability per unit volume) of finding an electron in a given region is given by the square of the wave function (ψ^2). Technically, since the wave function can contain imaginary numbers, squaring ψ cancels out any imaginary numbers and makes the probability density positive. A plot of the wave function squared (ψ^2) represents an orbital, a probability distribution map of the electron's position. This result can be extended to give information about MOs.

The Schrödinger equation cannot be solved exactly in most cases. To get a better understanding of what the equation involves, consider a simple model system for which an exact solution is possible: the particle in a box quantum model. In classical mechanics, a particle trapped in a large box can move at any speed inside the box and is no more likely to be in one position than another (Figure 1.19a). In quantum mechanics, however, the particle can only occupy certain energy levels and is more likely to be in certain positions than others, depending on its energy level (Figure 1.19b).

In terms of quantum mechanics, a particle of mass m moves inside a one-dimensional box along a line of length L between two potential energy barriers, which are the ends of the box. Therefore, the potential energy V inside the box is zero because the particle is free to move. Inside the box, where the potential energy term $V(x) = 0$, the Schrödinger equation simplifies to

$$-\frac{h^2}{8\pi^2 m}\frac{\partial^2 \psi(x)}{\partial x^2} = E\psi \tag{1.21}$$

From Eq. (1.21) it is appreciated that the energy is:

$$E = \frac{n^2 h^2}{8mL^2} \quad n = 1, 2, 3, 4, \ldots \tag{1.22}$$

One way of deriving the above formula is as follows: The kinetic energy of a particle of mass m is related to its velocity u by the equation $E_k = 1/2\ mu^2$. Thus, noting that the linear momentum is $p = mu$, and using the de Broglie relation, one can relate this energy to the wavelength of the particle:

$$E_k = \frac{1}{2}mu^2 = \frac{1}{2}\frac{\overbrace{(mu)^2}^{P}}{m} = \frac{\overbrace{(p)^2}^{h/\lambda}}{2m} = \frac{h/\lambda}{2m} = \frac{h^2}{2m\lambda^2} \tag{1.23}$$

de Broglie relationship

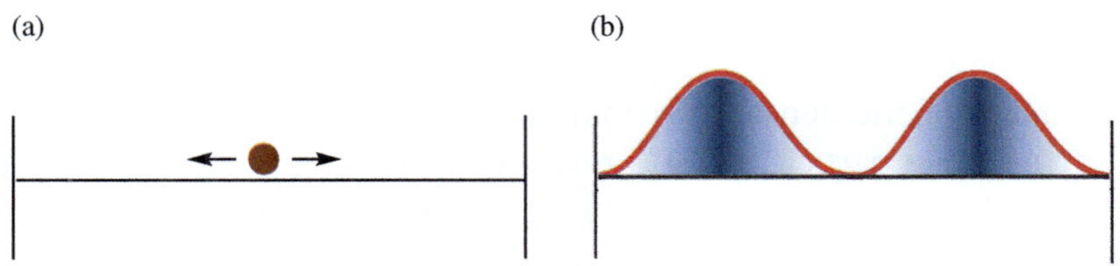

<table>
<tr><td>(a)</td><td>(b)</td></tr>
</table>

FIGURE 1.19 (a) A classical mechanics view of a particle in a box its position and its momentum are known. (b) In quantum mechanics, the particle occupies certain energy levels and is more likely to be in certain positions than others.

Assuming that the potential energy of the particle is zero throughout the interior of the box, the total energy is equal to the kinetic energy. Here we see that, like the guitar string (see as mentioned earlier), only integer multiples of half a wavelength can enter the box (see Figure 1.20; the waves have one, two, three valleys or crests, etc., where each valley or crest is half a wavelength). This means that the wavelengths that are possible for a particle to enter a box of length L must satisfy the condition that:

$$L = \frac{1}{2}\lambda, \frac{2}{2}\lambda, \frac{3}{2}\lambda \ldots = n \times \frac{1}{2}\lambda \text{ with } n = 1, 2, 3, \ldots \tag{1.24}$$

Therefore, the allowed wavelengths (λ) are:

$$\lambda = \frac{2L}{n} \text{ with } n = 1, 2, 3, \ldots \tag{1.25}$$

Substituting λ into Eq. (1.23), we obtain the formula 1.22:

$$E = \frac{h^2}{2m \left(2L/n\right)^2} = \frac{n^2 h^2}{8m L^2} \tag{1.26}$$

Note that the energy of the particle is quantised because n presents only integer values. The values of n are called quantum numbers (see the following). Equation (1.26) shows that the distance between neighbouring energy levels decreases as L (the length of the box) or m (the mass of the particle) increases. Put another way, the difference between consecutive energy levels is extremely small and essentially continuous. Therefore, until very small systems were studied, no one realised that energy was quantised!

The wave functions ψ obtained from the Schrödinger equation for the particle in the box are given by Eq. (1.27):

$$\psi_n(x) = \left(\frac{2}{L}\right)^{1/2} \sin\left(\frac{n\pi x}{L}\right) n = 1, 2, 3\ldots \tag{1.27}$$

Figure 1.20a shows the wave functions of the first five energy levels, while the probability density for these levels (proportional to the square of the wave function) is shown on the right (Figure 1.20b).

The shapes of the density distributions generated from the square of the wave functions of a particle in a box reveal valuable information. For example, at energy level 2, one can notice a node in the middle where the probability of finding the particle is zero. At this level, the particle has an energy of $h^2/2m L^2$ and is most likely to be found at the centres of the two hills. The probability densities are also highlighted by the shading of the bands underneath each wave function (Figure 1.19b).

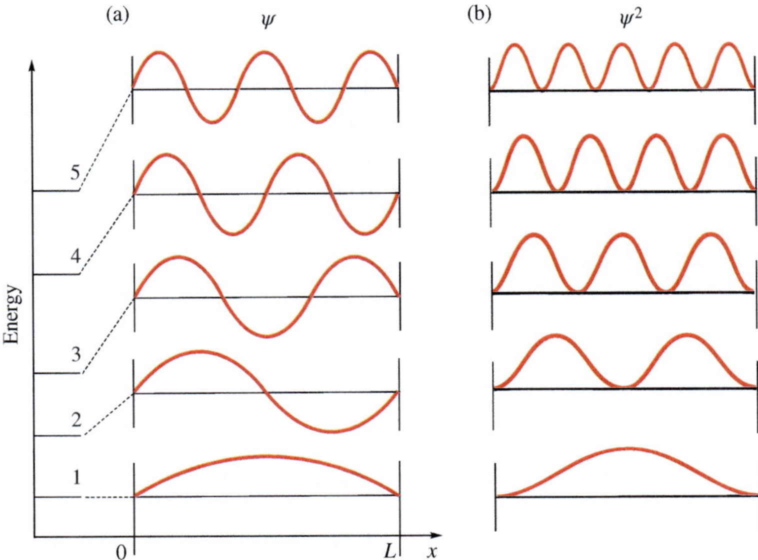

FIGURE 1.20   (a) Wave functions are obtained from the Schrödinger equation for a particle in a box for the first five energy levels. (b) The probability density for these levels (square of the wave function).

Some important mathematical properties for the wave function ψ are:

(a) ψ has boundary conditions that, like the boundary conditions on the guitar string, lead to a line spectrum. Some of the boundary conditions (or limits) of ψ are (i) at infinite distances $\psi = 0$ (the nucleus and the electron are separated). (ii) ψ is continuous and single-valued (i.e. presents a unique value for each region around the nucleus). Finally, (iii) the integration of ψ^2 over the whole space is equal to 1 (i.e. ψ is normalised) and (iv) the electron cannot move at infinite speed. (b) ψ is characterised by one or several nodes, that is surfaces where $\psi = 0$, implying a change of sign at its boundaries. (c) ψ presents amplitudes and phases. In addition, most importantly, (d) given their mathematical nature two or more wave functions ψ interact constructively or destructively.

The solution of the Schrödinger equation reproduces the hydrogen spectrum and explains the energy levels of the atom. It also allows the quantitative calculation of bond energies and bond lengths, molecular vibrational frequencies, etc. In short, the chemistry of the atom can be explained on the basis of quantum mechanics.

HYDROGENIC ORBITALS

Because the hydrogen atom has only one electron and therefore does not present electron–electron interactions, which would require explicit treatment, the Schrödinger equation can be solved exactly. The hydrogen electron can be thought of as a particle in a box, being confined to a small volume as a consequence of its attraction to the nucleus. Nevertheless, because of such attraction, the electron has potential energy, unlike the particle in the box. Furthermore, the electron can move in three dimensions, x, y and z, not just one.

Hydrogen has a positively charged nucleus Z, where $Z = 1$, and a negatively charged electron e at a distance r from the nucleus that is assumed to be static. Thus, by incorporating the kinetic energy of the electron and the potential energy of attraction of the nucleus/electron for the hydrogen atom, Eq. (1.18) is rewritten:

$$\left(-\frac{1}{2} \times \frac{h^2}{4\pi^2 m} \nabla^2 - \frac{Ze^2}{r}\right)\psi = E\psi \tag{1.28}$$

The Laplacian (∇^2) is a differential operator. In very general terms, the Laplacian of the wave function provides the kinetic energy, and it can be used to describe how a property is distributed and varies in a particular region; it can be expressed in Cartesian or spherical coordinates.

The solution of the Schrödinger equation gives rise to three quantum numbers that are necessary for the characterisation of each wave function. The fourth quantum number, spin angular momentum, was discovered experimentally by Stern and Gerlach in 1922. Table 1.1 lists the quantum numbers and includes a practical description of their meaning.

The orbital energy level or 'shell' is defined by the quantum number n and can be any positive integer between 1 and ∞. For a given energy level, there may be sub-shells, the number of which is defined by the quantum number ℓ. The term 'shell' reflects the fact that as n increases, the region of the highest probability density is like an almost concave shell of increasing radius. The average distance of an electron from the nucleus increases with the value of n.

TABLE 1.1   The Quantum Numbers and Their Meaning

Quantum number	Possible values	Characteristics
Principal quantum number, n	$1, 2, 3 \ldots \infty$	$n - 1$ nodal surfaces. It is related to orbital size and energy
Orbital angular momentum, ℓ	$0, 1, 2, n - 1$	ℓ non-spherical nocal surfaces, related to the orbital shape
Magnetic, $\mathbf{m}_\ell$	$-\ell \ldots +\ell$	Orientation of non-spherical nodes, related to its orientation in space
Spin angular momentum, $\mathbf{m}_s$	$\pm\frac{1}{2}$	Spin direction

TABLE 1.2   The Relationship and Meaning of Quantum Numbers n and ℓ

Value of n	Possible values of ℓ	Orbital name
1	0	1s
2	0, 1	2s, 2p
3	0, 1, 2	3s, 3p, 3d
4	0, 1, 2, 3	4s, 4p, 4d, 4f

TABLE 1.3   The Meaning of m_ℓ in the Orbital's Orientation

Values of n	2	2
Values of ℓ	0	1
Name	2s	2p
Possible values of m_ℓ	0	+1, 0, −1
Name	2s	$2p_x$, $2p_y$, $2p_z$ orientation of the orbitals

As shown in Table 1.2, the values that ℓ can take depend on the value of n: it can take any value from 0 to $n - 1 := 0, 1, 2, \ldots n - 1$. The different possible values of ℓ are labelled with letters instead of numbers, called s, p, d and f ('s' for sharp, 'p' for principal, 'd' for diffuse, and 'f' for fundamental). The value of ℓ provides information about the region of space in which an electron can be located; that is, it describes the shape of the orbital. As its name suggests, ℓ describes the orbital angular momentum of the electron within the orbital, which is the mass times the angular velocity at which the electron 'orbits' (in classical terms) around the nucleus. For example, electrons in an s orbital where ℓ is 0 present zero orbital angular momentum, which means that one must refrain from thinking of them as orbiting the nucleus, but simply as being distributed around it. An electron in a p orbital ($\ell = 1$) has non-zero orbital angular momentum, so one can think of it as located in two lobes around the nucleus. An electron in a d orbital ($\ell = 2$) or an f ($\ell = 3$) orbital has an increasing orbital angular momentum.

The third quantum number, m_ℓ, the magnetic quantum number, refers to the orientation of an orbital, which distinguishes the individual orbitals within a subshell. The possible values of m_ℓ are given suffixes to the letters defining the quantum number ℓ. These letters refer to the orientation of the orbitals along the x, y or z axes (Table 1.3). Remember that momentum is a vector property, so the orbital angular momentum ℓ has an orientation and $\boldsymbol{m_\ell}$ describes that orbital orientation. There are $(2\ell + 1)$ possible orientations for the orbital angular momentum vector, corresponding to the possible values of $\boldsymbol{m_\ell}$ for a given value of ℓ. Thus, for an electron in a p orbital ($\ell = 1$), there are three possible orientations for the orbital angular momentum vector $(2(1) + 1 = 3)$. These values are assigned as $-1, 0, +1$.

An electron can be thought of as rotating about an axis passing through it and therefore has spin angular momentum in addition to the orbital angular momentum discussed earlier. According to quantum mechanics, an electron adopts two spin states, represented by the arrows ↑ (up) and ↓ (down) or the Greek letters α and β. One can imagine an electron rotating anti-clockwise around its axis at a given velocity or clockwise at exactly the same velocity. These two spin states are distinguished by a fourth quantum number, the magnetic spin quantum number, m_s, which can take two values: $+\frac{1}{2}$ indicates an ↑ electron and $-\frac{1}{2}$ indicates an ↓ electron.

Quantum numbers provide a comprehensive description of the behaviour of electrons in terms of their energy, spatial distribution, orientation, and intrinsic spin, and define the unique set of quantum states that electrons can occupy within an atom. Figure 1.21 presents a diagram showing how the quantum numbers divide and subdivide.

The mathematical expressions of ψ are complex, see Table 1.4 (the equations are presented for the purpose of illustration, consult specialist texts if you are interested in more in-depth information).

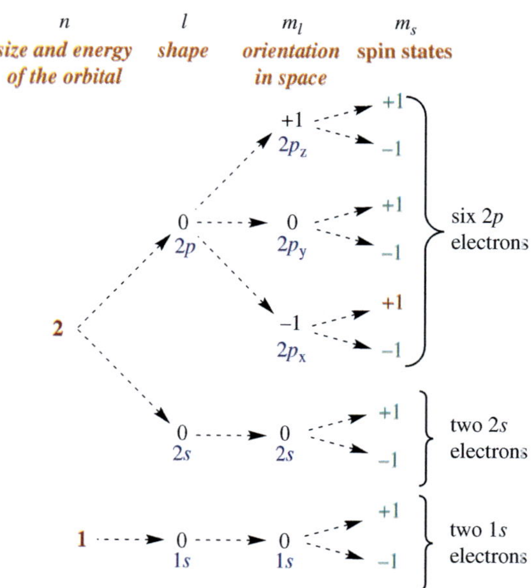

FIGURE 1.21   Diagram of the division and subdivision of the quantum numbers.

TABLE 1.4   Expressions of ψ for Orbitals $1s$, $2s$ and $2p$

$$\psi(1s) = \frac{1}{\sqrt{\pi}} \left(\frac{Z}{\sqrt{a_o}} \right)^{\frac{3}{2}} e^{\frac{-Zr}{2a_o}}$$

$$\psi(2s) = \frac{1}{4\sqrt{2\pi}} \left(\frac{Z}{a_o} \right)^{\frac{3}{2}} \left(2 - \frac{Zr}{a_o} \right) e^{\frac{-Zr}{2a_o}}$$

$$\psi(2\,p_x) = \frac{1}{4\sqrt{2\pi}} \left(\frac{Z}{a_o} \right)^{\frac{5}{2}} (x)\, e^{\frac{-Zr}{2a_o}}$$

$$\psi(2\,p_y) = \frac{1}{4\sqrt{2\pi}} \left(\frac{Z}{a_o} \right)^{\frac{5}{2}} (y)\, e^{\frac{-Zr}{2a_o}}$$

$$\psi(2\,p_z) = \frac{1}{4\sqrt{2\pi}} \left(\frac{Z}{a_o} \right)^{\frac{5}{2}} (z)\, e^{\frac{-Zr}{2a_o}}$$

where a_o is a collection of fundamental constants giving a conversion factor to atomic units of dimension; $a_o = 4\pi\,\varepsilon_o\,h^2 / m_e\,e^2 = 0.529$ Å (the Bohr radius), Z is the atomic number ($Z = 1$ for hydrogen) and r is the distance from the nucleus. Each ψ (atomic orbital) consists of three parts: (i) a normalisation constant so that the integration of the electron density over the whole space is equal to 1.0, (ii) a factor including the radical and angular nodes and (iii) an exponential term showing the decrease of ψ with the distance from the nucleus.

Therefore, it is convenient to use symbolic representations that correspond to three-dimensional images of the orbitals involved. For example, the $1s$ orbital contains no nodes but is always positive (or negative, depending on the label). One way of representing it is by means of a circle containing about 90% of the total electron density. Because a sphere adopts only one orientation, an s orbital presents only one value of m_ℓ: for any s orbital, $m_\ell = 0$. Figure 1.22 shows some of the plots used for the $1s$ orbit.

As can be seen, ψ is always positive for the $1s$ orbital. The probability density (ψ^2) decreases with increasing distance from the nucleus (r). It is worth noting that the electron density curve does not touch the axis, meaning that even though the probability of the electron being far from the nucleus is very small, it is never zero (Figure 1.22d). The values of ψ^2 are largest near the nucleus and then become smaller as one moves along a radius centred on the nucleus. This suggests that the electron does indeed spend some of its time in the region closest to the nucleus.

Nevertheless, integrating ψ^2 over the volume area shows that the maximum electron density is found at $r = a_o(0.529$ Å$)$. This is known as radial distribution. One does this by dividing the volume around the nucleus into thin, concentric, spherical layers. One then calculates the sum of the ψ^2 values in each layer to see which is most likely to contain the electron. In simple terms, we can say that we are plotting the number of points (probabilities) in each ring against the distance from the nucleus (Figure 1.23).

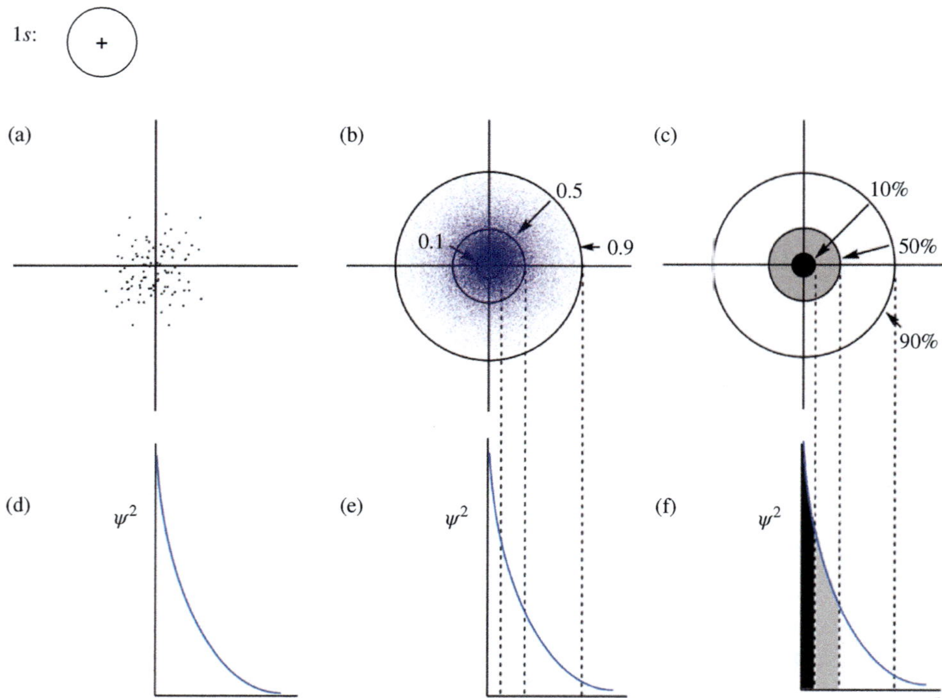

FIGURE 1.22   (a) Probability point diagram. It is an imaginary image of an electron changing position rapidly over time; this does not mean that an electron is a diffuse cloud of charge. (b) Density contours derived from representations in (a). (c) Representation of the volumes enclosing certain values of the probability of finding the electron. (d) Electron density plot from the nucleus. (e) Interpretation of electron density contours: they indicate where the electron spends most of the time. (f) Alternative representation of the volumes enclosing certain percentages of the electron density. r is given in Å.

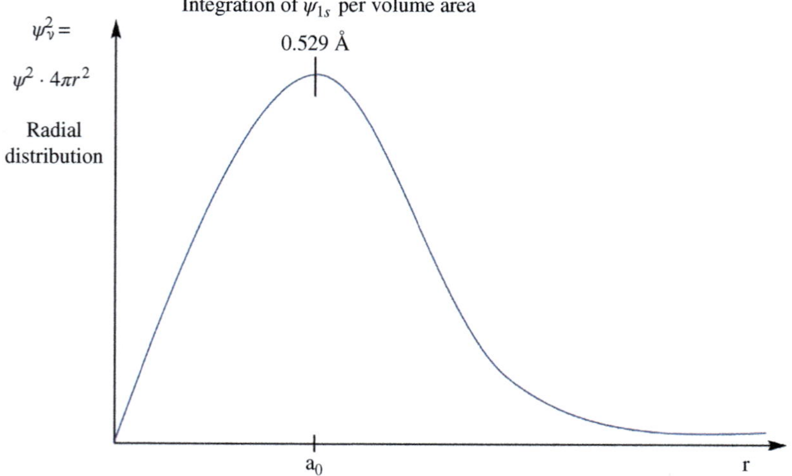

FIGURE 1.23   Probability of finding an electron in relation to the distance from the nucleus.

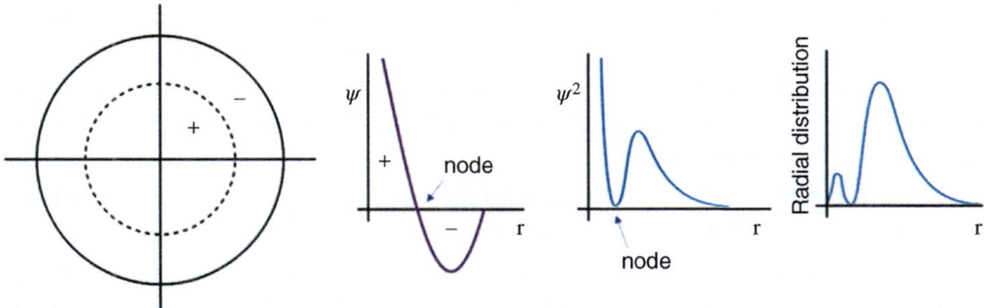

FIGURE 1.24 Representations of the 2*s* orbital.

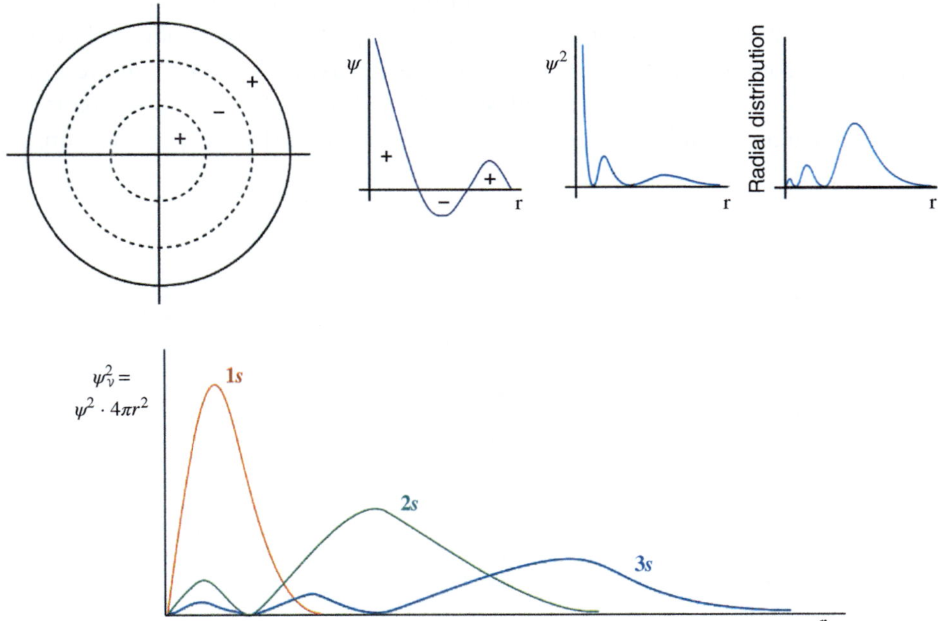

FIGURE 1.25 Top, representations of the 3*s* orbital. Down, radial distribution of orbitals 1*s*, 2*s* and 3*s*.

The 2*s* and 3*s* orbitals have both positive and negative ψ values on either side of a spherical nodal surface. Several representations of the 2*s* orbital are shown in Figure 1.24. The 2*s* orbital presents two regions of higher electron density. The radial probability distribution of the more distant region is larger than that of the closer one because the sum of ψ^2 is gathered over a much larger volume. There is a spherical node between the two regions, where the probability of finding the electron goes down to zero. This means that inside the 2*s* orbital, in contrast to the 1*s* orbital, there is a region in which there is no electron density at all. Since the 2*s* orbital is larger than the 1*s*, an electron in the 2*s* will spend more time away from the nucleus (in the larger of the two areas) than if occupying the 1*s* orbital.

There are three regions of high electron density and two nodes in the 3*s* orbital (Figure 1.25). Again, the highest radial probability appears in the region closest to the nucleus. This pattern of more nodes and a higher probability with increasing distance from the nucleus is continued with the 4*s* orbital, the 5*s* orbital and so on (Figure 1.25 down).

The probability density plot for *p* orbitals is no longer spherically symmetric. There are three 2*p* orbitals with quantum number $n = 2$, $\ell = 1$ which are isoenergetic. As shown in Figure 1.26, the 2*p* orbitals exhibit a particular orientation in space as well as two lobes. In this case, the nodal point is a plane between the two lobes, which is known as the nodal plane. These orbitals are distinguished by their third quantum number $\boldsymbol{m_\ell}$, which takes the values +1, 0, −1, corresponding to the *p* orbitals aligned along the x, y and z axes. As shown in Figure 1.27, the 3*p* orbital presents two nodes: one plane and one spherical node; the result is orbitals oriented on the three Cartesian axes. The 4*p* orbital presents two radial nodes apart from its nodal plane (not shown). Similar to the pattern of *s* orbitals, a 3*p* orbital is larger than a 2*p*, a 4*p* is larger than a 3*p*, and so on.

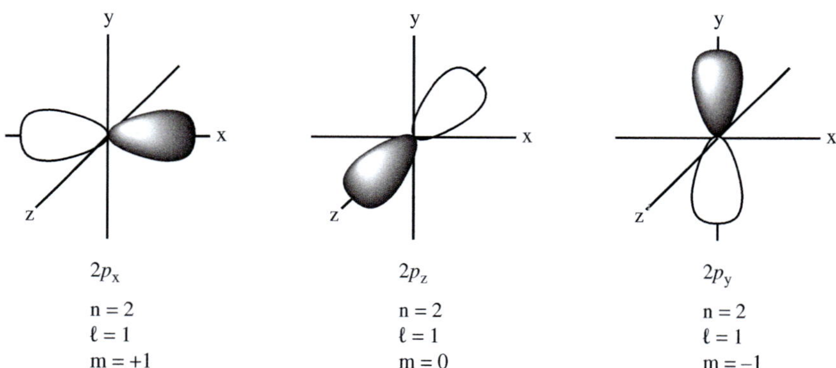

FIGURE 1.26   Representation of the 2*p* orbitals. Note: It is not possible to assign a specific $\boldsymbol{m_\ell}$ quantum number to a specific type of *p* orbital, but for didactic purposes, they have been assigned to show that when electrons are located in the same atom they must have a unique combination of the four quantum numbers.

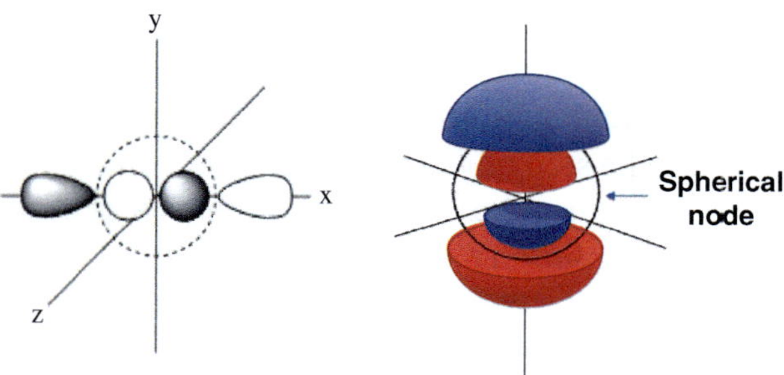

FIGURE 1.27   Representation of the 3p_x orbital: yz-plane nodes and a spherical node.

There are five 3*d* orbitals, which are isoenergetic and have two non-spherical nodes, either two orthogonal nodal planes or two nodal cones. The $3d_{xy}$, $3d_{xz}$ and $3d_{yz}$ orbitals present nodal planes along the xz and yz, xy and yz, and xy and xz planes, respectively. (Only $3d_{xy}$ is shown in Figure 1.28). The $3d_{x^2-y^2}$ orbital is obtained by rotating the $3d_{xy}$ orbit by 45°. The $3d_{z^2}$ orbit presents two nodal cones, which lie on the nucleus and are collinear with the z-axis (Figure 1.28).

The five 4*d* orbitals are constructed by superimposing a spherical node on the models described for the 3*d* orbitals (Figure 1.29).

The 4*f* orbitals constitute a set of seven orbitals with three nodes each. Six of these are orthogonal pairs whose shape can usually be obtained from the *d* orbitals by superimposing a nodal plane.

The energy of each hydrogen orbital is determined by the principal quantum number, *n*, according to Eq. (1.29).

$$E_n = -R \times \frac{Z^2 h}{n^2} \tag{1.29}$$

where R is the Rydberg constant, that is 313.6 kcal mol^{-1}, Z is the charge of the nucleus, $n = 1, 2, 3$ and the negative sign indicates that the energy of the orbital is lower than that of the separated proton and electron. In other words, as the

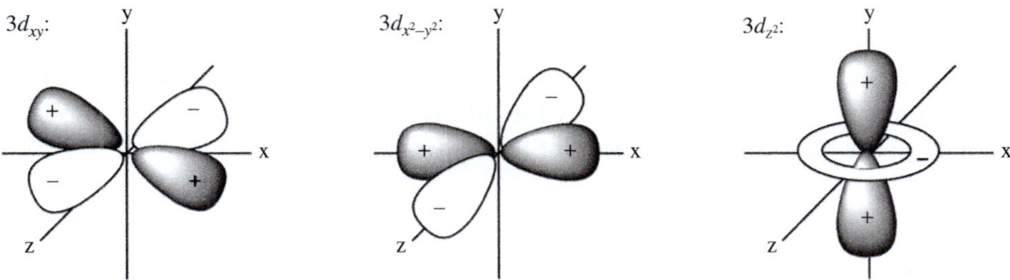

FIGURE 1.28 Symbolic representations of 3*d* orbitals.

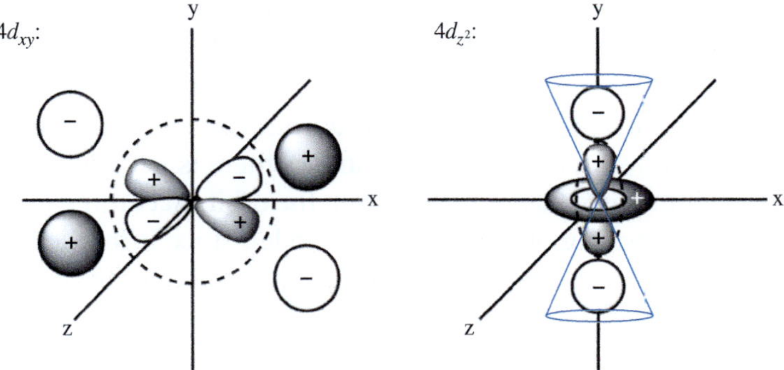

FIGURE 1.29 Symbolic representations of 4*d* orbitals.

electron approaches the nucleus, there is a favourable electrostatic interaction, so the energy decreases and becomes more negative.

According to Eq. (1.28), the order of the energy levels is $1s < 2s < 3s = 3p = 3d < 4s = 4p = 4d = 4f$. However, for atoms larger than hydrogen, electronic repulsion changes this energy order, so that experimentally it has been found that $1s < 2s < 2p < 3s < 3p < 4s < 3d < 4p < 5s < 4d < 5p < 6s < 4f < 5d < 6p < 7s$.

The quantum numbers n and ℓ together determine the spatial size of the orbital, i.e. the average distance $\bar{r}$ of the electron from the nucleus (Eq. 1.30, where $a_0 = 0.529\ \text{Å}$).

$$\bar{r} = a_0 \frac{n^2}{Z} \left[\frac{3}{2} - \frac{\ell(\ell + 1)}{2n^2} \right] \tag{1.30}$$

In the quantum mechanical theory of the atom, the three quantum numbers n, ℓ and m determine the energy levels of the hydrogen atom and their shapes (nodal patterns). Determining the energy associated with each orbital allows the precise assignment of the spectrum of the hydrogen atom. Consider, for example, a hydrogen atom with its electron in the $2p$ orbital: this atom is 78.4 kcal mol^{-1} more stable than the separated nucleus and electron. However, when this atom emits energy and falls to a lower energy state, it must be into the $1s$ orbital, which is 313.6 kcal mol^{-1} lower in energy than the 0-reference level. This means that the $2p \rightarrow 1s$ transition is accompanied by the emission of 235.2 kcal mol^{-1}, which corresponds to one of the lines observed in the hydrogen spectrum (Figure 1.15). In fact, all the energy shifts observed in the hydrogen spectrum are explained by the differences in the estimated levels with respect to the quantum description of the hydrogen atom.

WHY DOESN'T THE ELECTRON FALL INTO THE NUCLEUS? BOHR'S LEGACY AND THE QUANTUM MECHANICAL MODEL

Given that the atomic electron and the atomic nucleus present opposite charges, the question arises, why doesn't the electron fall into the nucleus? As discussed in this chapter, the answer 'because the electron orbits the nucleus like a planet orbits the sun' is unacceptable because it would imply the emission (loss) of energy and a spiral trajectory of the electron until it collides with the nucleus. An electron cannot move in a circular orbit unless there is a force holding it in that orbit; if there is no force, it will escape from the orbit. According to classical theory, this is a shortcoming of the planetary model, since it implies that two forces balance each other to keep the electron in its atomic orbit: electrostatic attraction, which implies a tendency for the electron to fall into the nucleus, and the opposite tendency for the electron to move away from the nucleus as a consequence of its kinetic energy. If this is the case, the electron should be at rest or moving in a straight line at a constant speed. However, Bohr's model was successful in terms of its consistency with some atomic spectroscopic observations, indeed, we still use the terms 'ground state' and 'excited state' and retain the central idea that the energy of an atom exists in discrete levels and changes when the atom absorbs or emits a photon of a particular energy.

In the quantum mechanical model of the atom, the factor that causes the tendency of the electron and nucleus to separate arises from the wave-like properties of the electron as a particle. Consider again the iconic experiment of an 'electron in a box', which has a certain energy associated with it. Because of its wave nature, the electron presents a certain wavelength, which is determined by the dimensions of the box in which it is contained. The smaller the box, the smaller the wavelength must be, so its energy increases. This effect is called *confinement energy*.

Therefore, when the electron is confined to a certain volume, it also acquires confinement energy, which corresponds to its tendency to escape from confinement. In other words, the confinement energy is the factor that causes the electron in an atom to tend to move away from the nucleus.

When a nucleus is placed in a box containing an electron, the electrostatic attraction lowers the potential energy, and the total energy is a combination of the potential energy and the confinement energy (kinetic energy).

In short, since the nucleus and electron have opposite charges, they naturally tend to attract each other and bind. However, the electron is not just a particle; it also behaves like a wave. It is best to think of the electron as a diffuse cloud of matter and electric charge surrounding the nucleus. As the electron and the nucleus move closer together, the electron becomes more and more confined to a given volume and its wavelength becomes shorter and shorter in order to 'fit' into that space. Because its wavelength is shorter, the electron's energy increases, and eventually it will be large enough to dominate the electrostatic potential energy that attracts the nucleus and the electron. An equilibrium is reached in which the atom is stable and the electron neither falls into the nucleus nor separates from it.

In chemistry, quantum mechanics plays a crucial role in understanding and predicting the behaviour of molecules and their constituents. It explains phenomena such as atomic structure, chemical bonding, molecular spectroscopy, electronic structure and energy levels. Quantum mechanics is instrumental in understanding the properties of MOs and, as we will see in the following chapters.

FURTHER READING

A. Streitwieser, C. H. Heathcock, *Organic Chemistry*, MacMillan, New York, **1981**.

E. V. Anslyn, D. A. Dougherty, *Modern Physical Organic Chemistry*, University Science Books, Sausalito, California, **2005**.

F. A. Carroll, *Perspectives on Structure and Mechanism in Organic Chemistry*, 2nd ed., Wiley, New Jersey, **2010**.

M. Silberberg, P. Amateis, *Chemistry: The Molecular Nature of Matter and Change*, 8th ed., McGraw Hill, New York, **2017**.

P. A. Cox, *Introduction to Quantum Theory and Atomic Structure*, Oxford University Press, Oxford, UK, **1996**.

R. Chang, J. W. Thoman Jr., *Physical Chemistry for the Chemical Sciences*, Royal Society of Chemistry, London, UK, **2014**.

EXERCISES

1.1 The energy of a photon is 7.50×10^{-16}. Calculate the frequency and the wavelength of the light.

1.2 Calculate the wavelengths of the light with the following frequencies: 5.75×10^{14} Hz; 5.15×10^{14} Hz; 4.27×10^{14} Hz. To which colour of the visible spectrum does each frequency belong?

1.3 Show that to heat 1 mole of water from 0 to 100 °C requires 2/3 mole quanta of infrared light with $v = 3 \times 10^{13}$ cycles s^{-1}.

1.4 What is the energy of a single photon of blue light of frequency 6.4×10^{14} Hz and the energy per mole of photons of the same frequency?

1.5 Calculate the wavelength of a baseball that weighs 142 g travelling at 150 km h^{-1}.

1.6 Derive all possible sets of quantum numbers for $n = 2$ and describe the meaning of these sets of numbers.

1.7 Using Eqs. (1.28) and (1.29) determine the energy and average size of the 1s, 2s, 3s, 2p, 3p and 3d orbitals.

CHAPTER 2

Hybrid Orbitals

INTRODUCTION

One of the most relevant characteristics of atoms is their ability to combine and produce molecules. Indeed, millions of chemical compounds are known to be formed from the combination of a few types of atoms, such as carbon. These molecules, as it is well known, present quite different physical and chemical behaviour, so that, it is clear that it is not the specific atoms but rather the way they are bound in molecules that gives rise to the distinctive properties of chemical compounds.

The concept of 'molecule' was introduced in 1811 by A. Avogadro, who concluded that a molecule is the smallest constituent in a substance or material that still conserves its characteristic properties.

It became well established that molecules are formed by atoms, which are connected by chemical bonds, whose nature and strength vary from molecule to molecule. As a matter of fact, the main types of bonds are classified as ionic, covalent, hydrogen bonds, and so on. This book will be the most concerned with covalent bonding, which is the most common in organic chemistry and which is adequately understood in terms of valence bond theory (VB theory) or molecular orbital theory (MO theory).

The ϕ wave functions describing the atomic orbitals (see Chapter 1) present the mathematical properties of common wave equations. In this way, the mathematical manipulation of two or more ϕ allows the prediction of the properties of the resulting atomic or MOs and is in line with the observed characteristics of molecules, in particular, their structural properties (bond lengths, bond angles) and their reactivity behaviour. This ability of the quantum mechanical model to predict interatomic bonding and chemical reactivity constitutes a transcendental achievement in the advancement of the chemical sciences. From this point on, we will use phi (ϕ) to describe atomic orbitals and psi (Ψ) to describe MOs.

In this regard, VB theory describes covalent bonding in terms of atomic orbitals. It was originally developed in the late 1920s by Walter Heitler, Fritz London, John Slater and Linus Pauling. The VB theory is a quantum mechanical description of bonding processes that take into account the overlap of the atomic orbitals involved in forming a chemical bond. According to VB theory, a covalent bond is the result of two atomic orbitals overlapping in such a way that a pair of electrons occupies the overlapping zone. From a quantum mechanical point of view, two orbitals overlap and form a bond when their wave functions combine *in phase*.

In general terms, VB theory states that:

- The bonding electrons must have opposite (paired) spins. In this regard, according to the Pauli Exclusion Principle, the overlap zone between two bonding atomic orbitals must be occupied by a maximum of two electrons whose spins are paired.

- The strength of a bond depends on the attractive interaction between the bound nuclei and their shared electrons, so that the greater the overlap of the ϕ wave functions associated with the atomic orbitals, the closer the nuclei are to the electrons and the stronger the bond. Obviously, the size of the overlapping bonding zone will depend on the shape of the orbitals.

- The formation of hybrid orbitals. When one considers the formation of a bond in a diatomic molecule such as HF (hydrogen fluoride), one finds that it is the overlap between the s orbital in hydrogen and a p orbital in fluorine that gives rise to a linear HF molecule. Initially, the carbon atom has two of its four valence electrons already paired, so these two-paired electrons are not able to participate in bond formation. Under this premise, the carbon atom should form only two bonds at an angle of 90°. In methane, however, carbon forms *four equal C—H bonds with a tetrahedral geometry*. How can one explain this observation? To explain this fact, Linus Pauling proposed that during bonding, the valence orbitals of isolated atoms mix to form new *hybrid* orbitals. These new orbitals can in turn form bonds with other atomic orbitals that match experimentally observed molecular geometries (e.g. methane). From a quantum mechanical point of view, it is possible to combine the wave functions of the valence orbitals according to mathematical principles to produce a set of new hybrid atomic orbitals. If the wave functions coincide in their positive or negative mathematical phases, the amplitudes increase in magnitude (add up constructively). By contrast, if the overlapping wave functions present opposite signs, the overall amplitude is reduced and may even cancel out completely. This model is known as orbital hybridisation theory and is explained in the following.

HYBRIDISATION THEORY

In organic chemistry, one of the most important hybridisation processes is, of course that involving the carbon atom. Let us then recall that one carbon atom contains six electrons in its neutral ground state. These six electrons would be arranged as follows: two electrons fit in the $1s$ orbital, two electrons fit in the $2s$ orbital, and finally, one electron fits in the $2p_x$ orbital and the last one in the $2p_y$ orbital $\left(1s^2\ 2s^2\ 2p_x{}^1\ 2p_y{}^1 2p_z{}^0\right)$. This electron distribution leaves the $2p_z$ orbital empty, as shown in Figure 2.1a. According to the hybridisation theory, the carbon atom re-distributes its four valence electrons so that each one occupies a second-level orbital $\left(1s^2\ 2s^1\ 2p_x{}^1\ 2p_y{}^1 2p_z{}^1\right)$. This leads to the formation of four sp^3 hybrid orbitals, each with 25% s-character and 75% p-character. The identical four new sp^3 orbitals are similar to typical p orbitals in that they also present a planar node; nevertheless, one of the lobes is larger than the other one (Figure 2.1b). This is because the contribution of the spherical $2s$ orbital combines in-phase (adds constructively) to one lobe of the $2p$ orbital and subtracts from the other lobe. The four sp^3 hybrid orbitals are arranged tetrahedrally

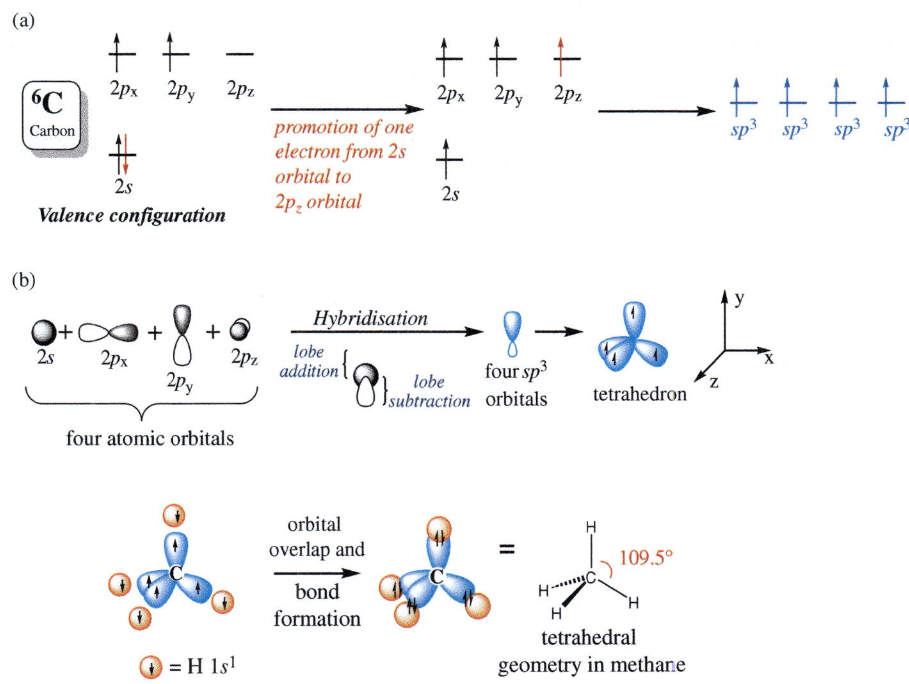

FIGURE 2.1   (a) Hybridisation of orbitals $2s$ and $2p$ to form four sp^3 orbitals. (b) Tetrahedral geometry of sp^3 orbitals. The original orbitals are shown in black colour, the hybrid orbitals are depicted in blue colour.

to minimise electronic repulsion and to facilitate overlap with other orbitals of other atoms in the formation of bonds. Figure 2.1b presents a graphical representation of the hybridisation process that gives rise to sp^3 orbitals. This theoretical model helps explain the tetrahedral shape of the methane (CH_4) molecule (Figure 2.1b), in which each carbon-hydrogen bond is located at a 109.5° angle to the other C—H bonds.

On the other hand, starting from the initial configuration of carbon $\left(1s^2\ 2s^2\ 2p_x^1\ 2p_y^1\ 2p_z^0\right)$, the 2s orbital and two 2p orbitals can also combine to produce three sp^2 hybrid orbitals, presenting each 33.33% s-character and 66.66% p-character within a geometrical arrangement such that each of them points to the corner of an equilateral triangle, with their axes 120° apart (Figure 2.2). This leaves one unchanged p orbital ($2p_z$) perpendicular to the three sp^2 hybrid orbitals. The three sp^2 hybrid orbitals of a carbon atom, as well as the unchanged $2p_z$, can overlap with the orbitals of other atoms and form bonds, as we will discuss in Chapter 3 (Figure 2.2b).

Similarly, the combination of one 2s orbital with one 2p orbital gives rise to two new sp-hybrid orbitals, each with 50% s and 50% p-character and oriented 180° apart, as shown in Figure 2.3. The $2p_y$ and $2p_z$ orbitals remain now unchanged. Here, the sp hybrid orbitals are able to form bonds with other atomic orbitals, maintaining a linear geometry, while the p orbitals can in turn participate in the formation of two additional bonds. This process corresponds to the formation of a triple bond between two carbon atoms (see Chapter 3).

In short, hybrid orbitals satisfy the following characteristics: (i) the number of newly formed hybrid orbitals is equal to the number of atomic orbitals that participate in the hybridisation process; (ii) the electronic nature of the hybrid

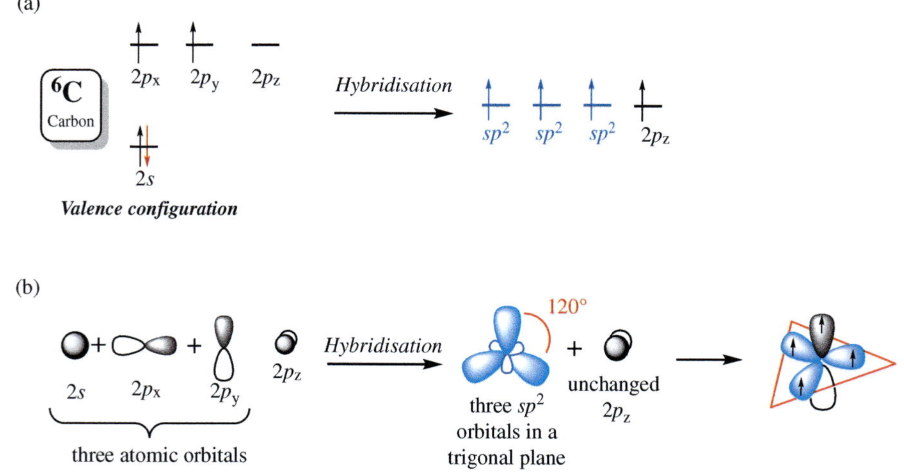

FIGURE 2.2   (a) Hybridisation of orbitals 2s and 2p to form three sp^2 orbitals. (b) Trigonal geometry of sp^2 orbitals. The original orbitals are shown in black colour, the hybrid sp^2 orbitals are depicted in blue colour.

FIGURE 2.3   (a) Hybridisation of one $2s$ orbital and one $2p$ orbital to form two sp orbitals. (b) Digonal (lineal) geometry of sp orbitals. The original orbitals are shown in black colour, the hybrid sp orbitals are depicted in blue colour.

orbitals depends on the type and number of the atomic orbitals from which they are formed; and (iii) the shape and orientation of the hybrid orbitals minimise electron–electron repulsion and maximise overlap with the atomic orbital with which they combine to form a bond.

The hybrid orbitals correspond to new ϕ' wave functions that contain the mathematical information that dictates its subsequent chemical behaviour. As mentioned earlier, mathematically, the bonding process involves linear combinations of atomic wave functions (orbitals) that provide equal energy (degenerate) hybrid orbitals whose position is determined by the orientation of the p-wave functions (p orbitals) used in the hybridisation process.

Importantly, the theory of hybridisation can be applied to all atoms and is not limited to carbon. For instance, below is the hybridisation scheme for the atomic orbitals in oxygen, nitrogen and boron (relevant atoms present in biomolecules and in other important organic compounds) as they form bonds (for example with hydrogen). In the case of nitrogen, its five valence electrons in orbitals $\left(2s^2\ 2p_x^{\ 1}\ 2p_y^{\ 1}2p_z^{\ 1}\right)$ can combine to afford four sp^3 hybrid orbitals (Figure 2.4a). These hybrid orbitals are oriented in an approximate tetrahedral arrangement. However, unlike carbon, a sp^3 hybrid orbital is occupied with a lone pair of electrons. The remaining unpaired electrons correspond then to the bonding electrons that can combine with other atomic orbitals and form bonds, for example, three N—H bonds in ammonia (Figure 2.4a).

In the case of oxygen with its six valence electrons, the combination of its valence atomic orbitals $\left(2s^2\ 2p_x^{\ 2}\ 2p_y^{\ 1}2p_z^{\ 1}\right)$ produces four sp^3 hybrid orbitals, two of them occupied with lone pairs of electrons and two half-occupied with one unpaired electron each, that can combine with other atomic orbitals and form bonds, for example, two O—H bonds in water (Figure 2.4b). The lone

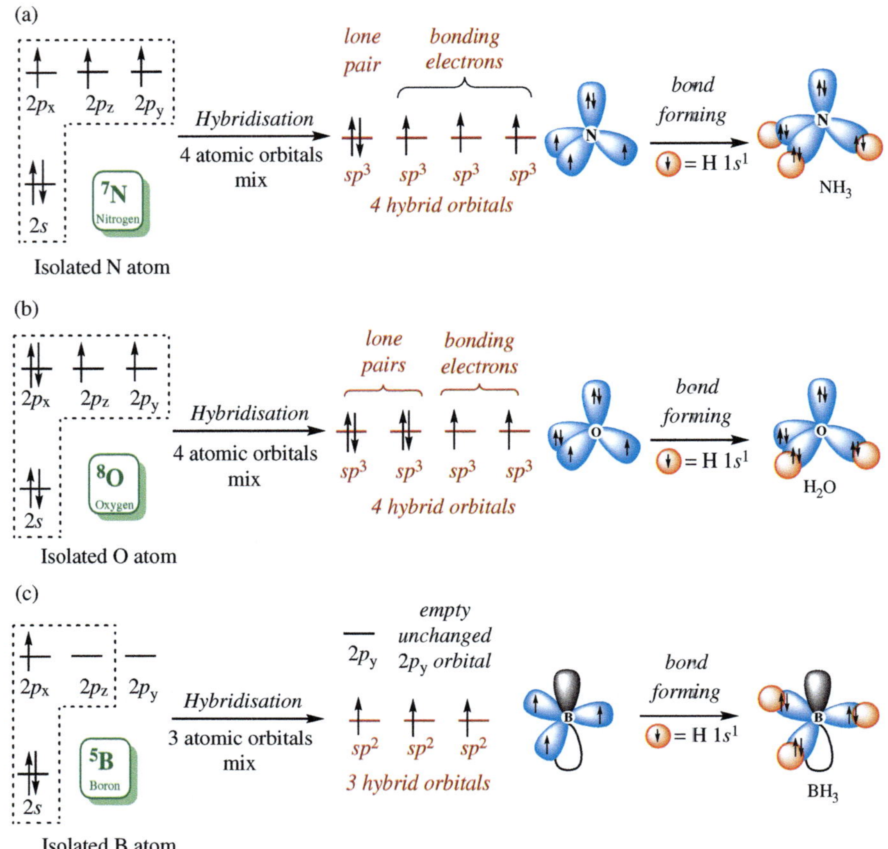

FIGURE 2.4   Hybridisation of the valence atomic orbitals in (a) nitrogen (N), (b) oxygen (O) and (c) boron (B).

electron pairs have an influence on both the structure of derivatives and their reactivity, as will be discussed in the following.

Boron is a particular case of an atom. In this case, when forming bonds with hydrogen atoms as in borane BH_3, three of the four original atomic orbitals combine to afford three sp^2 hybrid orbitals as well as an unchanged $2p_y$ orbital (Figure 2.4c). It can be appreciated that the lower energy orbitals are the first to be occupied by electrons; that is, the sp^2 orbitals with their higher s-character, are of lower energy than sp^3 orbitals and therefore boron in BH_3 prefers to adopt a sp^2 hybridisation with an empty $2p_y$ orbital. In other words, if an atom presents an empty orbital, it better be an orbital of the highest possible energy, since being empty will not affect the stability of the molecule. In the case of oxygen and nitrogen, the lone pairs of electrons (for example in water or ammonia) exhibit a preference to occupy the lower energy sp^3 orbital rather than a pure p orbital. In general, lone electron pairs on second-row elements tend to occupy hybrid orbitals with s-character.

Hybrid orbitals with different ratios of s and p orbitals can be constructed according to Eq. (2.1) where Φ_h is a sp^λ hybrid orbital that presents $1/(\lambda + 1)$ s-character and $\lambda/(\lambda + 1)$ p-character.

$$\Phi_h = \frac{1}{\sqrt{\lambda + 1}}\,\phi_s + \frac{\lambda}{\sqrt{\lambda + 1}}\,\phi_p \tag{2.1}$$

The most important hybrid orbitals in organic chemistry present sp, sp^2 and sp^3 hybridisation, which results from the combination of the s orbital with one, two or three p orbitals (Table 2.1).

Although less important in organic chemistry, atoms can also mix d-type atomic orbitals together with s or p orbitals to form planar square (dsp^2) or trigonal bipyramidal (dsp^3) geometries, among others. The representation of the most common hybrid orbitals is shown in Figure 2.5.

TABLE 2.1 Most Important Hybrid Orbitals and Some of Their Structural Properties

Percent of s character (%)	λ (Eq. 2.1)	Orbital	Geometry	Internal angle
50	1	sp	Digonal	180°
33.33	2	sp^2	Trigonal	120°
25	3	sp^3	Tetrahedral	109.5°

Orbital	Geometry	Representation
sp	lineal	
sp^2	trigonal	
sp^3	tetrahedral	
dsp^2	planar	
dsp^3	trigonal bipyramid	

FIGURE 2.5 Geometry and graphical representation of the most common hybrid orbitals.

WAVEFUNCTIONS ASSOCIATED WITH HYBRID ORBITALS

A hybrid orbital Φ_h corresponds to the linear combination of atomic orbitals ϕ. In particular, any hybrid orbital Φ_h derived from combinations of s and p orbitals is expressed as

$$\Phi_h = C_s\phi_s + C_{p_x}\phi_{p_x} + C_{p_y}\phi_{p_y} + C_{p_z}\phi_{p_z} \tag{2.2}$$

where the coefficients C correspond to the degree of participation of each atomic orbital. As it is the case in the initial atomic orbitals, hybrid orbitals are normalised; that is, their integration over all space is equal to 1, and this means that the probability of finding the electron in such an orbital is effectively real.

Considering that the direction of a vector in an x, y, z coordinate system is conveniently expressed by its directional cosines (Figure 2.6), then the hybrid orbital can also be expressed as follows:

$$\Phi_h = C_s\phi_s + C_p(\alpha\phi_{p_x} + \beta\phi_{p_y} + \gamma\phi_{p_z}) \tag{2.3}$$

Since Φ_h is normalised, then:

$$C_s^2 + C_p^2 = 1 \tag{2.4}$$

Naturally, the ratio $C_s{:}C_p$ determines the s-character present in the hybrid orbital. Setting $C_s^2 = n$ and $C_p^2 = m$, Φ_h can also be expressed as a hybrid $s^n p^m$.

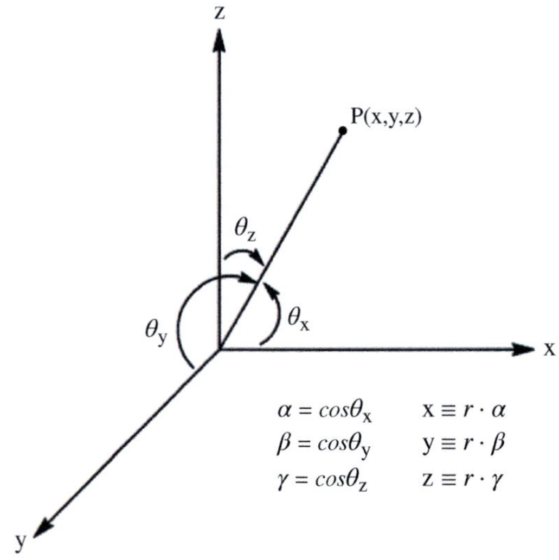

FIGURE 2.6　Coordinate system of a vector in x, y, z-axes and its directional cosines.

PROCEDURE TO BUILD A HYBRID ORBITAL

The following steps must be followed:

1. Write the corresponding equation indicating its correct orientation by means of α, β, γ angles.
2. Choose the coefficients C_s and C_p to obtain the desired hybridisation. For a $s^n p^m$ orbital,

$$C_s = \sqrt{n} \ \text{ and } \ C_p = \sqrt{m}$$

3. Normalise the orbital dividing by $\sqrt{C_s^2 + C_p^2}$

Thus, to define a sp^2 orbital, centred at the origin of an x, y, z system and oriented in the $+z$ direction:

1. Since θ_x and $\theta_y = 90°$, then $\cos\theta_x = \cos\theta_y = 0$; i.e. $\alpha = \beta = 0$:

$$\Phi_h = C_s\phi_s + C_p(\gamma\phi_{p_z}) \tag{2.5}$$

   where $\cos 0 = 1$, then $\gamma = 1$.
2. The ratio $C_s : C_p = \sqrt{n} : \sqrt{m} = \sqrt{1} : \sqrt{2} = 1 : \sqrt{2}$ Thus,

$$\Phi_h = \phi_s + \sqrt{2}\ \phi_{p_z} \tag{2.6}$$

3. Finally, normalisation is achieved by dividing Eq. (2.6) by $\sqrt{C_s^2 + C_p^2} = \sqrt{1+2} = \sqrt{3}$.
   That is to say:

$$\Phi_h = \frac{1}{\sqrt{3}}\ \phi_s + \frac{\sqrt{2}}{\sqrt{3}}\ \phi_{p_z} \tag{2.7}$$

Following the same procedure, it can be shown that the hybrid orbitals that result from the combination of the atomic orbitals s, p_x and p_y are:

$$\Phi_1 = \frac{1}{\sqrt{3}}\left(\phi_s + \sqrt{2}\ \phi_{p_x}\right) \tag{2.8}$$

$$\Phi_2 = \frac{1}{\sqrt{3}}\left(\phi_s - \frac{1}{\sqrt{2}}\phi_{p_x} + \frac{\sqrt{3}}{\sqrt{2}}\phi_{p_y}\right) \tag{2.9}$$

$$\Phi_3 = \frac{1}{\sqrt{3}}\left(\phi_s - \frac{1}{\sqrt{2}}\phi_{p_x} - \frac{\sqrt{3}}{\sqrt{2}}\phi_{p_y}\right) \tag{2.10}$$

where the orientation of Φ_1, Φ_2 and Φ_3 was chosen as indicated in Figure 2.7.

$$\Phi_2 = \frac{1}{\sqrt{3}}\left(\phi_s - \frac{1}{\sqrt{2}}\phi_{px} + \frac{\sqrt{3}}{\sqrt{2}}\phi_{py}\right)$$

$$\Phi_1 = \frac{1}{\sqrt{3}}\left(\phi_s + \sqrt{2}\phi_{px}\right)$$

$$\Phi_3 = \frac{1}{\sqrt{3}}\left(\phi_s - \frac{1}{\sqrt{2}}\phi_{px} - \frac{\sqrt{3}}{\sqrt{2}}\phi_{py}\right)$$

FIGURE 2.7   Spatial orientation of Φ_1, Φ_2 and Φ_3.

ORTHOGONALITY OF WAVE FUNCTIONS (ORBITALS)

As it was previously indicated, the wave functions Φ_h must satisfy the condition of normalisation, which effectively limits the orbital's space. On the other hand, hybrid orbitals must be independent entities even when they are isoenergetic (degenerate orbitals) in order to warrant their identity; this is known as the orthogonality condition.

Two wave functions or orbitals are said to be orthogonal when the integration of their product is equal to zero (Eq. 2.11).

$$\int \Phi_1 \Phi_2 \, dv = 0 \tag{2.11}$$

Normally, the orbitals in an atom are orthogonal to each other. For example, the product of orbitals 2.8 and 2.9 is equal to:

$$\frac{1}{3}\Big[\int s^2 \, dv = \frac{1}{\sqrt{2}}\int s p_x \, dv + \frac{\sqrt{3}}{\sqrt{2}}\int s p_y \, dv$$
$$+ \sqrt{2}\int p_x s \, dv - \int p_x^2 \, dv + \sqrt{3}\int p_x p_y \, dv\Big] \tag{2.12}$$
$$= \frac{1}{3}(1 - 0 + 0 + 0 - 1 + 0) = 0$$

It is important to mention that the product of the s and p atomic orbitals is equal to zero since the positive (in-phase) overlap cancels out the negative (out-of-phase) overlap.

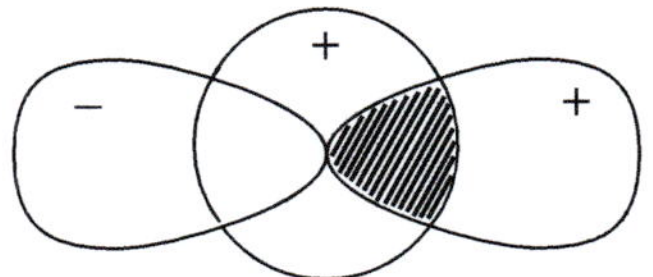

THE BENT BOND OR TAU MODEL

In addition to the very useful sigma/pi (σ/π) model of the double bond, an alternative description views the double bond as composed of two bent bonds or, as they are also called, tau (τ) bonds. For each atom of the double bond, the orbitals present five parts p-character and one part s-character. Hence, two equivalent τ bonds may be formed (Figure 2.8).

FIGURE 2.8   Model of a tau (τ) bond using sp^5 orbitals to formulate the double bond in ethylene.

The τ bond model has the significant property of conveying information about both the s and p orbital components of a double bond in a pictorially instructive manner. Consider, for instance, cyclopropane: for purely geometric reasons, the internuclear C—C—C angle is 60°. Whereas the natural bond angle for C_{sp^3} orbitals is 109.5°, for hybrid orbitals having more p-character, the natural angle is smaller. Thus, the C—C bond orbitals in cyclopropane have between sp^4 and sp^5 hybridisation and give rise to bent bonds (Figure 2.8).

EFFECTS OF HYBRIDISATION

The hybridisation of atomic orbitals is reflected in the structural properties of the molecules to which they give rise, such as bond angles and bond lengths. For the organic chemist, it is also important to note that the type of hybridisation is reflected in other properties such as internuclear coupling constants in NMR spectroscopy, the electronegativity of functional groups, and the acidity of the C—H bonds, among others. Here are some specific examples.

Coupling Constants in C—H bonds and s-Character

The atomic nucleus is a particle of subatomic size. As such, it can be described by a number of quantum properties. One of these is called nuclear spin, so nuclei that interact with magnetic fields (H_o) present a preferred direction of the nuclear spin (a simple analogy is that they behave like small magnets). The C—H bond consists of two nuclei with spin $+\frac{1}{2}$ or $-\frac{1}{2}$ (or ↑ and ↓) and two electrons that also have spin $= +\frac{1}{2}$ or $-\frac{1}{2}$ (or ↑ and ↓) (see Chapter 1). According to the Pauli exclusion principle, the electrons that form the C—H bond must have antiparallel spins to occupy the same orbital. When an external magnetic field (H_o) is applied to the C—H nuclear pair, the most stable electronic configuration (lowest in energy) is the one in which one of the nuclear spins is oriented antiparallel to H_o (Figure 2.9a) and in which all the remaining spins (nuclear and electronic) are arranged antiparallel to their neighbours. However, a slightly

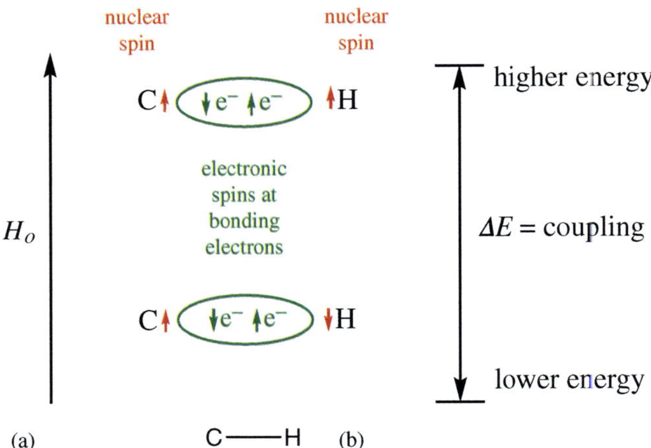

FIGURE 2.9   Effect of an external magnetic field (H_o) applied to the C—H nuclear pair.

higher-energy electronic configuration is one in which both nuclei are oriented parallel to the H_o (Figure 2.9 top).

Since the energy difference between the high-energy parallel and the low-energy antiparallel configurations is small, the ratio of molecules at these levels is close to 50:50, nevertheless, information about the orientation of the nuclear spins is transmitted through the electrons in the bond. Since the s orbitals are the only ones that have an electronic density in the nucleus, then the magnitude of the interaction (coupling) is directly proportional to the percentage of s-character in the bond that joins the coupled nuclei, according to Eq. (2.13).

$$^1J_{CH} = (500 \text{ Hz}) \text{ % } s\text{-character}$$

or

$$s\text{-character } (\%) = \rho = \frac{^1J_{C/H}}{500} \tag{2.13}$$

The reason for this effect is that s orbitals are the only ones possessing electron density at the nucleus, so that they interact more directly with the nucleus than p orbitals do. Thus, the greater a bond's s-character, the more spin information is transmitted between the bonded nuclei.

In this manner, the value of $^1J_{C/H}$ expected for sp^3, sp^2 and sp hybridisations is 125, 167 and 250 Hz, respectively. Table 2.2 presents other examples taken from the literature.

As can be seen in Table 2.2, normal alkanes present $^1J_{C/H}$ coupling constants close to 125 Hz, whereas cyclopropane and cyclobutane exhibit higher C—H coupling constants because the p-character is concentrated in the endocyclic C—C bonds. In these cyclic compounds, as a consequence of bond angle torsion, the carbon atoms cannot adopt the normal 109.5° bond angle associated with sp^3 hybridised carbons. Increasing the p-character in the C—C bond allows the bond angle to be smaller and therefore reduces bond angle strain. Consequently, the C—H bonds

TABLE 2.2 Coupling Constants $^1J_{C/H}$ for Several Organic Molecules of Interest

Bond	$^1J_{C/H}$	Bond	$^1J_{C/H}$
$CH_3-H(sp^3)$	125	$C_6H_5-H(sp^3)$	159
$CH_3CH_2-H(sp^3)$	126	$C_8H_7-H(sp^3)$	155
$(CH_3)_3C\,CH_2-H(sp^3)$	124	$CH_3-C\equiv C-H(sp^2)$	248
$C_6H_5\,CH_2-H$	122	$C_6H_5-C\equiv C-H(sp^2)$	251
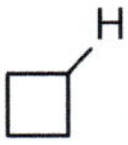 cyclobutane–H (sp^3)	134	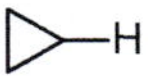 cyclopropane–H (sp^3)	161
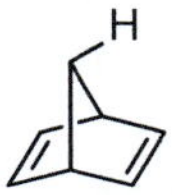 norbornene (sp^3)	135	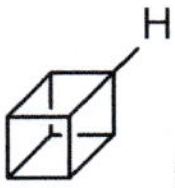 bicyclic (sp^3)	160

Note: The hybridisation shown corresponds to the carbon highlighted in bold type.

increase their *s*-character, reducing their length and increasing the magnitude of the $^1J_{C/H}$ coupling constant. In this context, olefinic C—H bonds exhibit couplings close to the ideal value of 167 Hz for sp^2 hybridisation, whereas acetylenes present $^1J_{C/H} = 250$ Hz as anticipated for sp hybridised C—H bonds.

Bond Angles in Molecules with Unshared Electron Pairs

Electrons that form part of internuclear bonds are stabilised by the electrostatic attraction with the neighbouring nuclei, and this attraction compensates for the electron–electron repulsion that is generated when they occupy the same orbital. As discussed earlier in this chapter, in the case of lone pairs of electrons, only a single positive nucleus can stabilise the negatively charged electrons and, as a consequence, the atomic hybridisation changes to concentrate the maximum possible *s*-character at the atomic orbital supporting the two unshared electrons. In this way, the orbital presents greater electron density in the region closest to the nucleus and the electrostatic attraction is increased. Of course, if the orbital containing the pair of unshared electrons presents greater *s*-character, then the hybridisation in the remaining bonds will be closer to that of *p* orbitals, with their smaller bond angles (Figure 2.10).

Similarly, N—H bonds in ammonia have increased *p*-character relative to a pure sp^3 orbital. The case of the compound PH_3 is particularly interesting because its atomic orbitals are reluctant to hybridise (the third-row valence orbitals of phosphorus are $3s^2\ 3p_x^1\ 3p_y^1\ 3p_z^1$). Thus, the three 1*s* hydrogen orbitals overlap with the three essentially pure 3*p* phosphorus orbitals, leaving

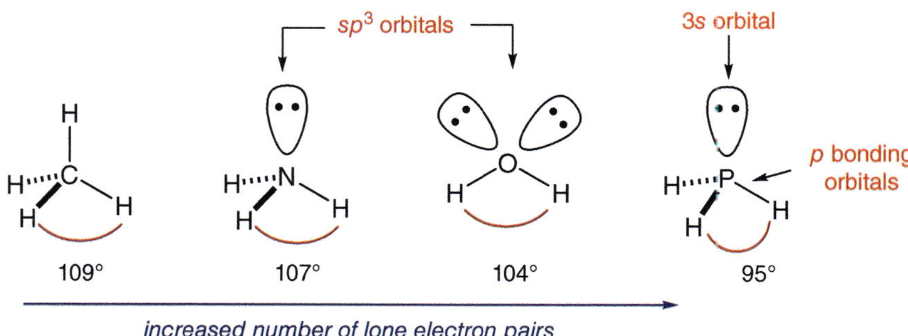

FIGURE 2.10 Bond angles in molecules with unshared electron pairs.

the lone pair of electrons at the $3s$ orbital. This almost pure s lone pair presents lower energy than an sp^3 lone pair in ammonia and is therefore less reactive, which helps explain why ammonia NH_3 is more basic than phosphine PH_3.

It is possible to quantify the s- and p-character of a bonding orbital (hybridisation index) by the simple Eq. (2.14). The experimentally observed angle is used in this equation to solve for i. The hybridisation is then defined as sp^i.

$$1 + i \cos \theta = 0 \tag{2.14}$$

The bond angle between two N—H bonds in the molecule of ammonia is 107.1°. Therefore, substituting this value in Eq. (2.14) gives Eq. (2.15). Thus, in the molecule of ammonia, the bonding orbitals of the nitrogen atom are $sp^{3.4}$ hybridised. Therefore, in the hybrid orbitals of ammonia, the N—H bonds present 22.7% of s-character, which is lower than the value (25%) in ideal sp^3 orbitals.

$$i_{NH_3} = \frac{-1}{\cos(107.1)} = 3.40 \tag{2.15}$$

By contrast, the O—H bonds in water involve sp^4 hybrid orbitals, which means that in a molecule of water, the orbitals that form the O—H bonds are 80% p and 20% s.

Bond Lengths

As seen in Table 2.3, C—C bonds with greater s-character are notably shorter. As the s-character of a hybrid orbital increases, its front lobe becomes smaller. This reduces the distance from the nucleus to the point where there is a maximum probability of finding an electron (that is, the highest electron density). As a result, bond length decreases as the hybridisation changes from sp^3 to sp^2 to sp. Table 2.3 compiles the covalent radii for carbon atoms with different

TABLE 2.3 Covalent Radii (Å) for C—C Bonds of Different Hybridisation

Hibridación en C′	C—C′	C=C′	C ≡ C′
sp^3	0.767		
sp^2	0.737	0.669	
sp	0.687	0.642	0.604

$$sp^3\text{--}sp^2 \qquad sp^2\text{--}sp^2$$

$$\text{calculated (Table 2.3)} \quad \dfrac{0.767 + 0.737}{} = 1.504 \text{ Å} \qquad \dfrac{0.669 + 0.669}{} = 1.338 \text{ Å}$$

$$\text{observed} \quad = 1.501 \text{ Å} \qquad = 1.336 \text{ Å}$$

FIGURE 2.11   Calculated and experimental C—C bond lengths in the propene molecule.

hybridisation and allows understanding the bond length tendency observed in alkanes, alkenes and alkynes.

As an illustrative example, Figure 2.11 presents the calculated and experimental bond lengths for the C—C bonds in propene.

Thermodynamic Acidity of Organic Compounds

By the same arguments presented earlier, carbanions are more stable when the electron pair associated with the conjugate base of an organic acid occupies an orbital with greater s-character. Thus, acetylene is several orders of magnitude more acidic than ethylene, since the removed protons are linked to sp and sp^2 hybridised orbitals, respectively (Figure 2.12).

Likewise, cyclopropane is much more acidic than propane and biphenylene is selectively metallated at the α position. Indeed, the exocyclic bonds in cyclopropane present more s-character than the C—H bonds in normal alkanes, whereas the C_α—H bond is more acidic than C_β—H in biphenylene due to the fact that the former is directly linked to the more electronegative C—C bond (higher s-character) (Figure 2.13).

Effect of Hybridisation on C—C Bond Dissociation Energies

Earlier discussion in this Chapter referred to the shortening of C—C bonds when the amount of s-character in such bonds increases. In this regard, it is well known that bond force constants are stronger for shorter bonds. It is therefore

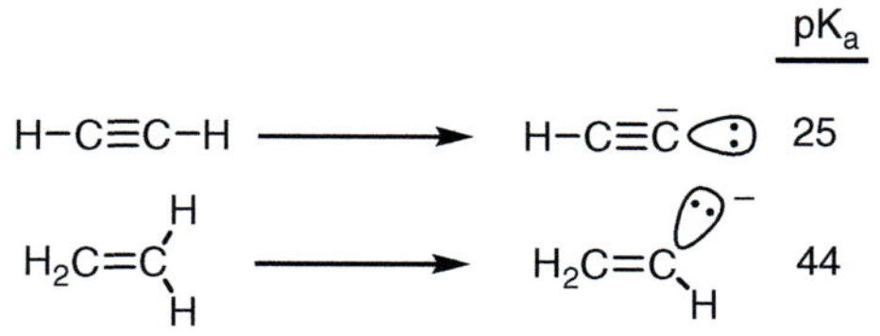

FIGURE 2.12 Difference in the acidic character between acetylene and ethene.

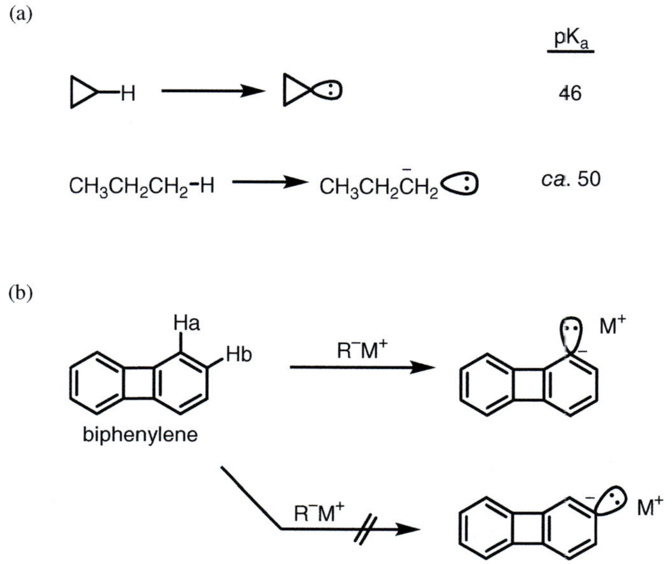

(a)

FIGURE 2.13 (a) Difference in the acidic character between cyclopropane and propane. (b) Difference in the acidic character between aromatic hydrogens at the α or β position in biphenylene.

TABLE 2.4 Relationship Between Bond Hybridisation, Bond Force Constants and Bond Dissociation Energies

C—C bond	Hybridisation	Bond force constant (dynes cm^{-1})	Bond dissociation energy (kcal mol^{-1})
$CH_3—CH_3$	$sp^3—sp^3$	4.5	83
$CH_3—CN$	$sp^3—sp$	5.3	103
NC—CN	sp—sp	6.7	112

not surprising that an increase in *s*-character for the hybridisation of C—C bonds is accompanied by an increase in the bond's force constant and then in its dissociation energy. Some examples collected are collected in Table 2.4.

FURTHER READING

A. Streitwieser Jr., C. H. Heathcock, *Introduction to Organic Chemistry*, 2nd ed., MacMillan, New York, **1981**.

D. E. Lewis, *Advanced Organic Chemistry*, Oxford University Press, Oxford, UK, **2016**.

D. J. Klein, N. Trinajstić, *J. Chem. Ed.* **1990**, 67, 633.

E. D. Becker, *High Resolution NMR*, 2nd ed., Academic Press, New York, **1980**.

E. D. Becker, *High Resolution NMR. Theory and Chemical Applications*, 2nd ed., Academic Press, New York, **1980**, Chapter 5.

E. G. Lewars, *Computational Chemistry Introduction to the Theory and Applications of Molecular and Quantum Mechanics*, Springer, Germany, **2016**.

E. V. Anslyn, D. A. Dougherty, *Modern Physical Organic Chemistry*, University Science Books, Sausalito, California, **2005**.

E. W. Della, E. Cotsaris, P. T. Hine, *J. Am. Chem. Soc.* **1981**, *103*, 4131.

F. A. Carey, R. J. Sundberg, *Advanced Organic Chemistry*, 3rd ed., Plenum Press, New York, **1990**.

H. E. Zimmerman, *Quantum Mechanics for Organic Chemists*, Academic Press, New York, **1975**.

L. Pauling, *J. Am. Chem. Soc.* **1931**, *53*, 1367.

L. Pauling, *The Nature of the Chemical Bond*, 3rd ed., Cornell University Press, Ithaca, **1960**.

M. B. Smith, J. March, *March's Advanced Organic Chemistry. Reactions, Mechanisms, and Structure*, 5th ed., Wiley-Interscience, New York, **2001**.

P. A. Rock, D. A. McQuarrie, *General Chemistry*, 2nd ed., Freeman, New York, **1987**.

W. K. Whitten, K. D. Gaipey, R. E. Davis, *General Chemistry with Qualitative Analysis*, 3rd ed., Saunders, San Francisco, **1988**.

EXERCISES

2.1 What kind of orbitals are the following?

$$\Phi_1 = \frac{1}{\sqrt{6}}\phi_s + \frac{1}{\sqrt{3}}\phi_{p_x} + \frac{1}{\sqrt{2}}\phi_{p_z}$$

$$\Phi_2 = \frac{1}{\sqrt{6}}\phi_s + \frac{1}{\sqrt{3}}\phi_{p_x} - \frac{1}{\sqrt{2}}\phi_{p_z}$$

2.2 Obtain the normalised orbitals Φ_1, Φ_2 and Φ_3 knowing that they are $s + p_x + p_y$ hybrids.

2.3 Write mathematical expressions for the *sp* hybrid orbitals aligned in the x coordinate. Show that they are mutually orthogonal.

2.4 From the infrared spectroscopy, it has been established that isoelectronic molecules such as CO_2, NO_2, NO_2^+ and N_3^- present frequencies of torsion line to 600 cm^{-1}. Propose electronic structures for them.

2.5 Explain the increasing order in the magnitude of the coupling constants $^1J_{C/F}$ in the following compounds:

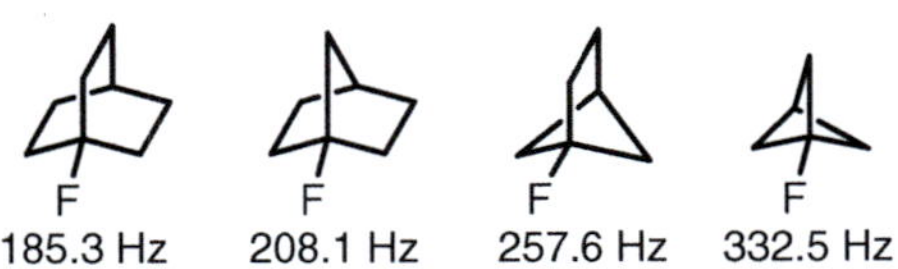

2.6 Explain the displacement in the C-13 NMR spectra to almost 4 ppm lower δ in **A** compared to **B**.

2.7 Lewis acids such as BF_3 form stronger complexes with amines relative to amides or nitriles. Suggest an explanation in terms of hybridisation.

2.8 Explain why cyclopropylamine is less basic than other aliphatic amines (e.g. propylamine).

2.9 Explain the following acidity constants:

Amines	$K_a = 10^{-33} - 10^{-35}$
Amides	$K_a = 10^{-14} - 10^{-16}$
Imides	$K_a = 10^{-9} - 10^{-10}$

CHAPTER 3

Bond Formation from Atomic Orbitals

INTRODUCTION

The first and second chapters in this book deal with the development of the quantum mechanical model for atomic orbitals (AOs) and the concept of hybridisation. On this basis, the formation of molecular orbitals (MOs) is discussed in the present chapter. As it will be seen in the following chapters, the physical and chemical properties of any molecule are properly explained in terms of its MOs. In the Lewis and valence bond models of molecular structure, electrons are located on atoms or between pairs of bonded atoms. In MO theory, all valence electrons are delocalised throughout the molecule and are not confined to individual bonds. In other words, the MO approach seeks to obtain suitable wave functions for the whole molecule.

MIXING OF s ORBITALS

Let us start with a simple diatomic molecule such as the molecule of hydrogen, H_2. The H—H bond results from the overlap of the AOs involved, in this case, two $1s$ orbitals. As we have seen previously, the wave functions describing the AOs can be combined mathematically and such combination will be additive for two wave functions of the same sign (*in phase* combination). If this is the case, a higher electron density is obtained in the region of the H—H bond. The $1s$ orbitals for two non-bonded hydrogens (Figure 3.1, left side) are apart from each other, so that they do not occupy a common region of space between the atoms. However, as they approach each other, there will be a region of overlap. If the overlap is sufficiently efficient, this leads to the formation of an MO (Figure 3.1, right side). Any electron occupying this MO will be attracted to both hydrogen nuclei and will therefore present lower energy relative to the situation where the electron is confined to

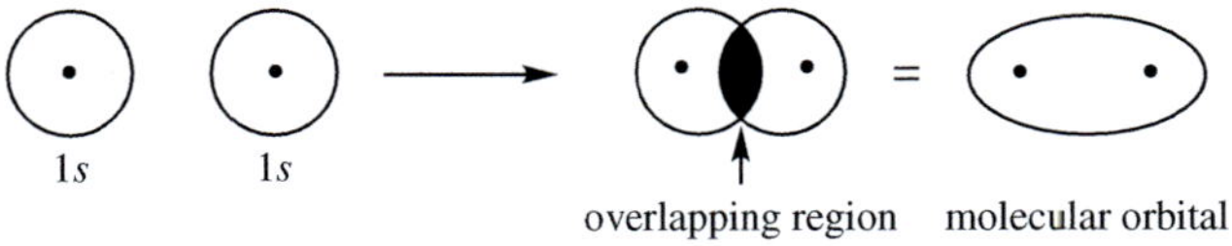

FIGURE 3.1   Overlap of two 1s orbitals in the H_2 molecule. The dots represent the hydrogen atom nuclei.

one AO. Although the entropy associated with the isolated atoms decreases, the potential energy in the molecule also decreases: the nucleus–nucleus repulsion is reduced, whereas the nucleus–electron attraction becomes dominant.

MOs correspond to linear combinations of atomic orbitals (LCAO) and the MO in Eq. (3.1) is the *in-phase* linear combination of two hydrogen atomic orbitals.

$$\psi = \phi_{H1s} + \phi_{H1s} \tag{3.1}$$

The magnitude of the overlap is given by Eq. (3.2). The overlap integral between two atomic orbitals is determined by multiplying the two AO wavefunctions with each other (in this case two ϕ_{H1s} AOs) and then taking the integral of this product. In other words, the overlap integral is the area under a plot of the product of the two atomic orbitals (AOs) (Eq. 3.2).

$$S = \int \phi_{H1s} \phi_{H1s}\, dv \tag{3.2}$$

Integral S corresponds to the overlap integral, which can take values from $+1$ to -1. When the orbitals are separated so that there is no overlapping region between them, the overlap integral between these two orbitals is 0 (Figure 3.2a). As the orbitals come closer together, they begin to present regions of overlap and so S becomes larger than 0 (Figure 3.2b). As the overlap integral becomes more significant, the MO formed will be associated to a lower energy level relative to the isolated AOs, to the point of minimum possible energy (Figure 3.2c). This *in-phase* combination forms a bonding MO.

A bonding MO presents lower energy than the AOs that form it for several reasons: (i) the electron density is mainly located between the nuclei, the nucleus–nucleus repulsion decreases while the nuclear–electron attraction increases. (ii) Any electron occupying such an MO is attracted to both nuclei, so it presents lower energy than the atomic orbitals before bonding. (iii) Two electrons in an MO can locate their negative charges within a larger volume, reducing electron–electron repulsion. They also present lower kinetic energy, just like a particle confined in a larger box (Chapter 1).

It is worth noting that the overlap integral S cannot be exactly zero even when the nuclei are separated, since the surface boundaries of the orbitals do not represent a 100% probability of finding the electron. On the other hand, the overlap integral S cannot be exactly 1, since as the internuclear distance decreases, the repulsive nucleus–nucleus and electron–electron interactions

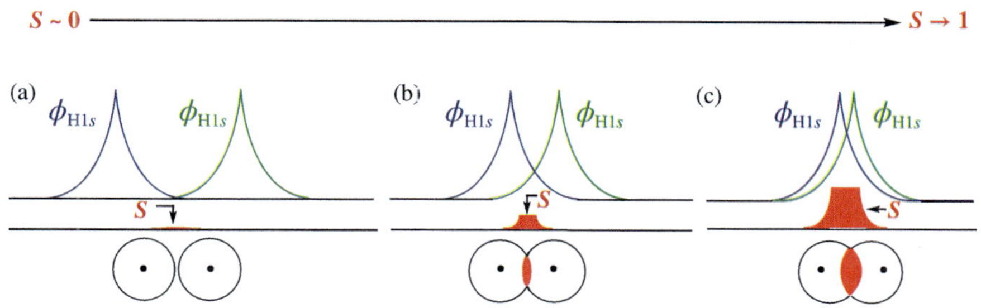

FIGURE 3.2 Schematic illustration of the meaning of orbital overlap and the overlap integral S. Mathematically, S is the area under the curve. The dots represent nuclei.

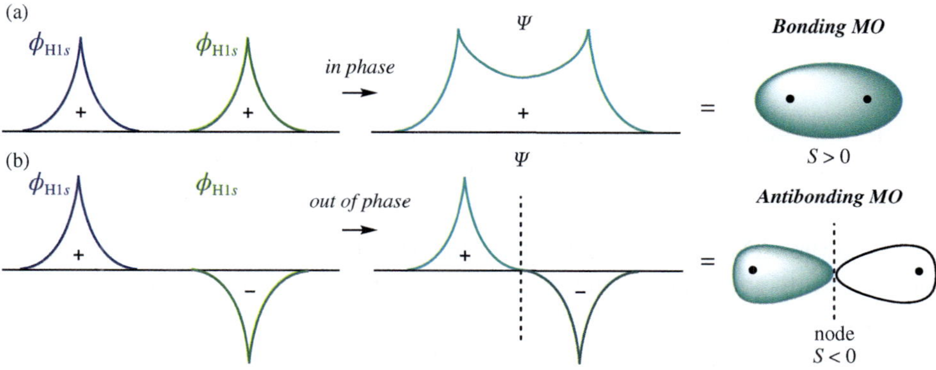

FIGURE 3.3 *In-phase* and *out-of-phase* overlap. The dots represent the hydrogen nuclei.

also increase, so that bond lengths of minimum energy result from the balance where the overlap is maximum and the internuclear and interelectron repulsion is minimum (Figure 3.3a). Therefore, it is correct to say that $0 < S < 1$.

Negative values of the overlap integral S result when the combined wave functions are of opposite sign (Figure 3.3b). In this case, the resulting wave function Ψ presents a zero amplitude in the internuclear region. This is called destructive or *out-of-phase* overlap. In fact, this situation is energetically unfavourable relative to isolated AOs, and actually when $S < 0$ we are dealing with an anti-bonding MO. An anti-bonding MO is higher in energy relative to the AOs that form it. With most of the electron density outside the internuclear region and a node between the nuclei, the nucleus–nucleus repulsion increases. When an electron occupies this orbital, it is largely excluded from the internuclear region and presents higher energy by comparison with the case when that electron is located in one isolated AO.

In the same way that the combination of n atomic orbitals forms n hybrid orbitals (Chapter 2), the combination of n atomic orbitals gives rise to the same number n of MOs. Thus, taking for example the bond formed by two hydrogen atoms (two $1s$ atomic orbitals), two MOs are formed: the bonding orbital $\Psi_1 = \phi_{1s} + \phi_{1s}$ and the anti-bonding orbital $\Psi_2 = \phi_{1s} - \phi_{1s}$. An energy-level MO diagram is a practical way of illustrating the relationship between the energy of

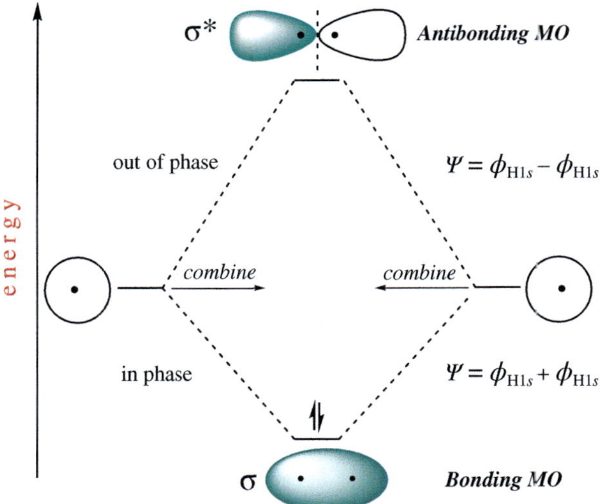

FIGURE 3.4   Energy level molecular orbital diagram for the *in phase* and *out of phase* combination of two 1s orbitals. The dots represent the hydrogen nuclei.

MOs and the AOs from which they are formed (Figure 3.4). The vertical scale corresponds to the system's energy. The two original AOs are shown on the left and right sides, and the MOs resulting from their *in-phase* and *out-of-phase* combination are shown in the middle. Note that the bonding orbital presents lower energy than the original 1s orbitals, whereas the anti-bonding orbital presents higher energy.

To construct the energy level MO diagram, one first places all possible MOs in a qualitative manner with a basis on the starting valence atomic orbitals AOs. Then, using the principle of atomic construction, one accommodates the valence electrons in the available MOs, according to the procedure: (i) The electrons are first placed in the lowest energy MOs and subsequently in orbitals of increasing energy. (ii) According to the Pauli Exclusion Principle, there can be up to two electrons in each of the MOs. When two electrons are in a MO, they must be paired (that is, they present opposite spin). (iii) When there is more than one MO with the same available energy, the electrons occupy them individually and adopt parallel spins (Hund's rule).

The MOs are assigned labels for proper identification, providing pertinent information about the symmetry properties of their wave functions. Both the bonding and anti-bonding MOs derived from 1s atomic orbitals are cylindrically symmetric about an imaginary line connecting the nuclei (along the z-axis). This means that if one looks end-to-end at the MOs in Figure 3.5, one can rotate them around the axis between the two atoms at any angle and the result will be indistinguishable from the initial view. MOs with this property are said to present σ symmetry. Anti-bonding orbitals are marked with a superscript star σ* (Figure 3.5).

As it can be seen in Figure 3.6, the formation of the hydrogen molecule H_2 from two isolated hydrogen atoms is energetically rather favourable because

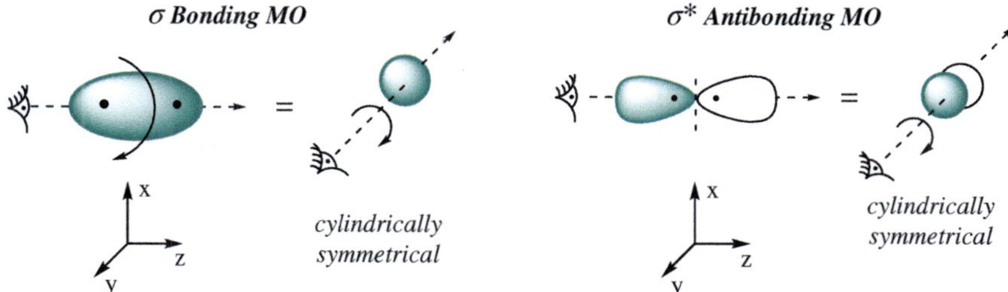

FIGURE 3.5 Symmetry of the sigma and sigma* orbitals. Rotation around the internuclear axis does not result in modified wave functions, thus the MO orbital is symmetric with respect the internuclear axis. The dots represent nuclei.

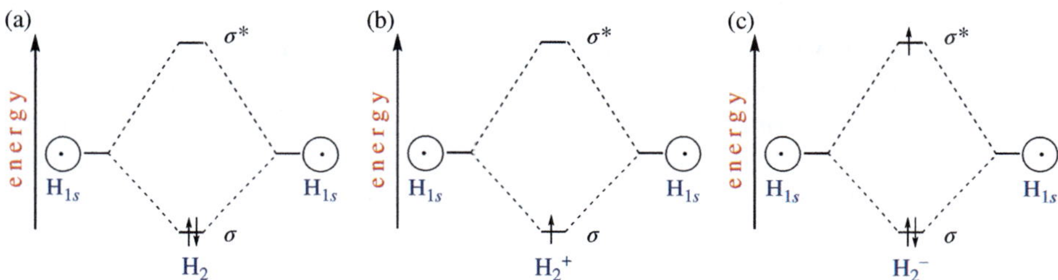

FIGURE 3.6 MO diagram for H_2, H_2^+ and H_2^- molecules.

upon bond formation the two electrons will occupy the lowest energy MO (σ); the resulting configuration is written σ^2 (Figure 3.6a). By contrast, the H_2^+ ion is less stable than the H_2 molecule because the H_2^+ molecule presents only one electron in the bonding level (σ^1, Figure 3.6b). One can therefore expect H_2 to be more strongly bonded than H_2^+, and indeed the experimental dissociation energy of H_2 is 103 kcal mol^{-1}, whereas that of the ion H_2^+ is only 61 kcal mol^{-1}. Similarly, in the H_2^- anion, the increase in energy caused by the electron occupying the $\sigma*$ level cancels out the decrease in energy of one of the electrons in the σ orbital ($\sigma^2\sigma*^1$ configuration, Figure 3.6c). This analysis is experimentally confirmed by the observation that the bond lengths in H_2^+ and H_2^- are longer than in the neutral molecule H_2, i.e. the bonds are weaker in the ionic species.

In the hypothetical molecule He_2, the available four electrons fill the bonding and anti-bonding orbitals ($\sigma^2\sigma*^2$). Consequently, the stabilisation gained by the electron pair in the bonding MO is cancelled by the destabilisation of the electron pair in the anti-bonding MO (Figure 3.7a). In fact, the anti-bonding MO increases the molecule's energy more than the binding MO lowers it. Experimentally, the available evidence points to the non-existence of a covalent molecule of He_2. In this regard, one early triumph of MO theory was its ability to predict the existence of He_2^+ and the unlikely existence of He_2 (Figure 3.7b). The three electrons in He_2^+ are distributed so that a pair of electrons occupy the σ MO and a single electron is located in the $\sigma*$ MO ($\sigma^2\sigma*^1$ configuration).

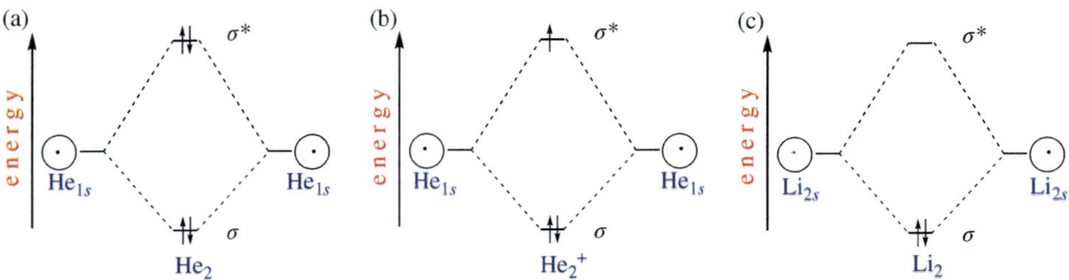

FIGURE 3.7   Energy level MO diagrams for the hypothetical He_2 molecule, ionic He_2^+ and Li_2 species.

Despite the weak bond present in He_2^+, this ion does exist. In fact, this species has often been observed experimentally.

In the case of MO formation from higher period atoms, one often ignores the inner electrons. Indeed, generally only the outer (valence) AOs interact efficiently to form MOs. For example, in the case of Li_2, the two $2s$ valence electrons with opposite spins fill the binding MO (σ), leaving the anti-bonding σ^* MO empty (Figure 3.7c). Li_2 has also been observed experimentally.

MIXING OF p ORBITALS

In a diatomic molecule, the z-axis is conventionally taken to correspond to the internuclear axis, so that the three p orbitals are no longer equivalent. In particular, the $2p$ orbitals can combine for bond formation in two different ways: one p orbital of each atom can overlap *end-to-end*, but the other two p orbitals of each atom must combine *side-by-side* (Figure 3.8).

Just as the two MOs in the molecule of H_2 result from the *in-phase* and *out-of-phase* mixing of two s orbitals, two MOs also result from the mixing of p_z orbitals in an *end-to-end* mode. The *in-phase* combination results in a rich electron density zone between the nuclei. Filling this orbital with electrons will lead to an attraction between the atoms and a bond will be formed (Figure 3.9, bottom). By contrast, the *out-of-phase* combination results in the formation of a nodal surface in the internuclear region, so that the electron density at this point is zero and the internuclear repulsion gives the orbital a higher energy (Figure 3.9, top). Both MOs have cylindrical symmetry and are therefore called σ and σ^* orbitals. By contrast, a bond formed by filling a MO arising from the interaction of two end-to-end p orbitals is called a σ_p bond (to distinguish it from the σ bonds formed by s orbitals).

When the $2p$ AOs $2p_x$ or $2p_y$ approach each other with an orientation perpendicular to the bond axis, *in-phase* and *out-of-phase* give the bonding and anti-bonding MOs shown in Figure 3.10. Their symmetry is not cylindrical. Rotating the orbitals 180° around the axis between the nuclei gives the opposite phase MOs. When an orbital presents this symmetry property, it is labelled π for

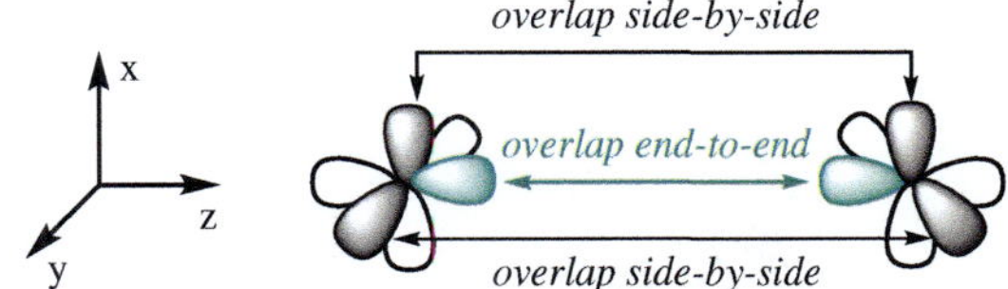

FIGURE 3.8 Combination modes of *p*-orbitals.

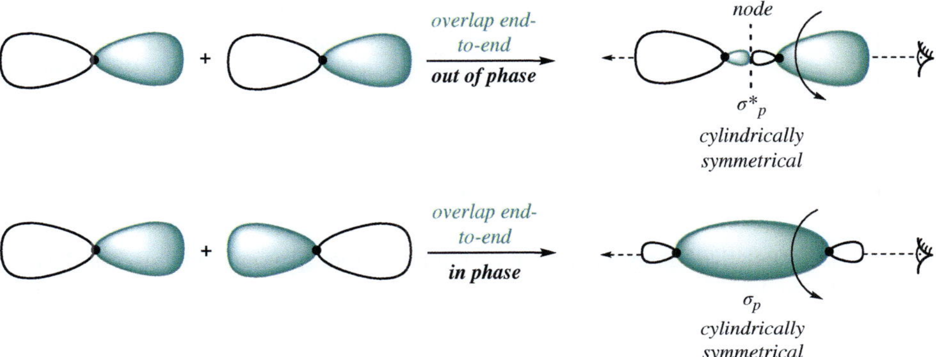

FIGURE 3.9 The mixing of p_z orbitals in an *end-to-end* mode and their cylindrical symmetry.

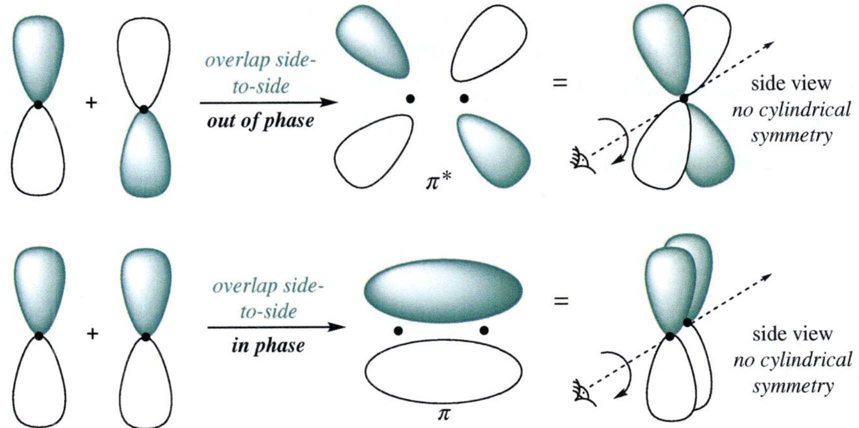

FIGURE 3.10 The combination of AOs ($2p_x$ or $2p_y$) with an orientation perpendicular to the bond axis. Lateral view showing that there is no cylindrical symmetry about the interatomic axis. The dots represent nuclei.

the bonding orbital and π^* for the anti-bonding orbital. The π-bonded MO concentrates the electron density in the region between the two nuclei, but unlike a σ-bonded MO, the concentration is above and below the internuclear axis rather than along it (Figure 3.10, bottom). Thus, a π-bonding MO is generally less strongly bonding than a σ bonding MO. The lateral interactions of the $2p_x$ and

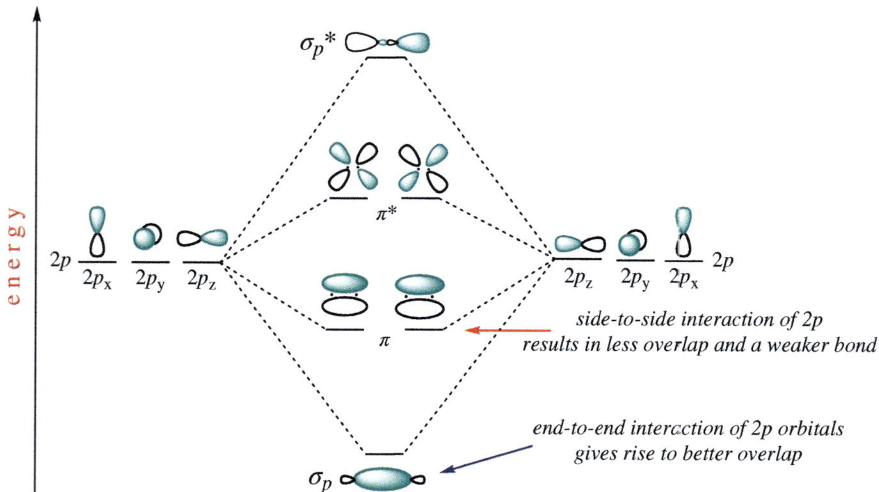

FIGURE 3.11 Relative energy diagram of MOs formed from $2p$ atomic orbitals. The dots represent the nuclei.

$2p_y$ orbitals generate a pair of mutually perpendicular and degenerate (that is, or equal energy) π orbitals and, in turn, a pair of mutually perpendicular degenerate π^* orbitals (Figure 3.10, top).

Two p-orbitals overlap more effectively by end-to-end rather than side-to-side combination. Therefore, σ_{2p} MOs are usually lower in energy than the π_{2p} MOs. By the same token, the destabilising effect of the σ^*_{2p} MO is greater than that of the π^*_{2p} MO. Thus, the energy order for MOs derived from $2p$ orbitals is typically $\sigma_{2p} < \pi_{2p} < \pi^*_{2p} < \sigma^*_{2p}$. Figure 3.11 shows a relative energy diagram of MOs formed from $2p$ AOs.

FACTORS AFFECTING THE MAGNITUDE OF ORBITAL INTERACTIONS

Symmetry

AOs must present the same symmetry to interact efficiently. For example, taking the internuclear axis at the z-direction, s and p_x orbitals cannot combine (Figure 3.12a). The constructive overlap is cancelled by an equal amount of destructive overlap, so these two AOs cannot interact to form MOs. In terms of symmetry, the s orbitals are symmetric with respect to the yz-plane, which means that when the wave function is reflected in this plane there is no change in the wave function sign. By contrast, reflection of the $2p_x$ orbital in this plane results in a change of sign, which implies its antisymmetric character. The overlap integral is necessarily zero because the two orbitals present different symmetry with respect to this plane. The same effect is found when trying to combine orbitals p_x with p_y, or p_x with p_z (Figure 3.12a). By contrast, Figure 3.12b shows AOs with the correct symmetry and therefore their overlap does lead to the formation of bonding MOs.

Energy Match from Different AOs

Figure 3.13 shows how the energy of the resulting MOs is affected as the difference in energy between the original AOs increases. When the AOs are of the same or similar energy, the anti-bonding MO and the bonding MO are significantly above and below the energies of the AOs from which they are formed (shown schematically as **A** and **B**, Figure 3.13a). In these cases, the contributions of these OAs will be equal in the bonding MOs, while they will have equal but opposite contributions in the anti-bonding MOs. By keeping the energy of orbital **A** constant while decreasing the energy of orbital **B**, it can be seen that as the energy difference between the two AOs increases, the bonding MO moves closer to the lower energy AO (**B**), while the anti-bonding MO moves closer in energy to the higher energy AO (**A**) (Figure 3.13b). The bonding MO becomes more dominated by the AO contribution of **B**, and the anti-bonding MO becomes more dominated by the contribution of the **A** atomic orbital. If the energy difference of the AOs is too large, the two MOs present essentially the same energy as the two original AOs (Figure 3.13c). In this situation there is

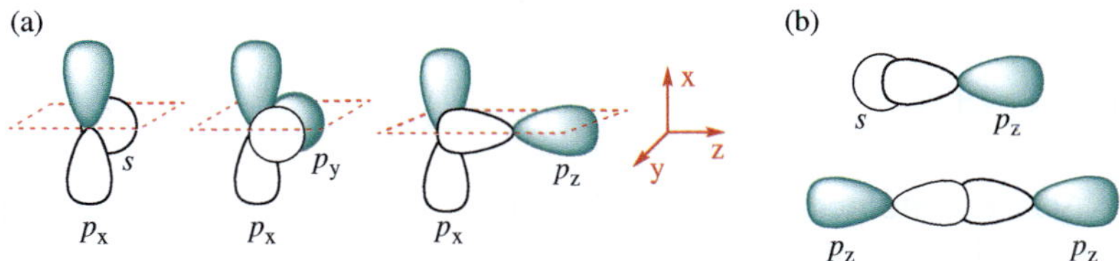

FIGURE 3.12   (a) The constructive overlap is cancelled out by the destructive overlap so that the overlap integral is zero. (b) AOs with the correct symmetry combine constructively during overlap, leading to the formation of bonding MOs.

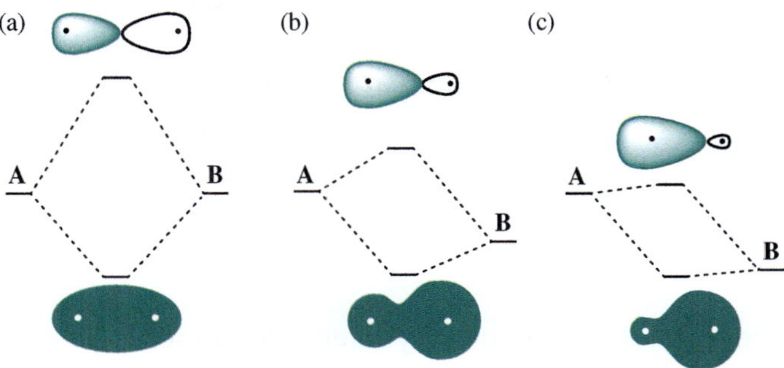

FIGURE 3.13   Effect of energy difference between AOs (**A** and **B**) upon formation of MOs. The dots represent nuclei.

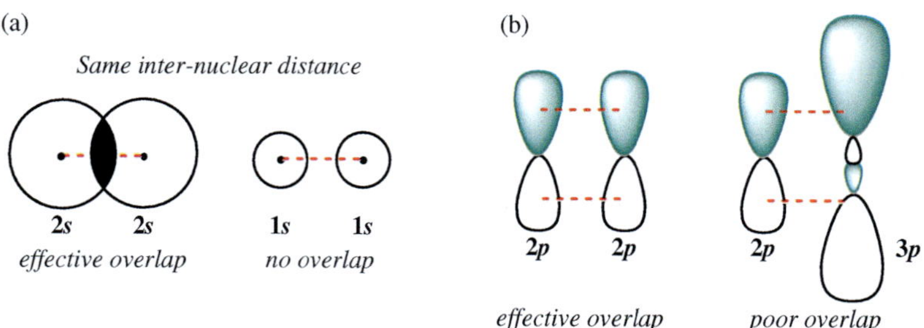

FIGURE 3.14 Influence of AO size on the relative efficiency of orbital overlap. The dots represent nuclei.

very little interaction between them; the bonding MO is a very slightly modified orbital of the more stable, less energetic AO (**B**), while the anti-bonding MO is an orbital very similar in shape to the less stable, more energetic AO (**A**).

AOs of the same kind present dissimilar energy when located in different atoms. The difference originates from the fact that the opposite charge attraction between the electron in a particular orbital and the atomic nucleus depends on its nuclear charge. For instance, a *2p* orbital in oxygen and a *2p* orbital in nitrogen do not present the same energy. The more electronegative an atom is, the more it attracts the atomic electrons; thus, the lower the energy of its AOs, the stronger the nucleus–electron electrostatic attraction will be and the more stabilised the electrons will be.

Size of the Orbitals

For two AOs to interact efficiently, having similar energies is not the only determining criterion. Another important factor is the size of the AOs. To achieve maximum overlap, the orbitals need to be of similar size. For example, two *2s* orbitals will overlap significantly at a typical internuclear distance for second-period elements. However, at that distance, two *1s* orbitals do not overlap and so they are not able to form MOs (Figure 3.14a). On the other hand, Figure 3.14b shows how a *2p* orbital overlaps more efficiently with another *2p* orbital rather than with a larger *3p* orbital. So, although the orbitals have the appropriate symmetry to interact, their size also influences the degree of effective overlap.

BONDING IN HOMO-DIATOMIC MOLECULES

Many properties of MOs are adequately interpreted in terms of electronic configurations. For example, according to Lewis or Valence-bond theory, molecular O_2 should be diamagnetic, which is contrary to the experimental evidence.

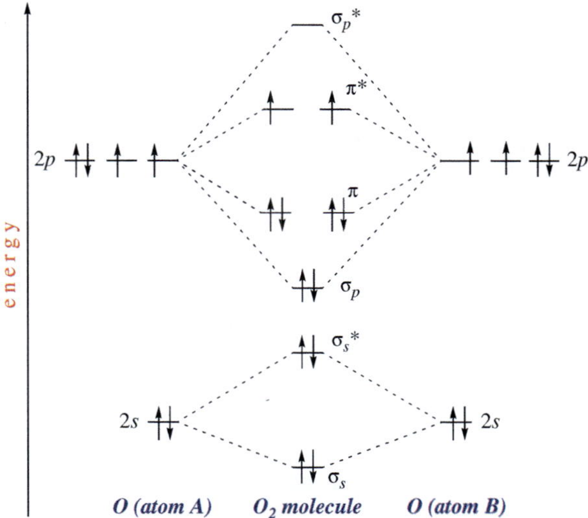

FIGURE 3.15 Energy level MO diagram for the oxygen molecule O_2.

MO theory helps us understand this issue. The electronic configuration of the oxygen atom is $[He]2s^2 2p^4$; thus, there are 12 electrons to feed an energy level MO diagram for the O_2 molecule. One obtains the ground state electronic configuration of O_2 by accommodating these 12 valence electrons (6 from each oxygen atom) in the MO diagram shown in Figure 3.15. The electrons are placed according to the MO energy levels; that is, the first ten electrons occupy the five orbitals with the lowest energy, from σ orbitals up to the π orbitals. There are two more electrons to accommodate, and the next available orbitals are the isoenergetic π^* orbitals. According to Hund's principle, each electron must occupy one of the two orbitals with parallel spins. This conclusion is a major achievement of MO theory: since the top two spins are not paired, their magnetic fields do not cancel, indicating that the molecule should be paramagnetic, exactly as observed. Therefore, the electronic configuration of O_2 is: $\sigma_{2s}{}^2 \sigma_{2s}{}^{*2} \sigma_{2p}{}^2 \pi_{2p}{}^4 \pi_{2p}{}^{*1} \pi'_{2p}{}^{*1}$.

In this context, the MO bond order corresponds to the number of electrons in bonding MOs minus the number of electrons in the anti-bonding MOs, multiplied by ½ (Eq. 3.3).

$$\tfrac{1}{2}[(\text{no. of } e^- \text{ in bonding MOs}) - (\text{no. of } e^- \text{ in anti-bonding MOs})] \quad (3.3)$$

Therefore, for the O_2 molecule, the bond order is $\tfrac{1}{2}(8-4)=2$.

MO theory also correctly predicts the non-existence of molecule Ne_2. The hypothetical Ne_2 molecule would present 16 valence electrons (eight from each neon atom). Placing these electrons in the energy diagram, it can be seen that all the MOs are full, giving a bond order of zero (Figure 3.16). Indeed, the Ne_2 molecule is too unstable to exist.

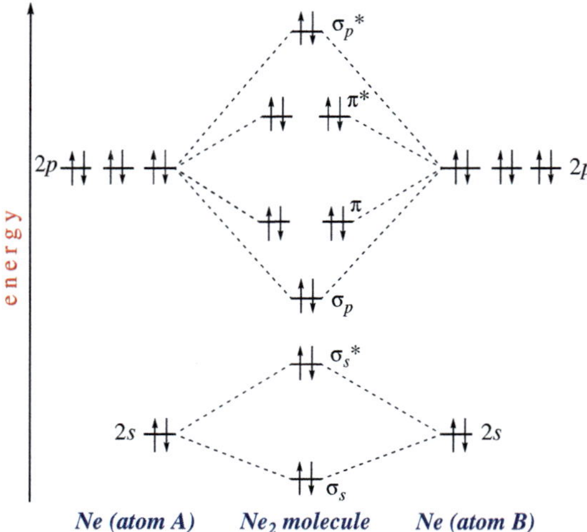

FIGURE 3.16   Energy level MO diagram for hypothetical Ne$_2$.

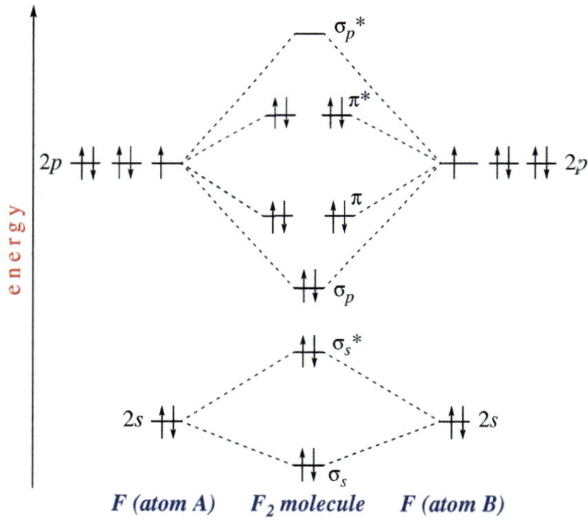

FIGURE 3.17   Energy level MO diagram for the F$_2$ molecule.

The electronic configuration of the fluorine atom is $[He]2s^2 2p^5$, so in the F$_2$ molecule there are seven valence electrons from each fluorine atom, for a total of 14 electrons. In this case, when constructing the MO diagram, it can be seen that the electrons are arranged in the paired form up to the π^* MO orbital, meaning that F$_2$ should be diamagnetic (Figure 3.17). The electrons in the σ_p orbital are the only ones that are not balanced by electrons in the corresponding unoccupied σ_p^*. These two electrons are responsible for holding the two fluorine atoms together. The F$_2$ molecule has an order of 1 and the F—F bond is then weaker and longer than the one in O$_2$. This agrees with the experimental results; the bond energy of

F_2 is 37 kcal mol^{-1}, with a F—F bond length of 1.41 Å, while for O_2 the bond energy is 117.8 kcal mol^{-1}, and the O=O bond length is 1.21 Å.

In principle, when considering the bonding of homo-diatomic molecules in the series from Li_2 to Ne_2, the contributions from $1s$ with $1s$ orbital overlap should be included in the development of the corresponding MO energy level diagrams. However, as the nuclear charge of an atom increases, the energy level of all its AOs diminishes. This is a consequence of the orbitals becoming smaller in size under the influence of the increased nuclear charge, which results in increased stabilisation of the electrons closer to the nucleus. Thus, in molecules such as Li_2 and the hypothetical He_2, the $1s$ orbital contracts to such an extent that its size is rather small, resulting in little effective overlap with the second atom's $1s$ orbital, so this interaction can be discarded. In addition, the atoms that lie to the right side of the periodic table's second period, such as O, F and Ne, are relatively small. Thus, when electrons pair up in their $2p$ orbitals, the strong electrostatic repulsion raises the energy of the $2p$ orbitals sufficiently above the energy of $2s$ orbitals, to the point that prevents orbital mixing with $1s$ AOs. Therefore, the participation of $1s$ AOs in the corresponding MOs can be ruled out for these molecules.

By contrast, boron, carbon and nitrogen atoms are relatively large, so electron–electron electrostatic repulsion is weaker. As a result, the energies of $2s$ and $2p$ AOs are comparable, and therefore the resulting MOs will reflect some combination with each other. Consider, for example, the two σ_{2s} and two σ_{2p} orbitals (Figure 3.18, left side); they are related by symmetry and are quite similar in energy so that they will interact to some extent with each other. When they

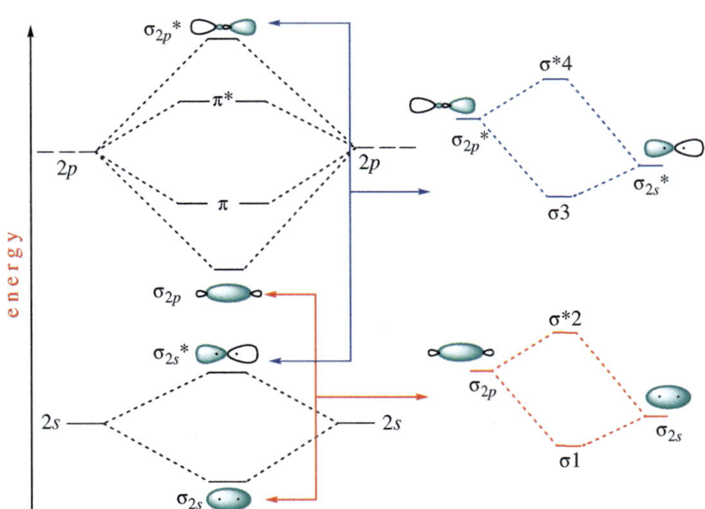

FIGURE 3.18 Mixing of the σ_{2s} and σ_{2p} orbitals lowers the energy of the σ_{2s} MO (now labelled $\sigma 1$) and raises the energy of σ_{2p} (now labelled $\sigma * 2$). The energy of the $\sigma * 2$ orbital is higher than that of the σ_{2p} orbital. Similarly, the mixing of the $\sigma_{2s}*$ and $\sigma_{2p}*$ orbitals result in a lowering of $\sigma_{2s}*$ (now labelled $\sigma 3$) and a raising of $\sigma_{2p}*$ (now labelled $\sigma * 4$).

mix, a new bonding MO as well as an anti-bonding MO are created (Figure 3.18, right side). As it was discussed earlier, when two orbitals of different energy are allowed to interact, the new bonding MO (labelled as $\sigma 1$) is closer in energy and properties to the lower energy AOs, in this case, the σ_{2s} orbital. On the other hand, the new anti-bonding MO ($2*\sigma$) is closer in energy to the higher energy AOs (σ_{2p}). The $\sigma*_{2s}$ and $\sigma*_{2p}$ anti-bonding orbitals exhibit also the correct symmetry to interact with each other (Figure 3.18, right side). This interaction modifies and lowers the energy of the $\sigma*_{2s}$ anti-bonding orbital, giving rise to the $\sigma 3$ orbital, and modifies and raises the energy of the $\sigma*_{2p}$ ($\sigma*4$) orbital. As a result, it is no longer possible to think of σ orbitals made up exclusively of $2s$ or $2p_z$ orbitals; rather, all four orbitals (two σ_{2s} and two σ_{2p}) must be used to form the four MOs. The overall effect is to decrease the energy of the σ_{2s} and $\sigma*_{2s}$ MOs and to increase the energy of the σ_{2p} and $\sigma*_{2p}$ MOs.

A salient consequence is the reversal in energy order for the σ_{2p} and π_{2p} MOs: $\pi_{2p} < \sigma_{2p} < \pi*_{2p} < \sigma*_{2p}$ (Figure 3.19).

With the updated arrangement of relative MO energies, one can construct adequate MO energy diagrams for the boron B_2, carbon C_2 and nitrogen N_2 homo-diatomic molecules. The boron molecule B_2 presents six valence electrons, four of them occupying the σ_{2s} and $\sigma*_{2s}$ MOs (Figure 3.20a). The remaining two electrons occupy the two π_{2p} MOs, one electron in each orbital and adopting parallel spins (Hund's rule). With four electrons occupying bonding MOs and two electrons occupying anti-bonding MOs, the bond order for boron B_2 is equal to 1. As expected from a molecule presenting two lone electrons in

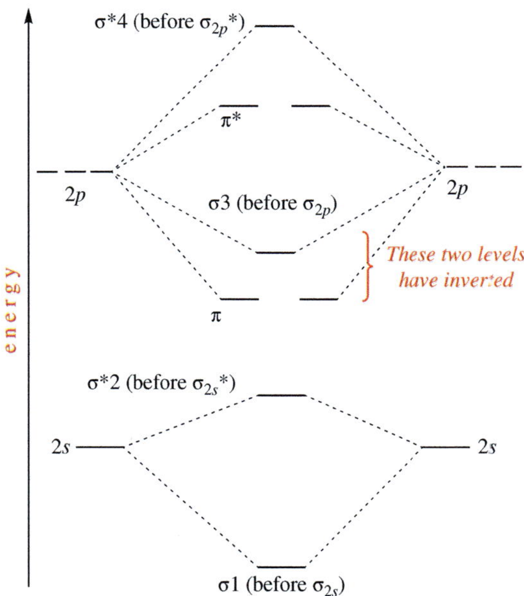

FIGURE 3.19   Energy level MO diagram as a result of mixing the orbitals σ_{2s} and σ_{2p} with orbitals $\sigma_{2p}*$ and $\sigma_{2p}*$.

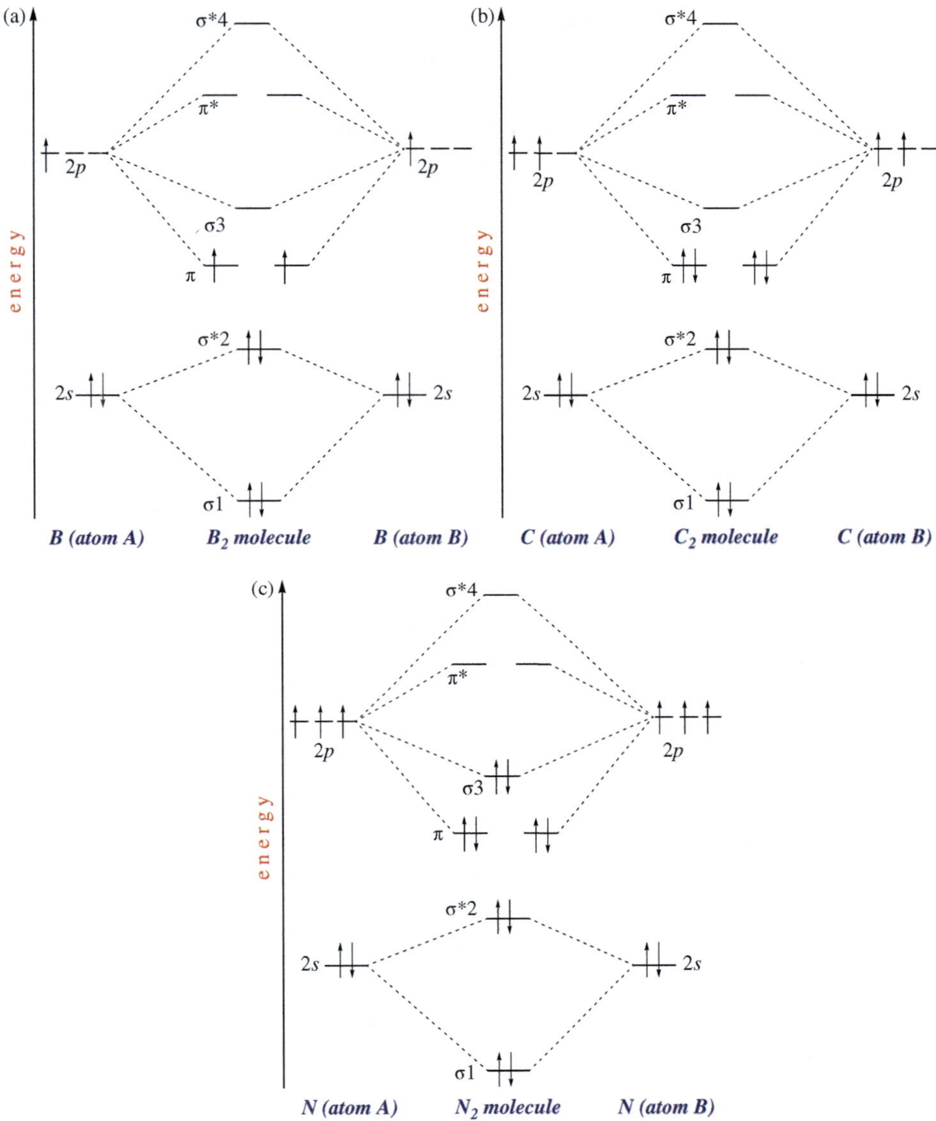

FIGURE 3.20 MO energy diagrams for (a) boron B_2, (b) carbon C_2 and (c) nitrogen N_2.

the upper energy semi-occupied MOs, B_2 is paramagnetic with a bond energy of 69.3 kcal mol^{-1} and a bond length of 1.59 Å. B_2 is not an isolable compound, but it has been detected spectroscopically.

In the case of the carbon homo-diatomic molecule C_2 with eight valence electrons, the two highest energy MOs are labelled as π_{2p} (Figure 3.20b). With two additional bonding electrons relative to B_2, the bond order of C_2 is equal to 2 and consequently, the C=C bond is stronger and shorter with bond energy equal to 148 kcal mol^{-1} and a bond length equal to 1.31 Å. C_2 is diamagnetic because all its electrons are paired.

In the case of nitrogen N_2 with ten valence electrons filling the bonding MOs (Figure 3.20c), the bond order is equal to 3, which is consistent with a triple bond $N \equiv N$ representation in the Lewis structure. Accordingly, the bond energy is equal to $226.1\,kcal\,mol^{-1}$, which is in line with a much stronger bond relative to the $C=C$ double bond in carbon C_2. Furthermore, the N_2 molecule is, as predicted, diamagnetic.

BONDING IN HETERO-DIATOMIC MOLECULES

Because of intrinsic differences in electronegativity, the interatomic bond in heteronuclear diatomic molecules is polar, with the electrons shared unequally by the two bonded atoms. Equation (3.1) can be rewritten as a linear combination of atoms **A** and **B** (Eq. 3.4):

$$\psi = C_A \phi_A + C_B \phi_B \tag{3.4}$$

where the coefficients C_A and C_B indicate the contribution of the corresponding atomic orbital to the MO. Taking the squared wavefunctions to be interpreted as electron density probabilities, a larger $C_A{}^2$ will indicate that the MO ψ is more similar to the atomic orbital of **A**, ϕ_A, and that the electron density will be higher near **A**. When $C_B{}^2$ is larger, then the MO is more similar to the atomic orbital of **B**, ϕ_B, and the electron density is higher near the **B** atom. In general, as explained earlier, the atom with the lowest atomic orbital energy dominates the corresponding MO and the electron density at the bonding orbitals is higher at this atom.

To find the appropriate ground state electron configuration of diatomic heteronuclear molecules, the same approach used for the analysis of diatomic homonuclear molecules is followed. Nevertheless, the MO energy diagrams must be modified to take into account differences in electronegativity among the bonded atoms. The MO diagram for nitric oxide (NO) is shown in Figure 3.21.

Because the AOs of the more electronegative oxygen atom are lower in energy by comparison with those in the less electronegative nitrogen atom, the energy level MO diagram of NO is dissymmetric. The MO energy diagram must accommodate 15 electrons (seven from nitrogen and eight from oxygen). Note that each bonding orbital is closer in energy to that present in the dominant contributing orbital. Similarly, each anti-bonding orbital is closer in energy to the dominant contributing orbital in N relative to the contributing orbital in O. This leads to a dissymmetry of the MOs: all the bonding orbitals show higher contributions from the oxygen AOs, while the anti-bonding orbitals show higher contributions from the nitrogen AOs. As it would be expected from a comparison of the electronegativities of N and O, the diagram includes eight electrons in the bonding orbitals and three electrons in the anti-bonding orbitals, and the overall electron distribution is polarised towards oxygen. As the valence electrons fill

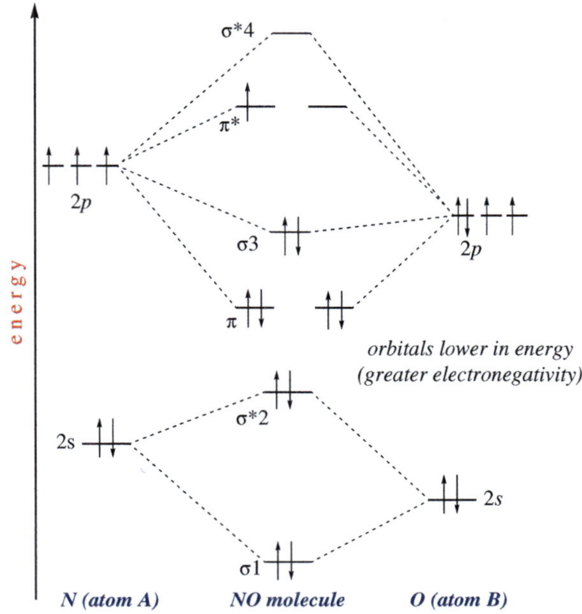

FIGURE 3.21   The energy level MO diagram for nitric oxide (NO).

MOs, the lone electron occupies one of the π^*_{2p} orbitals. The nitrogen $2p$ orbitals contribute more to this orbital, so the lone electron spends more time near N. There are eight bonding electrons and three anti-bonding electrons, giving a bond order of $\frac{1}{2}(8 - 3) = 2.5$, which is in line with the experimental bond dissociation energy of $150\,\text{kcal}\,\text{mol}^{-1}$ and a bond length of $1.14\,\text{Å}$. The single unpaired anti-bonding electron makes the molecule chemically very reactive and paramagnetic. As anticipated, when this electron is removed to form the positively charged NO^+ ion, the bond order increases to 3.0, and the bond length in the NO^+ ion shortens to $1.06\,\text{Å}$.

Figure 3.22 shows the MO diagram for the HF molecule. Given the large electronegativity difference between fluorine and hydrogen, one can expect the bonding orbital to be mainly composed by the fluorine AO and the anti-bonding orbital to show a dominant contribution by the hydrogen AO. The hydrogen $1s$ orbital is energetically well above fluorine's $2s$ orbital and, to a lesser extent, the $2p$ orbital. As a consequence, even though symmetry does allow two s-type orbitals to overlap, there is virtually no interaction between the hydrogen $1s$ orbital and the fluorine $2s$ orbital. Nevertheless, owing to the small energy difference and the symmetry match, the hydrogen $1s$ orbital can mix with the fluorine $2p_z$ orbital. The result is the formation of a bonding level and an anti-bonding level. Note that the $2p_x$ and $2p_y$ orbitals at fluorine cannot interact with the $1s$ orbital of hydrogen because they no longer have the same symmetry. The two filled $2p_x$ and $2p_y$ orbitals at F are non-bonding MOs they present energies that are more similar to those of the isolated AOs. Since the σ^*

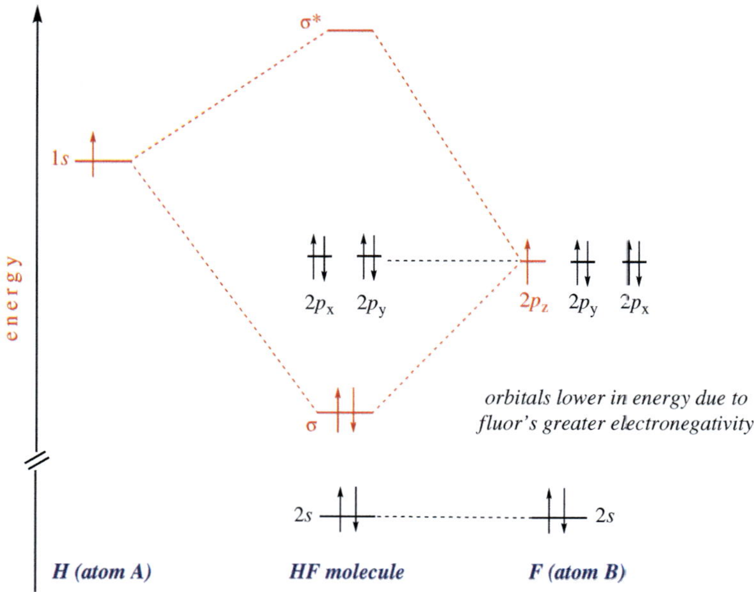

FIGURE 3.22   Energy level MO diagram for HF.

orbital is largely a $1s$ orbital of the hydrogen's AO, and the σ MO itself is largely a $2p_z$ atomic orbital located at F, this is consistent with considerable electron transfer from H to F. This results in a highly polarised molecule, with F carrying a partial negative charge and H a partial positive charge.

In the case of the LiH molecule, H is more electronegative than Li. One can ignore the $1s$ orbital on lithium as it is too low in energy to interact significantly with any other AO. Thus, there are two AOs that can interact to form LiH, the hydrogen $1s$ orbital and the lithium $2s$ orbital. The interaction of these two orbitals gives rise to a bonding σ MO and an anti-bonding σ^* MO. The σ MO is low in energy whereas σ^* MO is above the $2s$ orbital at Li. The σ bonding MO consists of an unequal mixture of the two bonding AOs, with the largest contribution from the $1s$ AO at H, as it is closer in energy to this MO. By contrast, the $2s$ AO at Li turns out to be the largest contributor to the σ^* anti-bonding MO. Therefore, most of the electron density in the LiH ground state molecule will be located at hydrogen. As shown in Figure 3.23, LiH has only two valence electrons located at the σ MO orbital. With both electrons located in the σ bonding orbital and no electrons located in the anti-bonding σ^* orbital, the molecule is predicted to be stable with respect to the dissociated atoms. Although LiH is usually found as a crystalline solid, it also exists in the gas phase and has been studied experimentally. The dissociation energy of LiH is $56\,\text{kcal mol}^{-1}$, its bond length is 1.60 Å and presents a large dipole moment of 5.9 D.

An energy level diagram for carbon monoxide (CO) must consider the arrangement of ten valence electrons, four from carbon ($[He]2s^22p^2$), and six

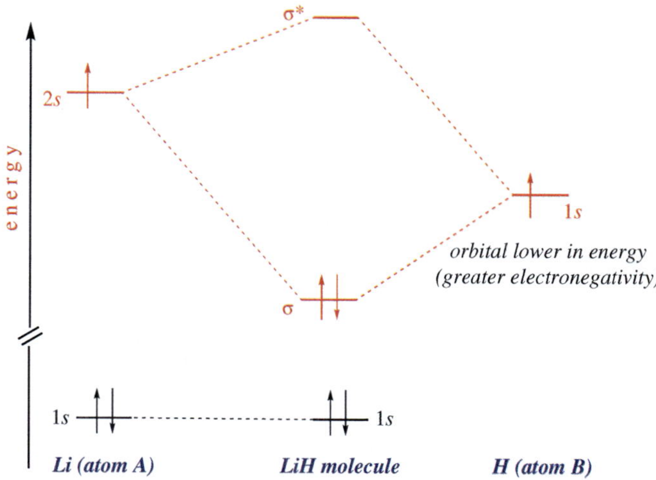

FIGURE 3.23   MO energy diagram for LiH.

from oxygen ($[He]2s^2 2p^4$). The energy level MO diagram is shown in Figure 3.24a. Because of the significantly large nuclear charge of oxygen (being more electronegative than carbon), its $2s$ and $2p$ orbitals are lower in energy relative to the $2s$ and $2p$ orbitals of carbon. The shape of the diagram is therefore slightly dissymmetric. The arrangement of the energy levels in the diagram is similar to the energy level diagram for N_2 (Figure 3.20c), which is also a 10-electron system (they are isoelectronic molecules). The difference is in the absolute values of the different energy levels. There are six electrons occupying the $\sigma 2$ and π bonding MOs. The $\sigma 2$ level and the isoenergetic (degenerate) pair of π MOs are clearly bonding MOs and have a larger contribution by the oxygen atom. Unfortunately, the MO diagrams do not show explicitly the individual contributions of the carbon and oxygen AOs to the four MOs. On one hand, two electrons occupy the $\sigma 1$ MO with a large contribution from the oxygen AO, making it mainly a non-bonding orbital mostly located on oxygen. On the other hand, in the $\sigma 3$ MO carbon is the major contributor, and it corresponds essentially to a carbon-based lone pair. The π^* orbitals are unoccupied.

From the point of view of coordination chemistry, in CO the filled $\sigma 3$ and the empty π^* MOs are critical for its chemical properties since they are involved in bonding interactions with various d-block metal orbitals. When acting as a ligand for metal centres, CO can form a σ-metal-carbon bond by donating an electron lone pair from the carbon $\sigma 3$ MO. CO can also accept π electron density from filled d-metallic orbitals into the CO π^* anti-bonding orbital. CO can therefore be considered as a σ Lewis base and a π Lewis acid capable of stabilising low oxidation state metals. This type of π-bonding is referred to as π back-bonding. Figure 3.24a shows the MO diagram for CO formation built from the energy level MO diagram of N_2, Figure 3.24b shows a MO diagram where the contribution of

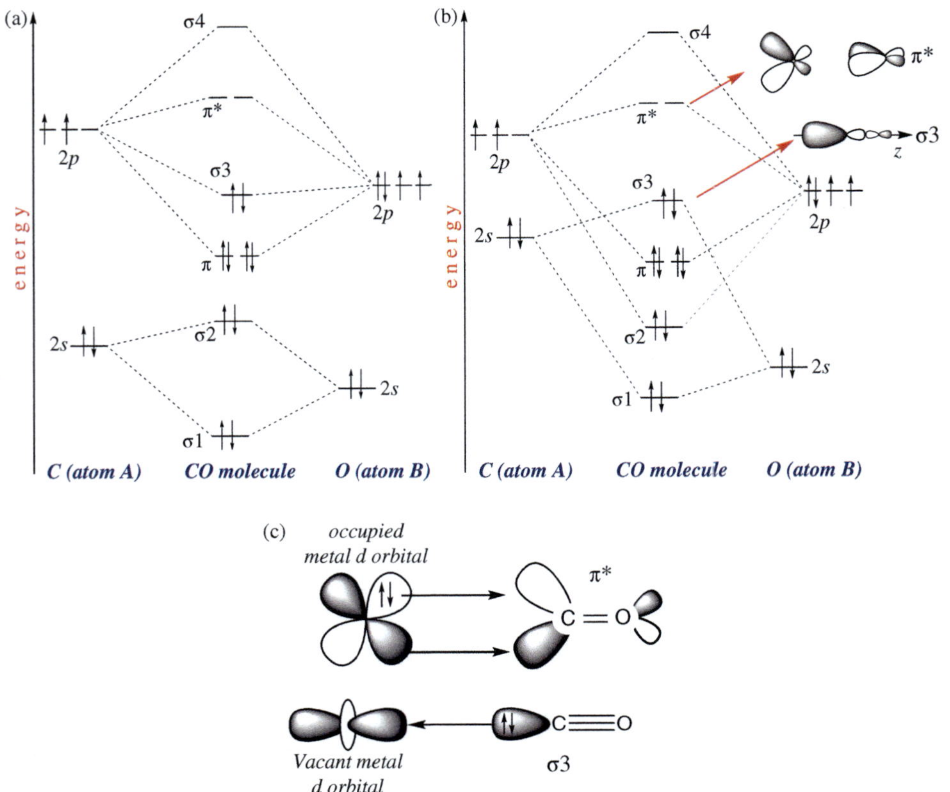

FIGURE 3.24   (a) and (b) Energy level MO diagram of CO and (c) metal–CO coordination.

the AOs to all MOs are better appreciated, and Figure 3.24c a MO interpretation of π back-bonding.

BONDING IN TRIATOMIC MOLECULES

The interpretation of MOs as LCAO can of course be applied to more complex polyatomic molecules. The aim in this section is to present in graphical manner the principles of MO theory when applied to several simple triatomic molecules, giving qualitative information about which orbitals can overlap constructively in the formation of these molecules, and then to understand their properties in a general way. The reader should be aware that the formal determination of the shape and energy of the correct MOs of more complex molecules requires the application of quantum mechanical calculations and experimental methods. In this regard, the identification of symmetry elements is crucial in MO theory. The assignment of symmetry to the orbitals of interest allows to predict which of the orbitals are likely to overlap during the course of a reaction; nevertheless, symmetry analysis does not provide information about the shape of the resulting molecule. In this sense, Valence Shell Electron Pair Repulsion (VSEPR) is the simplest method for predicting molecular shape.[1] Once the molecular shape has been determined, an analysis of the symmetry elements in the involved orbitals allows to determine which linear combination of AOs is consistent with a particular structural shape. For example, in the molecule of BeH_2 VSEPR analysis predicts a linear arrangement of the three bonded atoms, and one must establish whether the overlap of two hydrogenic $1s$ orbitals with the $2s$ and $2p$ orbitals of Be is in line with such geometry.

With four AOs to combine, a possible first step would be to work with two hydrogen atoms and analyse potential interactions with the AOs in beryllium. The starting assumption is that the wave functions for the two hydrogen $1s$ orbitals have a positive value around their nuclei that will interact with the beryllium atom in BeH_2. Orbital combination affords in-phase ($\sigma1$) and out-of-phase ($\sigma2$) orbital combinations with a small energy difference between the two combinations of the H orbitals (Figure 3.25, right side), ready to interact with the beryllium AOs (Figure 3.25, left side).

Analysis of the Symmetry Elements Present in Each Molecular Orbital

Taking the internuclear axis as the z-axis, both H orbital combinations are cylindrically symmetric relative to the yz plane, so they are labelled as 'S'. However, the symmetry element in the xy-plane, which runs vertically, is different for these orbitals. The $\sigma1$ orbital is symmetric (S) in relation to the xy-plane, but the $\sigma2$ orbital is antisymmetric (A). On the other hand, analysis of the AOs in beryllium shows that the $2s$ orbital is symmetric with respect to both symmetry elements (SS), while the $2p_x$ and $2p_y$ orbitals are AS and SS, respectively, whereas the $2p_z$ orbital is SA. It should be remembered that only orbitals presenting the same symmetry with respect to the pertinent symmetry elements in the molecular structure and in the orbitals of the components can be combined. The Be $2s$ orbital (SS) is therefore able to interact with the in-phase $\sigma1$ orbital (SS) to give the σ_s bonding MO and the σ_s* anti-bonding MO. On the other hand, the $2p_z$ orbital (SA) is able to interact with the out-of-phase $\sigma2$ orbital (SA) to form the σ_p and σ_p* MOs. The $2p_x$ and $2p_y$ orbitals do not have the correct symmetry to interact with any of the available orbitals and remain unaffected, so they become non-bonding orbitals in the Be_2H molecule (Figure 3.25).

It can be noticed that Figure 3.25 depicts the shape of the MOs in a pictorial way, showing that the energy of the orbital increases as the number of nodes increases. Unlike Lewis theory, which describes two equivalent Be—H bonds,

---

[1] The premise of VSEPR theory is that bonding and lone pairs repel each other and will therefore adopt the geometry that places the electron pairs as far apart as possible. The reader is advised to consult the bibliography provided at the end of this chapter for a better understanding of the subject.

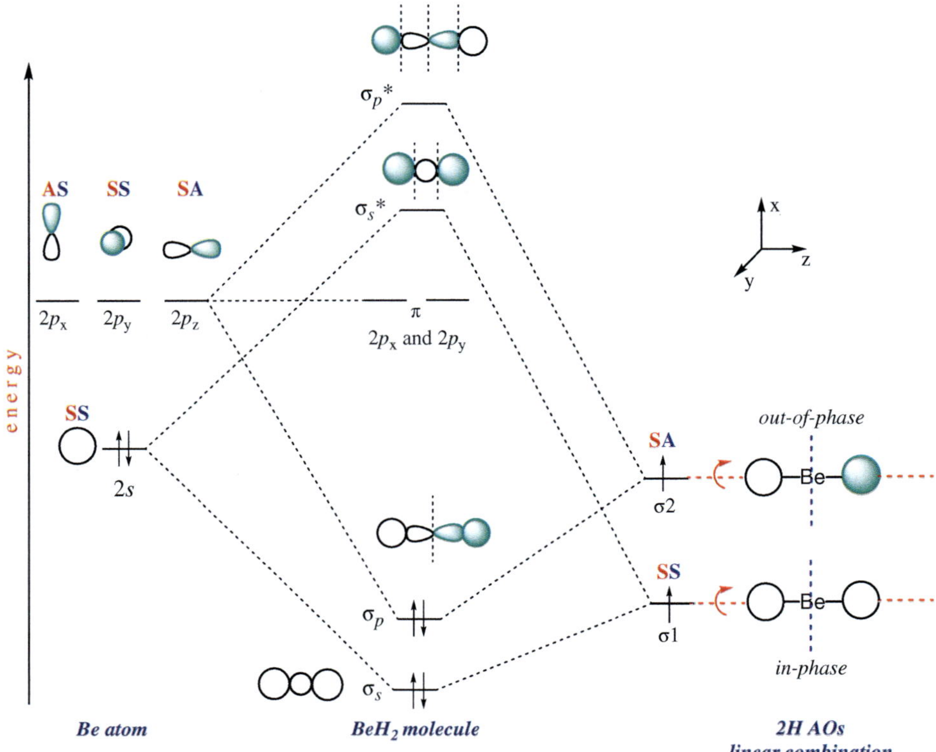

FIGURE 3.25   Energy level MO diagram for BeH$_2$.

MO theory predicts that four electrons are distributed in two sigma orbitals, giving a total of two bonds; nevertheless, these bonds are MOs delocalised on three atoms, and each MO presents a particular energy.

Spectroscopic data indicate that H_2O is 'bent', with an H—O—H bond angle of 104.5° (correctly predicted by the VSEPR method). In this geometry, the symmetry elements in the water molecule are the rotation around the x-axis (which passes through the oxygen), reflection in the mirror plane of the xy-plane, and reflection in the plane of the sheet (xz-plane).[2] As already done earlier with the BeH$_2$ molecule, it is convenient to start by analysing the interactions of the peripheral 1s hydrogens, which give rise to one in-phase and one out-of-phase combination in a bent arrangement (Figure 3.26). The next step consists in the analysis of the symmetry elements present in all the orbitals of the water molecule. For instance, regarding the out-of-phase $\sigma2$ orbital, rotation by 180° around the x-axis it is appreciated that the orbital phases are interchanged, so it is classified as antisymmetric (A). On the other hand, reflection on the xz-axis does not affect the $\sigma2$ orbital, which is then identified as symmetrical (S) with respect

---

[2] In group theory, the axis that passes through the oxygen of the H_2O molecule is the principal axis, so by convention the z-axis is taken along this direction. In this book we will arbitrarily leave the xyz-axes fixed as indicated throughout the text (with the xz-plane as the page plane).

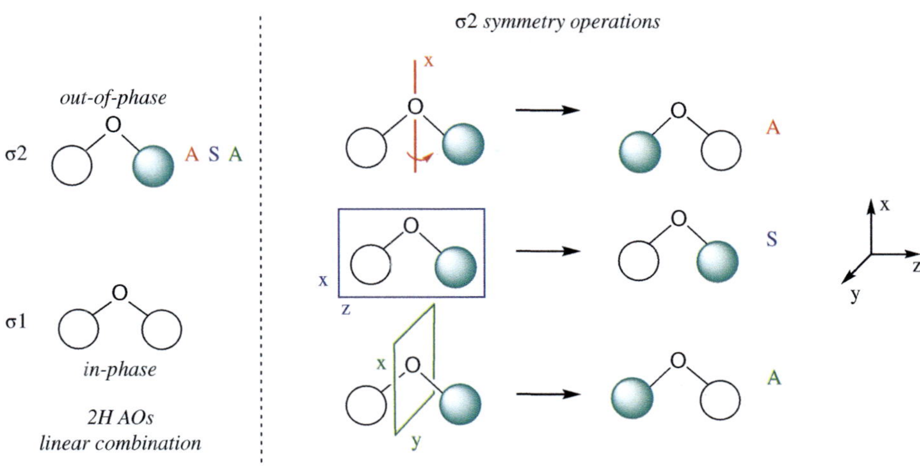

FIGURE 3.26 Symmetry of operations (rotation or planes) performed on the AOs in the water molecule.

to this operation. Finally, the $\sigma2$ orbital is antisymmetric (A) with respect to the reflection on the xy-axis that crosses the molecule passing through the oxygen atom. Thus, the $\sigma2$ orbital is labelled as ASA.

The rest of the orbitals are treated in the same way and the energy level diagram in the top of Figure 3.27 shows the symmetry of all the atomic and molecular orbitals in the water molecule. From the analysis of the diagram, it can be predicted that the $2p_y$ orbital of oxygen cannot overlap with any orbital as consequence of its AAS symmetry with respect to the three symmetry operations mentioned earlier, which does not match with any other AO. Therefore, this $2p_y$ orbital becomes a non-bonding MO. As it was the case with BeH_2, the $2p_z$ orbital (ASA) can interact by symmetry with the out-of-phase $\sigma2$ orbital, giving rise to the bonding MO labelled as σ_z and the anti-bonding MO $\sigma_z{}^*$. It is important to note that both the $2s$ and $2p_x$ orbitals on oxygen can interact with the in-phase $\sigma1$ orbital as they present the same SSS symmetry. As it was already mentioned, the combination of three AOs affords three MOs, one strongly bonding, one strongly anti-bonding and one of an intermediate nature. These three MOs are identified as σ_{sx}, σ_{sx} and $\sigma_{sx}{}^*$. The bottom of Figure 3.27 shows the interactions between the $2s$, $2p_x$ oxygen orbitals and the $\sigma1$ in-phase MO orbital.

It is known experimentally that methane exists in a tetrahedral geometry. A convenient way to visualise the MOs in this molecule is to start with the central orbital (carbon) and subsequently proceed to examine the combinations of the peripheral atomic orbitals on the four hydrogens that participate in effective overlap. To do this, the methane molecule is conveniently placed inside a cube, with the carbon atom in the centre and the four hydrogen atoms in the four corners (Figure 3.28). This is reasonable because the opposite corners of a cube describe a perfect tetrahedron. The next step is to analyse each of the $2s$ and $2p$ AOs of the carbon in the cube. The $2s$ orbital on the carbon atom can overlap in phase with the four $1s$ orbitals of hydrogen (Figure 3.28a).

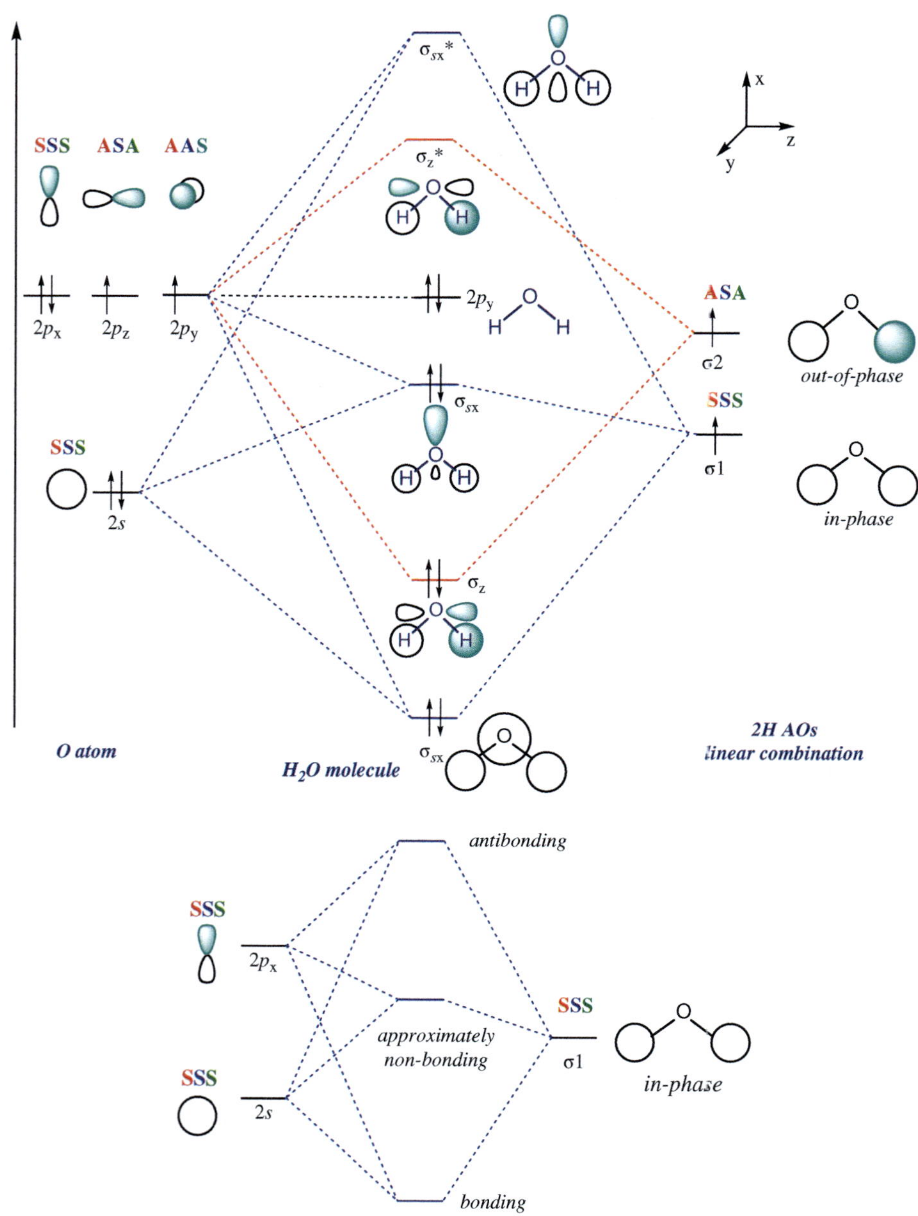

FIGURE 3.27 Top, energy level MO diagram for water. Bottom, interactions between the $2s$, $2p_x$ oxygen orbitals and the $\sigma 1$ in-phase orbital in the molecule of water.

This bonding combination corresponds to a MO delocalised over all five atoms. The corresponding anti-bonding combination is obtained by inverting the sign of the central atom's wave function (not shown). Each of the $2p$ orbitals points to opposite faces of the cube (Figure 3.28b). The four hydrogen $1s$ orbitals can also combine with the $2p$ orbitals on carbon as long as they are in phase with the central p-orbital lobes. This means that the hydrogen AOs on opposite sides of the cube present opposite phases relative to each other. This gives rise to the

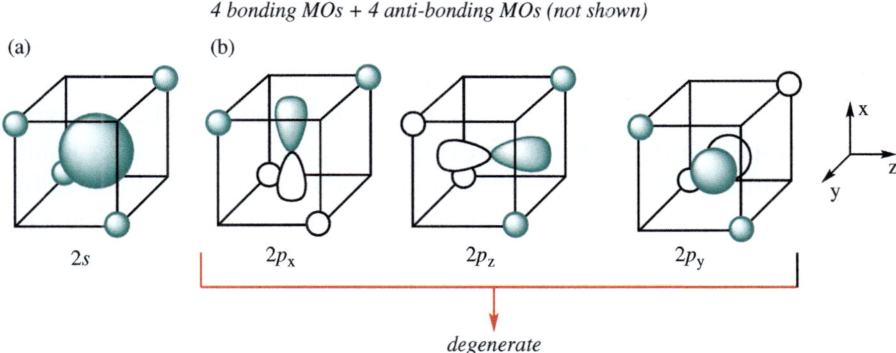

FIGURE 3.28   Molecular bonding orbitals for methane.

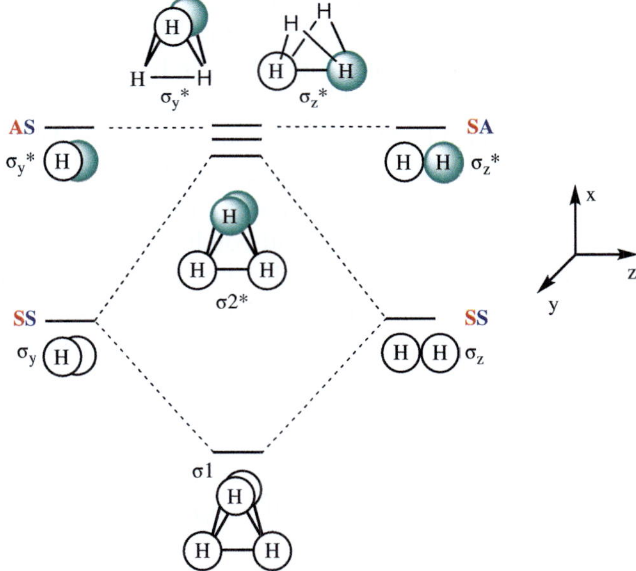

FIGURE 3.29   Energy level MO diagram for the tetrahedral lineal combination of orbitals in hypothetical H_4.

three MOs that are degenerate (isoenergetic). Together with the corresponding four anti-bonding orbitals, the total number of MOs is eight, in accordance with anticipation with the eight original AOs.

Although the above description is a simplistic way of constructing the MOs in methane, it is worth constructing a diagram from scratch for illustrative purposes. This is done by combining two 'hydrogen molecules' (or, more correctly, their wave functions) into a composite H_4 unit in a tetrahedral arrangement. For such hypothetical H_4 species, one possible combination would involve the two hydrogen 1s orbitals aligned along the z-axis, with their in-phase (σ_z) and out-of-phase (σ_z*) combinations (Figure 3.29, right side). Another anticipated AO combination takes place along the y-axis and presents in-phase (σ_y) and out-of-phase $(\sigma_y{*})$ configurations (Figure 3.29, left side). The pertinent symmetry elements are oriented on the xz- and yx-planes. The bonding orbital σ_z is symmetric with respect to both planes and is accordingly labelled SS. The anti-bonding orbital $\sigma_z{*}$ is symmetric with respect to the xz-plane but antisymmetric with respect to the xy-plane and is labelled as SA. On the other hand, the σ_y combination is symmetric with respect to both planes (SS). Finally, the anti-bonding orbital $\sigma_y{*}$ is antisymmetric with respect to the xz-plane but symmetric with respect to the xy-plane and is labelled AS. It can be appreciated that only the σ_z and σ_y orbitals present the appropriate symmetry for their combination to produce a bonding ($\sigma 1$) and an anti-bonding ($\sigma 2{*}$) orbital. The $\sigma_z{*}$ and $\sigma_y{*}$ orbitals remain unchanged and present an energy similar to the $\sigma 2{*}$ MO. In this way, four orbitals are built in a tetrahedral arrangement, which will be then combined with the valence orbitals of the carbon atom.

The orbital combinations created for the H_4 hypothetical component are then confronted with the AOs on the carbon atom (Figure 3.30). The energies of the H_4 orbitals are similar to those in an isolated $1s$ orbital in hydrogen because the four hydrogen atoms are separated by the carbon atom. One can then classify the orbitals according to their symmetry relative to the xz- and xy-planes. There are two SS orbitals on each side; however, the overlap integral of the interaction between the $2s$ orbital on carbon and the $\sigma 2{*}$ orbital is zero, since the same-phase bonding interaction cancels out with the opposite phase anti-bonding interaction. Nevertheless, there are still four possible interactions with the correct symmetry, which give rise to four bonding and four anti-bonding MOs. The bonding MOs are shown in Figure 3.30, where the lowest energy MO has all its interactions in phase and the

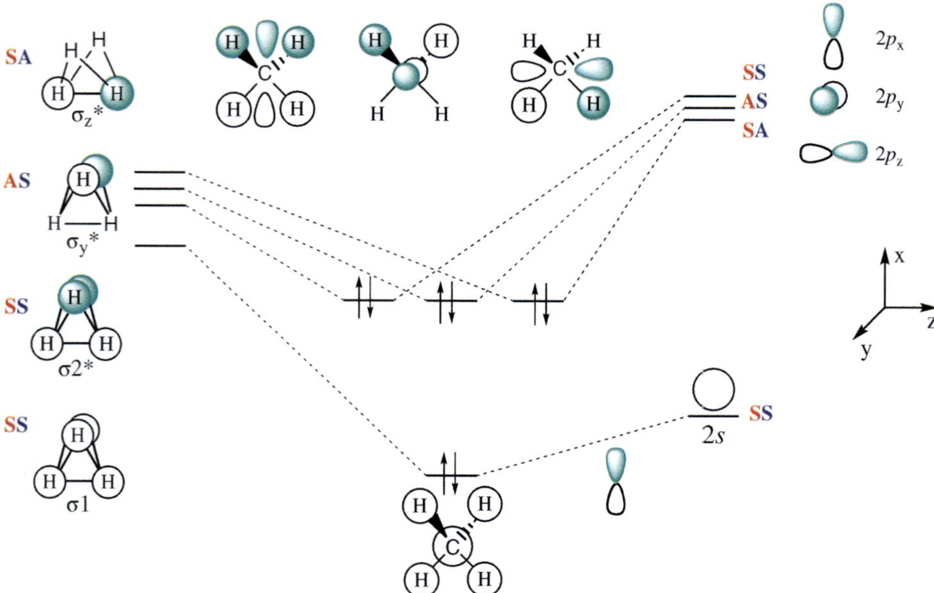

FIGURE 3.30   Energy level MO diagram for methane (anti-bonding MOs not shown for clarity).

other three MOs are degenerate (isoenergetic). The eight valence electrons occupy bonding MOs, making methane a very stable molecule. (The antibonding MOs are not shown in Figure 3.30).

MO theory suggests that methane presents one MO of one distinct type and three of another type, but that the electrons they contain are distributed among the five atoms. No hydrogen atom contains neither more nor fewer electrons than any other; they are all equivalent. One of the difficulties with these representations is that there is no single orbital that can be assigned to the C—H bonds. This is less intuitive to chemists, who find it much easier to visualise localised bonds. In this sense, chemists use Pauling's idea of hybridisation to overcome this difficulty. As we saw in Chapter 2, the AOs of the carbon atom can mix together, giving a set of hybrid orbitals with s and p characters in different proportions, which are then used to form MOs. Using the principles of hybridisation, the four sp^3 hybrid orbitals of methane, which point to the corners of a tetrahedron, can overlap with the $1s$ orbitals of hydrogen atoms (Figure 3.31a). Each hybrid AO presents a major lobe and a minor lobe aligned along a particular direction. To achieve the best overlap, an orbital must approach the major lobe head-on, making the hybrid AO more 'directional' than the AOs from which it is formed. Each overlap gives rise to one σ bonding MO and one σ^* anti-bonding MO. The eight valence electrons are placed in the four bonding MOs to form one σ C—H bond each. The anti-bonding MOs remain empty. The MOs are thus localised and can be associated with each bonding MO, whereas the pair of electrons in them form a single bond between the carbon and one of the hydrogens. Such a bond is called a two-centre, two-electron bond.

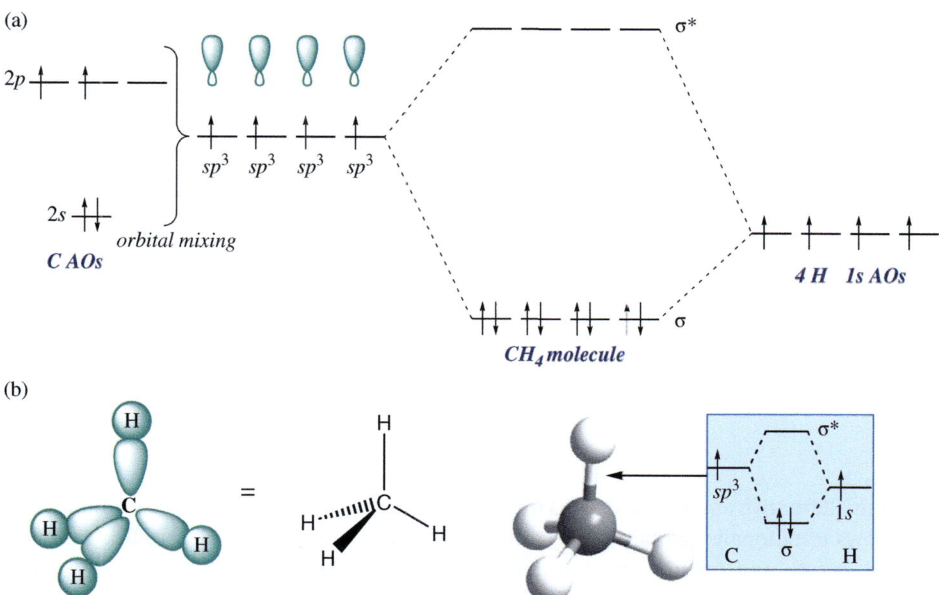

FIGURE 3.31 Energy level MO diagram of methane using hybridised atomic orbitals.

This picture has the advantage that the MOs present a direct relationship to the lines drawn in conventional structures (Figure 3.31b).[3] The MOs formed by overlapping sp^3 hybrid AOs present cylindrical symmetry about the σ bond axis, and the resulting bonds are therefore called σ bonds.

Hybridisation makes it easier to build larger structures. In the case of ethane, for example, each carbon uses three sp^3 AOs that overlap with three hydrogen atoms, leaving one sp^3 orbital on each carbon atom for C—C bonding (Figure 3.32).

Ethene is a planar molecule with C—C—H bond angles close to 120°, so a set of adequate orbitals that adopt this geometry is required. As discussed in Chapter 2, carbon's sp^2 hybridisation gives rise to this arrangement. These hybrids can be formed by combining the $2s$ and the two in-plane $2p_y$ and $2p_z$ AOs. The $2p_x$ AO is out of the plane and remains unchanged. So, for carbon, there are three sp^2 orbitals and one $2p_x$ to form MOs. The σ and $\sigma*$ MO orbitals are formed by the interaction of one sp^2 orbital in each carbon atom. Consequently, there are two $2p_x$ orbitals, one in each C, which interact side-by-side (Figure 3.33). The unhybridised $2p_x$ orbitals are more energetic than the sp^2 orbitals and interact less efficiently, so they give rise instead to π and $\pi*$ orbitals whose energies lie between those of the σ and $\sigma*$ orbitals. As it was seen previously, the MOs formed by the overlap of $2p_x$ AOs do not present cylindrical symmetry about the bond axis. The bonds resulting from the filling of these MOs are therefore called π bonds. Each carbon atom contributes two electrons to the π and σ MO, whereas the remaining two electrons are involved in the formation of the two C—H bonds.

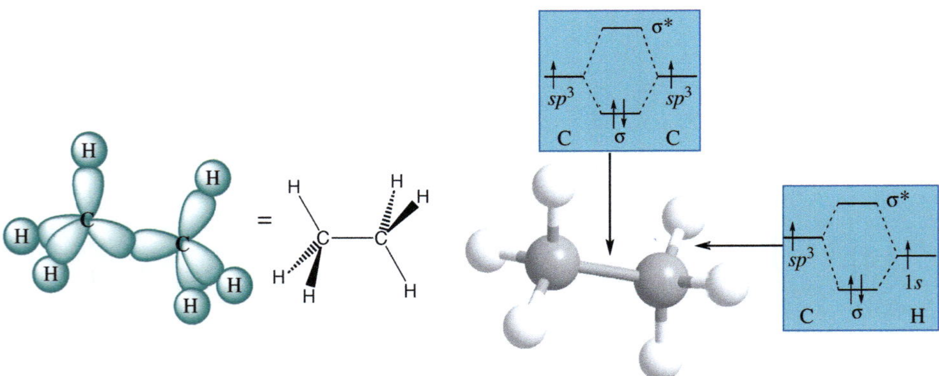

FIGURE 3.32   MO representation of the individual bonds present in the molecule of ethane.

---

[3] As with alternative approaches to MO theory, the concept of hybridisation is a useful model that allows to rationalise certain properties of chemical bonds in molecules, in this case the strength, length and orientation of covalent bonds, but fails in others, such as ionisation potentials. Hybridisation is not a physically real process, but a practical model.

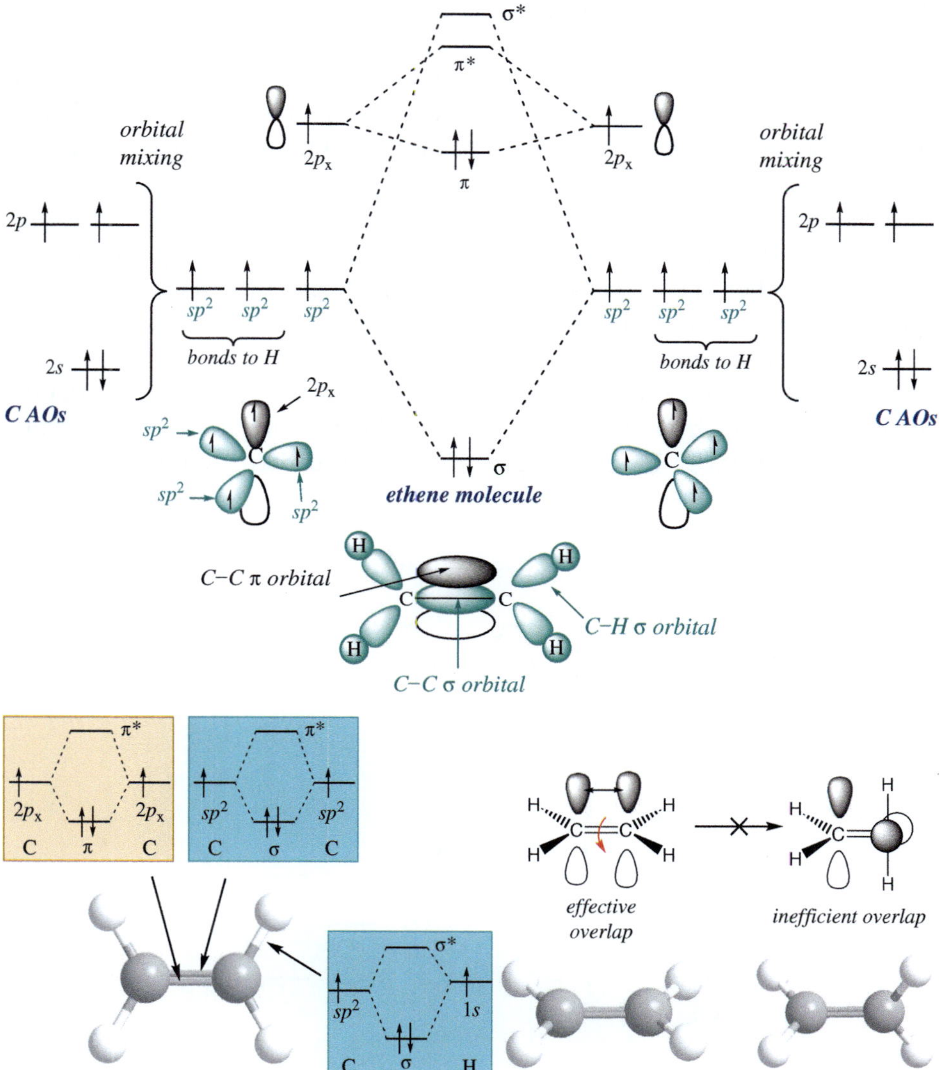

FIGURE 3.33 Energy level MO diagram of ethene using hybridised AOs.

One can appreciate the important differences between the C—C single bond in ethane and the C=C double bond in ethene. Rotation about the C—C bond in ethane is feasible since this bond is made up of electrons occupying a σ-type MO, which presents cylindrical symmetry around the bond axis. Importantly, rotation around the C—C bond does not affect the degree of overlap between the two sp^3 hybrid orbitals on carbon (Figure 3.32). By contrast, one of the bonds giving rise to the C=C double bond results from the electron occupation of a π-type MO, which is formed by the overlap of two out-of-plane $2p_x$ AOs, which do not present cylindrical symmetry. Consequently, rotation of the out-of-plane

CH_2 groups provokes that the lobes of the two $2p_x$ AOs will no longer be parallel and will not be able to overlap (Figure 3.33, bottom right side).

Acetylene is a linear C_2H_2 molecule with one triple carbon-carbon bond that is the simplest alkyne. To describe the bonding in acetylene, hybrid AOs on carbon arise from the hybridisation of the $2s$ and the $2p_y$ AOs and oriented at 180° to each other. This leaves unchanged $2p_z$ and $2p_x$ AOs that take part in the formation of two π bonds as the result of side-by-side overlap of the $2p_x$ and $2p_y$ AOs. On the other hand, the C—H σ bonds are formed by the overlap of the sp-hybrid orbitals at carbon with the $1s$ hydrogen orbital (Figure 3.34a). Figure 3.34b shows the MOs separately. Acetylene has ten valence electrons, all of which can be accommodated in bonding MOs.

The carbonyl group is one of the most important functionalities in organic chemistry. Formaldehyde (the simplest carbonyl-containing molecule) presents a planar geometry with H—C—O angles of 120°. The carbon atom makes use of

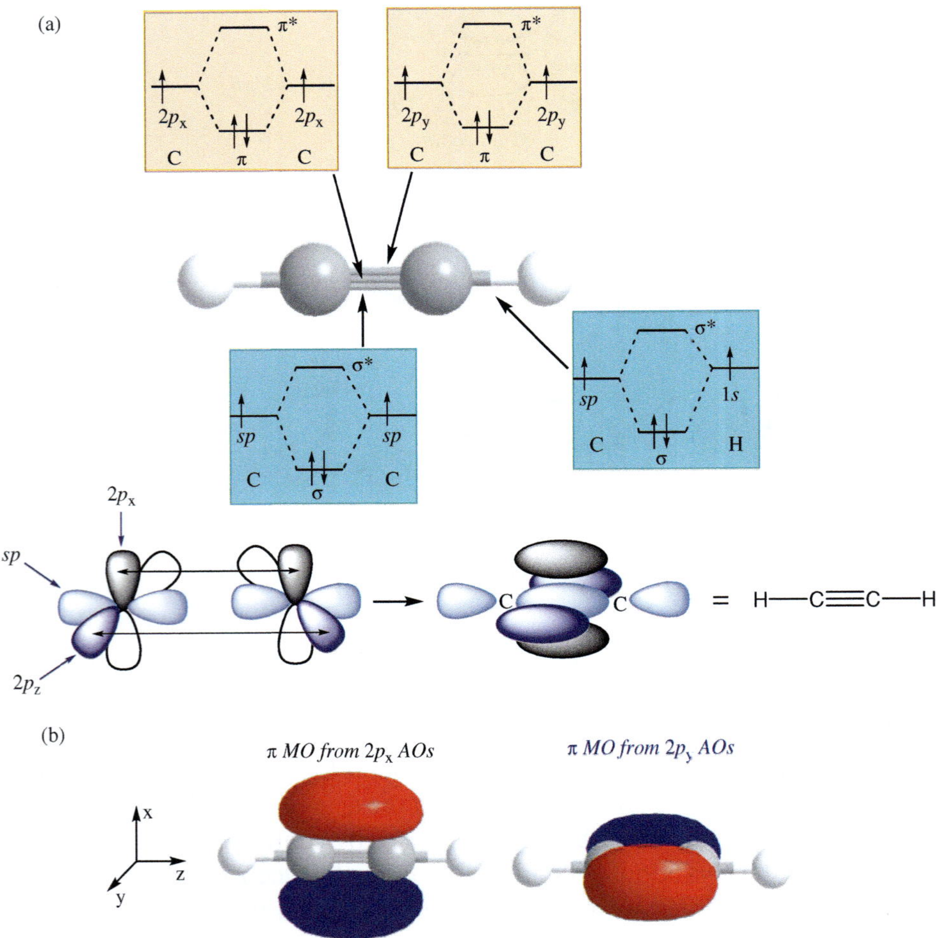

FIGURE 3.34   Energy level MO representation for the acetylene molecule.

three sp^2-hybridised orbitals to form three σ bonds, one with the oxygen atom and two with the hydrogen atoms. With regard to the oxygen atom, it forms a σ bond with carbon, occupies two sp^2-hybridised orbitals for its lone pairs, and the remaining $2p_x$ orbital combines with the unhybridised $2p_x$ orbital on carbon to form one π and one π^* MOs (Figure 3.35).

Owing to the higher electronegativity of oxygen relative to carbon, the π_{CO} orbital is energetically more similar to the $2p$ orbital in oxygen by comparison with the $2p$ orbital in carbon. Therefore, the π orbital presents a greater contribution from the oxygen $2p$ orbital. Consequently, the π orbital is distorted so that it is larger on the oxygen side than on the carbon side; concomitantly, the bonding electrons spend more time near the oxygen atom.

The polarisation of the C=O group is represented in Figure 3.35 by the dipole symbols δ^+ and δ^-. By contrast, the anti-bonding π^* orbital is closer in energy to the $2p$ orbital at carbon, and this is reflected in a greater contribution of this AO. The π^* anti-bonding orbital is therefore distorted towards the carbon terminus of the bond. Although the empty π^* orbital has no influence on the structure of the C=O bond, it does affect its reactivity: nucleophiles will attack the carbon atom in a carbonyl because it is more favourable to transfer electrons to the anti-bonding π^* orbital. In particular, the larger orbital coefficient at the carbon's π^* orbital allows for a better overlap with the nucleophile's reacting orbitals.

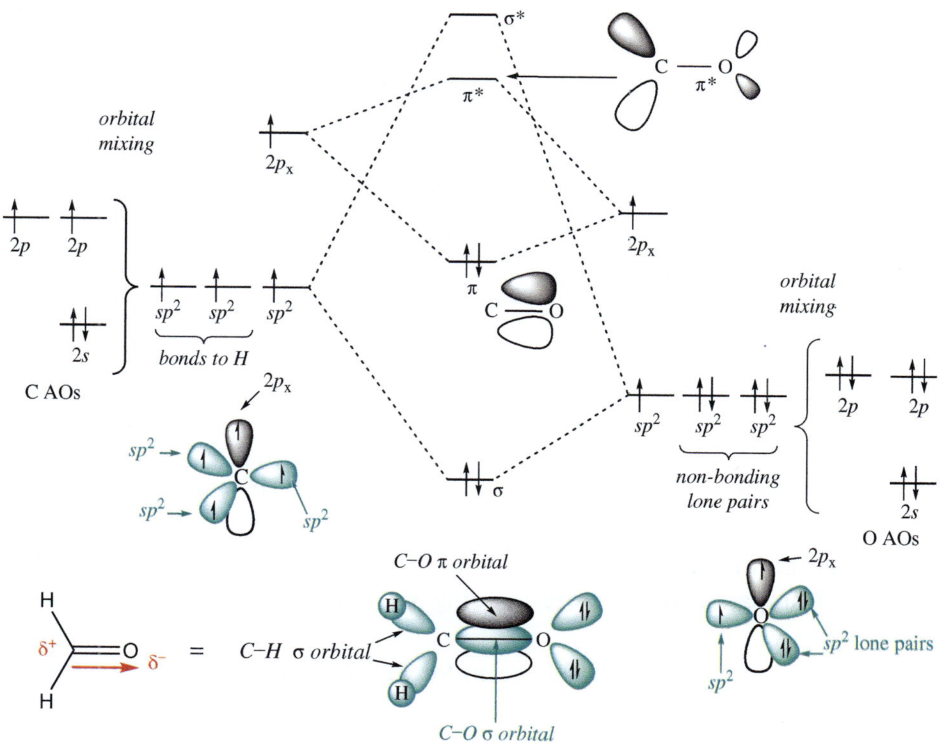

FIGURE 3.35 Energy level MO diagram of formaldehyde using hybridised AOs.

1,3-Butadiene
conjugated system

overlap of p orbitals

FIGURE 3.36   Conjugated π-system in 1,3-butadiene.

CONJUGATED SYSTEMS

Two or more conjugated C=C bonds can form delocalised π-systems with distinct MO energy levels and characteristic chemical behaviour. 1,3-Butadiene is a typical conjugated diene (Figure 3.36). The four carbon atoms are all sp^2 hybridised and each carbon presents an unhybridised p-orbital. The superposition of these four p AOs gives four MOs. Conjugated π-systems are conveniently studied using Hückel's theory, as is discussed in Chapter 4 of this book.

FURTHER READING

D. E. Lewis, *Advanced Organic Chemistry*, Oxford University Press, Oxford, UK, **2015**.

E. V. Anslyn, D. A. Dougherty, *Modern Physical Organic Chemistry*, University Science Books, Sausalito, California, **2005**.

I. Fleming, Molecular *Orbitals and Organic Chemical Reactions*, John Wiley & Sons, New York, **2010**.

J. Keeler, P. Wothers, *Chemical Structure and Reactivity: An Integrated Approach*, Oxford University Press, Oxford, UK, **2008**.

L. Pauling, *The Nature of the Chemical Bond*, 3rd. ed., Cornell University Press, Ithaca, New York, **1960**.

M. J. Winter, *Chemical Bonding*, 2nd ed., Oxford University Press, Oxford, UK, **2016**.

EXERCISES

3.1 Draw the energy level MO diagrams for O_2^{2-}, O_2^2, H_2^+ and Be_2. Calculate the bond order in these molecules and predict whether they are stable or not.

3.2 Draw the molecular orbital diagram for H_3^+ in a putative linear geometry. Compare with the corresponding MO diagram assuming a triangular geometry.

3.3 Build an energy level MO diagram for the N_2 molecule using hybridised orbitals.

3.4 Construct an energy level MO diagram for formaldehyde using unhybridised atomic orbitals.

CHAPTER 4

The Hückel Molecular Orbital Method (HMO)

THE LINEAR COMBINATION OF ATOMIC ORBITALS (LCAO) METHOD

A most useful approximation in molecular orbital theory considers the molecule as a set of molecular orbitals (MO), which are defined as two electron (of opposite spin) wave functions that encompass the entire molecule. According to this method, a given MO is defined as a linear combination of atomic orbitals (LCAO):

$$\Psi_i = C_{i1}\phi_1 + C_{i2}\phi_2 + \ldots + C_{in}\phi_n \tag{4.1}$$

or

$$\Psi_i = \sum_{j=1}^{n} C_{ij}\phi_j \tag{4.2}$$

where C_{i1} are the *coefficients* of the linear combination, indicative of the relative participation of each atomic orbital ϕ_j in the molecular orbital Ψ_i.

Consider, for example, a molecule formed by three atomic orbitals. According to the principle of orbital conservation, LCAO affords three molecular orbitals Ψ_1, Ψ_2 and Ψ_3.

$$nAO \Rightarrow nMO \tag{4.3}$$

$$\Psi_1 = C_{11}\phi_1 + C_{12}\phi_2 + C_{13}\phi_3 \tag{4.4}$$

$$\Psi_2 = C_{21}\phi_1 + C_{22}\phi_2 + C_{23}\phi_3 \tag{4.5}$$

$$\Psi_3 = C_{31}\phi_1 + C_{32}\phi_2 + C_{33}\phi_3 \tag{4.6}$$

Each MO (Ψ_i) is associated with its corresponding energy (E_i), which is obtained by solving Schrödinger's equation. A working concept behind the

LCAO method is that in the vicinity of each molecular atom its influence predominates, so that the molecular wave function in such regions of influence is approximately similar to the original atomic wave functions.

THE HÜCKEL MOLECULAR ORBITAL (HMO) METHOD

As already indicated, σ bonds are usually lower in energy than π bonds. Thus, it is understandable that those molecules possessing π bonds (that is, double and triple bonds, aromatic systems and so on) show physical properties and reactivity dictated mainly by those π bonds, and not so much by the σ bonds, which constitute mainly the structural molecular framework. With this consideration and keeping in mind the orthogonality principle which rules out effective overlap between σ and π orbitals, Hückel (1931) proposed to separate the σ from the π orbitals and use exclusively the latter in theoretical treatments. A further simplifying premise made by Hückel was the assumption of ideal planar π conjugated systems and C—C bonds of similar length. In addition, the overlap between non-adjacent atoms was considered negligible, so bonding between such atoms was discarded.

Thus, in the Hückel method n atomic π orbitals afford n molecular orbitals (MOs) Ψ_i, which according to Schroedinger's equation present unique values of energy:

$$\hat{H}\Psi_i = E_i\Psi_i \tag{4.7}$$

Because the introduced approximations make it impossible to determine the exact values of the energy of the MOs, the variation method is used to solve Eq. (4.7) in order to obtain the best average values of orbital energy. This requires one to multiply both sides of Eq. (4.7) by Ψ_i, and to integrate over all space, so that Eq. (4.7) becomes

$$\int \Psi_i \hat{H}\Psi_i\, dv = \int E_i \Psi_i^2\, dv \tag{4.8}$$

where E, the average energy, is a constant that comes out from the integral function so that

$$E_i = \frac{\int \Psi_i \hat{H}\,\Psi_i\, dv}{\int \Psi_i^2\, dv} \tag{4.9}$$

The wave function describing orbital Ψ_i must be normalised (Eq. 4.10); therefore, the denominator in Eq. (4.9) is equal to 1, and the evaluation of E_i will be simplified to Eq. (4.11).

$$\int \Psi_i^2\, dv = 1 \tag{4.10}$$

$$E_i = \int \Psi_i \hat{H}\, \Psi_i\, dv \tag{4.11}$$

Hückel's method proceeds now to find the energy values E_i associated with wave functions Ψ_i in the π system. For example, in a molecule containing three atoms

$$\Psi_1 = C_{11}\phi_1 + C_{12}\phi_2 + C_{13}\phi_3 \tag{4.4}$$

The only correct description for Eq. (4.4) is that in which the coefficients C_{11}, C_{12} and C_{13} minimise the energy of the MO Ψ_1.

Substitution of Ψ_1 (Eq. 4.4) into Eq. (4.9) yields expressions 4.12 and 4.13,

$$E_1\int(C_1\phi_1 + C_2\phi_2 + C_3\phi_3)^2\, dv = \int(C_1\phi_1 + C_2\phi_2 + C_3\phi_3) \\ \hat{H}(C_1\phi_1 + C_2\phi_2 + C_3\phi_3)\, dv \tag{4.12}$$

That is,

$$\begin{aligned}
E_1 = \big[&C_1^2 \int\phi_1^2\, dv + C_2^2 \int\phi_2^2\, dv + C_3^2 \int\phi_3^2\, dv + 2C_1 C_2 \int\phi_1\phi_2\, dv \\
&+ 2C_1 C_3 \int\phi_1\phi_3\, dv + 2C_2 C_3 \int\phi_2\phi_3\, dv = C_1^2 \int\phi_1\hat{H}\phi_1\, dv \\
&+ C_2^2 \int\phi_2\hat{H}\phi_2\, dv + C_3^2 \int\phi_3\hat{H}\phi_3\, dv + C_1 C_2 \int\phi_1\hat{H}\phi_2\, dv \\
&+ C_1 C_2 \int\phi_2\hat{H}\phi_1\, dv + C_1 C_3 \int\phi_1\hat{H}\phi_3\, dv + C_1 C_3 \int\phi_3\hat{H}\phi_1\, dv \\
&+ C_2 C_3 \int\phi_2\hat{H}\phi_3\, dv + C_2 C_3 \int\phi_3\hat{H}\phi_2\, dv
\end{aligned} \tag{4.13}$$

At this point it is convenient to introduce a number of abbreviations and to define their meaning:

$$\hat{H}_{rr} = \int\phi_r\hat{H}\phi_r\, dv \tag{4.14}$$

which is defined as the *Coulombic integral*, which corresponds to the energy of the electron in its isolated atomic orbital (AO).

$$\hat{H}_{rs} = \int\phi_r\hat{H}\phi_s\, dv \tag{4.15}$$

which is defined as the *resonance integral* and corresponds to the energy of the electron interacting with two nuclei; this term is equivalent to the stabilisation achieved in such a process,

$$S_{rr} = \int\phi_r^2\, dv \tag{4.16}$$

which is defined as the *normalisation integral* and corresponds to the probability of finding the electron in the orbital. Of course, this term is equal to unity when ϕ_r is normalised.

$$S_{rs} = \int\phi_r\phi_s\, dv \tag{4.17}$$

which is defined as the overlap integral and corresponds to the magnitude of overlap between the ϕ_r and ϕ_s orbitals.

Rewriting Eq. (4.13):

$$
\begin{aligned}
E_1 &= \left[C_1^2 S_{11} + C_2^2 S_{22} + C_3^2 S_{33} + 2C_1 C_2 S_{12} + 2C_1 C_3 S_{13} + 2C_2 C_3 S_{23} \right] \\
&= C_1^2 H_{11} + C_2^2 H_{22} + C_3^2 H_{33} + 2C_1 C_2 H_{12} + 2C_1 C_3 H_{13} + 2C_2 C_3 H_{23}
\end{aligned}
\tag{4.18}
$$

It should be noted that Eq. (4.18) incorporates all integrals as fixed quantities whose values will be determined by the molecular structure.

As already indicated, the energy E of the MO is a function of the variables C_1, C_2 and C_3, so in order to obtain the first of three equations with optimum values for these coefficients, one proceeds to obtain a partial derivative with respect to C_1, leaving C_2 and C_3 as constants. This procedure affords Eq. (4.19):

$$
\begin{aligned}
&\left(\frac{\delta E_1}{\delta C_1} \right) \left[C_1^2 S_{11} + C_2^2 S_{22} + C_3^2 S_{33} + 2C_1 C_2 S_{12} + 2C_1 C_3 S_{13} + 2C_2 C_3 S_{23} \right] \\
&+ E_1 [2C_1 S_{11} + 2C_2 S_{12} + 2C_3 S_{13}] = 2C_1 H_{11} + 2C_2 H_{12} + 2C_3 H_{13}
\end{aligned}
\tag{4.19}
$$

Because the correct C_1, C_2 and C_3 coefficients should correspond to an energy minimum (variation principle), then

$$
\frac{\delta E_1}{\delta C_1} = 0
\tag{4.20}
$$

Thus, the first term in Eq. (4.19) vanishes. Dividing by two and rewriting, one obtains Eq. (4.21).

$$
(H_{11} - ES_{11})C_1 + (H_{12} - ES_{12})C_2 + (H_{13} - ES_{13})C_3 = 0
\tag{4.21}
$$

Equation (4.21) is the first of three simultaneous equations; the second and third are obtained upon derivation with respect to C_2 and C_3.

$$
(H_{21} - ES_{21})C_1 + (H_{22} - ES_{22})C_2 + (H_{23} - ES_{23})C_3 = 0
\tag{4.22}
$$

$$
(H_{31} - ES_{31})C_1 + (H_{32} - ES_{32})C_2 + (H_{33} - ES_{33})C_3 = 0
\tag{4.23}
$$

Eqs. (4.21, 4.22 and 4.23) are equal to zero, and can be expressed in determinant form:

$$\begin{vmatrix} (H_{11} - ES_{11}) & (H_{12} - ES_{12}) & (H_{13} - ES_{13}) \\ (H_{21} - ES_{21}) & (H_{22} - ES_{22}) & (H_{23} - Es_{23}) \\ (H_{31} - ES_{31}) & (H_{32} - ES_{32}) & (H_{33} - ES_{33}) \end{vmatrix} = 0 \tag{4.24}$$

Recall that all subscripts in this determinant refer to the AOs participating in the resonance and Coulombic integrals. It can be appreciated that more complex systems will yield as many simultaneous equations, and therefore rows and columns for the determinant, as AOs are involved.

Consideration of the nature of the integrals represented in Eq. (4.24) allows for several simplifications. In particular,

$$H_{11} = H_{22} = H_{33} = H_{rr} = \alpha \tag{4.25}$$

where α corresponds to the energy of the electron in an *isolated* carbon p-orbital.

$$H_{12} = H_{21} = H_{23} = H_{rs} = \beta \tag{4.26}$$

where β corresponds to the energy of stabilisation gained by the electron in a carbon p-orbital interacting with two nuclei. Nevertheless,

$$H_{rs} = 0 \tag{4.27}$$

for non-vicinal atoms. On the other hand,

$$S_{11} = S_{22} = S_{33} = S_{rr} = 1 \tag{4.28}$$

as expected from the normalisation condition. However,

$$S_{12} = S_{21} = S_{23} = S_{rs} = 0 \tag{4.29}$$

that is, the interatomic overlap integral is neglected (although in fact for vicinal atoms this integral has usually a value close to 0.25).

As a result of the above simplifications (that is, Eqs. 4.25–4.29), the determinant for the allylic system can be expressed as

$$\begin{vmatrix} (\alpha - E) & \beta & 0 \\ \beta & (\alpha - E) & \beta \\ 0 & \beta & (\alpha - E) \end{vmatrix} = 0 \tag{4.30}$$

The algebra of determinants allows multiplication or division of all terms in a column of row by a constant. Dividing Eq. (4.30) by β:

$$\begin{vmatrix} \left(\dfrac{\alpha - E}{\beta}\right) & 1 & 0 \\ 1 & \left(\dfrac{\alpha - E}{\beta}\right) & 1 \\ 0 & 1 & \left(\dfrac{\alpha - E}{\beta}\right) \end{vmatrix} = 0 \tag{4.31}$$

Substitution of $x = (\alpha - E)/\beta$ affords a very simple determinant (Eq. 4.32), which is readily solved since it is only expressed in terms of x, 1 and 0.

$$\begin{vmatrix} x & 1 & 0 \\ 1 & x & 1 \\ 0 & 1 & x \end{vmatrix} = 0 \tag{4.32}$$

Thus,

$$x^3 - 2x = 0 \tag{4.33}$$

and

$$x = 0 \tag{4.34}$$

$$x = \pm\sqrt{2} \tag{4.35}$$

Recall that $x = (\alpha - E)/\beta$; therefore, $E = \alpha - x\beta$, so that x defines the energy of the system relative to α (the energy of the electron in an isolated orbital) as a function of β.

The above values of x obtained for the allylic system permit to arrange Ψ_1, Ψ_2 and Ψ_3 in terms of their associated energy, and relative to the level of reference α (Figure 4.1). Therefore, the energy associated to each orbital depends on the magnitude of x: negative values correspond to MOs more stable than the reference orbital at α, while orbitals with $x > 0$ are higher in energy.

Consideration of the number of π electrons present in the allylic system allows estimation of the total energy. For example, in the case of the allylic cation, radical or anion it is observed that all three species present the same bonding energy since they possess the same magnitude of $x\beta$ (Figure 4.2).

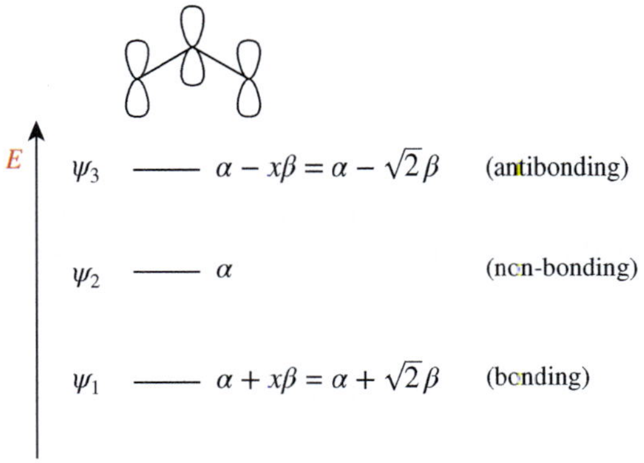

E

ψ_3 ——— $\alpha - x\beta = \alpha - \sqrt{2}\beta$ (antibonding)

ψ_2 ——— α (non-bonding)

ψ_1 ——— $\alpha + x\beta = \alpha + \sqrt{2}\beta$ (bonding)

FIGURE 4.1   Relative energies for the three MOs in the allylic system in terms of α and β.

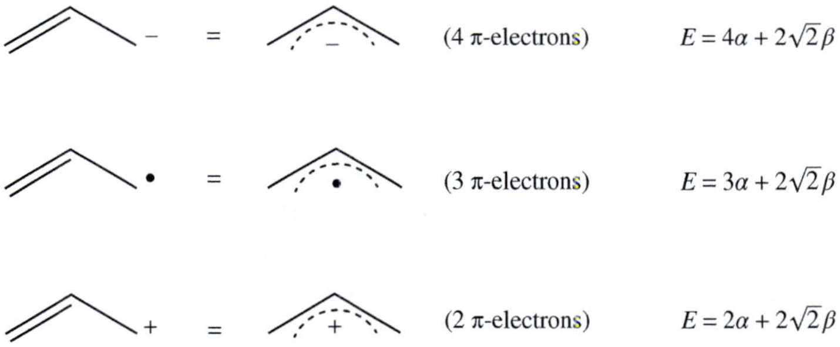

(4 π-electrons) $E = 4\alpha + 2\sqrt{2}\beta$

(3 π-electrons) $E = 3\alpha + 2\sqrt{2}\beta$

(2 π-electrons) $E = 2\alpha + 2\sqrt{2}\beta$

FIGURE 4.2   Hückel molecular orbital (HMO) energies for the allylic cation, radical and anion.

SIMPLIFIED PROCEDURE FOR THE APPLICATION OF HÜCKEL'S METHOD

This procedure consists of six steps:

1. A determinant is written with as many lines and columns as atoms are involved in the π system of interest.
2. Each column and line are labelled according to the appropriate AO to be combined: column number 1 for ϕ_1, column number 2 for ϕ_2, and so on.
3. Starting at the upper left corner, the diagonal positions, that is, the intersections ϕ_r/ϕ_s, when $r = s$, are filled with x.
4. For all other intersections ϕ_r/ϕ_s, a value of 0 is assigned whenever atoms r and s are not bonded, and a value of 1 when r and s are vicinal atoms.
5. The determinant is set equal to zero.
6. The determinant is solved for x (each value of x corresponds to the energy value associated to the MO).

APPLICATION OF HÜCKEL'S METHOD: SEVERAL EXAMPLES

Ethylene

The π-system in ethylene is the simplest because it is formed by only two AOs (Figure 4.3). Therefore, the corresponding determinant consists of two lines and two columns:

$$\begin{array}{c} \\ \phi_1 \\ \phi_2 \end{array} \begin{array}{cc} \phi_1 & \phi_2 \\ \left| \quad \right. & \left. \quad \right| \\ \left| \quad \right. & \left. \quad \right| \end{array} \tag{4.36}$$

Filling the diagonal with x:

$$\begin{array}{c} \\ \phi_1 \\ \phi_2 \end{array} \begin{array}{cc} \phi_1 & \phi_2 \\ \left| x \right. & \left. \quad \right| \\ \left| \quad \right. & \left. x \right| \end{array} = 0 \tag{4.37}$$

The non-diagonal elements correspond in this example to intersections between vicinal atoms ϕ_1/ϕ_2 and ϕ_2/ϕ_1; therefore, those positions are assigned values of 1, and the determinant is set equal to zero.

$$\begin{array}{c} \\ \phi_1 \\ \phi_2 \end{array} \begin{array}{cc} \phi_1 & \phi_2 \\ \left| x \right. & \left. 1 \right| \\ \left| 1 \right. & \left. x \right| \end{array} = 0 \tag{4.38}$$

Evaluation of the determinant affords Eq. (4.39); therefore, $x = \pm 1$.

$$x^2 - 1 = 0 \tag{4.39}$$

$$x^2 = 1 \tag{4.40}$$

$$x = \pm 1 \tag{4.41}$$

The energy of the two MOs in ethylene is then determined, $E = \alpha - x\beta$, and the corresponding orbital diagram is built (Figure 4.4). Finally, for neutral

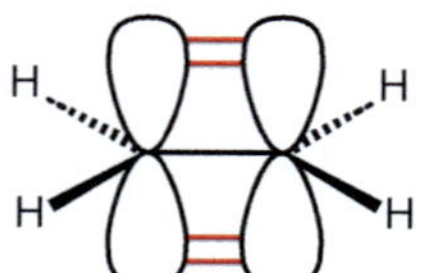

FIGURE 4.3   Ethylene's π-system.

ethylene, each atomic *p*-orbital involves one electron, so that two electrons occupy the low-energy bonding MO (Figure 4.4).

The Cyclopropenyl System

This cyclic π-system contains three AOs (Figure 4.5), which leads to a 3×3 determinant (Eq. 4.42). All ϕ_r/ϕ_s, $r \neq s$, correspond to vicinal atoms.

$$
\begin{array}{c}
\quad \phi_1 \quad \phi_2 \quad \phi_3 \\
\begin{array}{c} \phi_1 \\ \phi_2 \\ \phi_3 \end{array}
\begin{vmatrix} x & 1 & 1 \\ 1 & x & 1 \\ 1 & 1 & x \end{vmatrix} = 0
\end{array}
\tag{4.42}
$$

The solution of this determinant involves Eqs. (4.43–4.45).

$$
x^3 + 2 - 3x = 0 \tag{4.43}
$$

which factors into:

$$
(x + 2)(x - 1)(x - 1) = 0 \tag{4.44}
$$

$$
x = -2, +1, +1 \tag{4.45}
$$

With these values of x, the relative energies for the three MOs are obtained (Figure 4.6). It can be appreciated that according to the HMO method, the

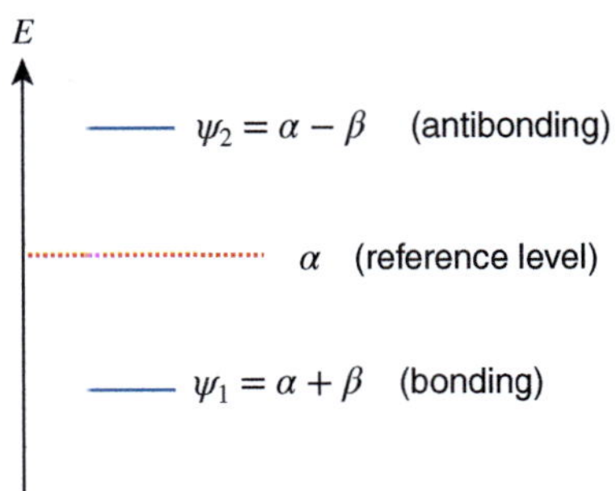

FIGURE 4.4   Relative energies for the bonding and anti-bonding MO orbitals in ethylene.

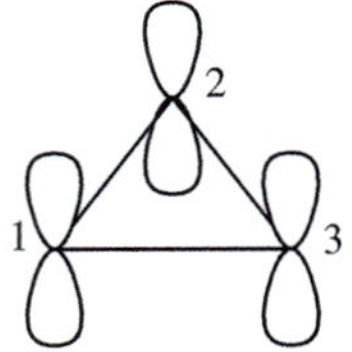

FIGURE 4.5   Cyclopropenyl π-system.

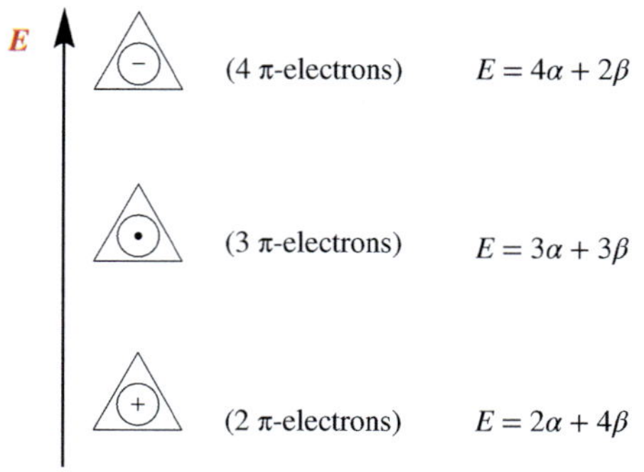

$$(4\ \pi\text{-electrons}) \qquad E = 4\alpha + 2\beta$$

$$(3\ \pi\text{-electrons}) \qquad E = 3\alpha + 3\beta$$

$$(2\ \pi\text{-electrons}) \qquad E = 2\alpha + 4\beta$$

FIGURE 4.6   Relative energies for the cyclopropenyl cation, radical and anion.

cyclopropenyl cation is more stable (it possesses greater bonding character) than the corresponding radical or anion.

Estimation of the Delocalisation Energy

The cyclopropenyl cation, radical and anion are π-delocalised species as consequence of their molecular framework, which forces the atomic p-orbitals into overlap. Such electronic delocalisation may lead to energy lowering of the π-system, which can be confirmed by comparison with localised reference fragments (Figure 4.7):

These results are in line with the concept of aromaticity as it applies to $(4n+2)\,\pi$-electrons delocalised in a planar ring. In addition, it should be noticed that the cyclopropenyl radical presents an unpaired electron in orbital Ψ_2; this *open-shell* nature is reflected in its increased reactivity.

Cyclobutadiene

For medium-size determinants it is convenient to apply the procedure of expansion of a determinant into cofactors. In the case of cyclobutadiene, the original 4×4 determinant (Eq. 4.46) may be expanded as illustrated in Eq. (4.47).

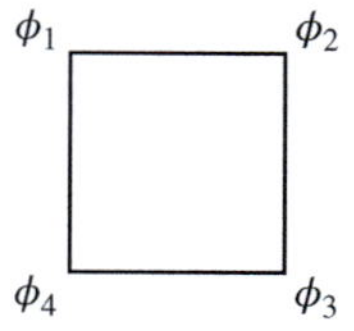

$$
\begin{array}{c|cccc}
 & \phi_1 & \phi_2 & \phi_3 & \phi_4 \\
\hline
\phi_1 & x & 1 & 0 & 1 \\
\phi_2 & 1 & x & 1 & 0 \\
\phi_3 & 0 & 1 & x & 1 \\
\phi_4 & 1 & 0 & 1 & x
\end{array} = 0
\tag{4.46}
$$

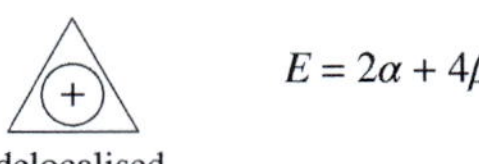

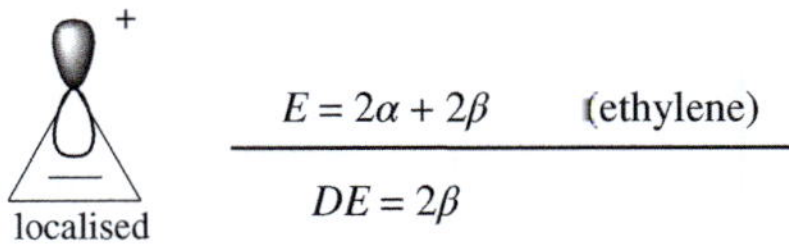

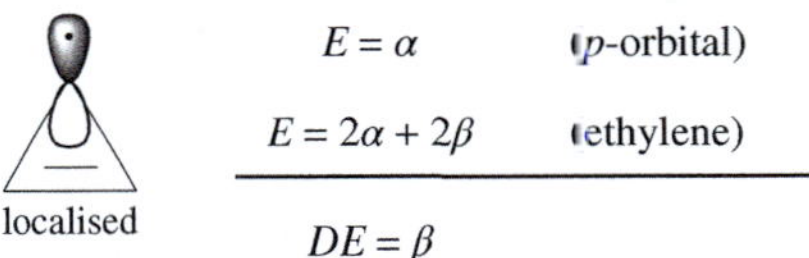

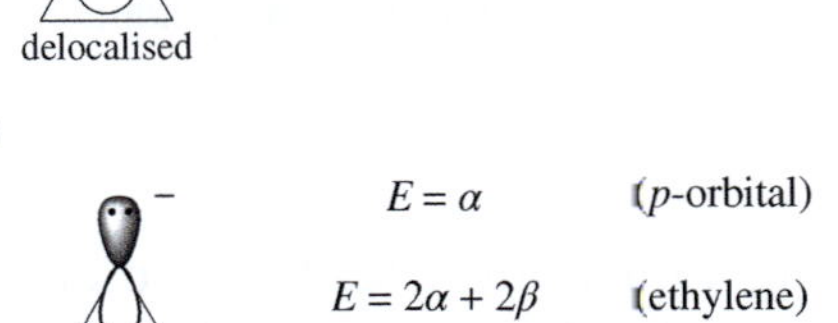

FIGURE 4.7 Delocalisation energies (DE) in cyclopropenyl cation, radical and anion.

$$x\begin{vmatrix} x & 1 & 0 \\ 1 & x & 1 \\ 0 & 1 & x \end{vmatrix} + 1 \cdot -1 \begin{vmatrix} 1 & 1 & 0 \\ 0 & x & 1 \\ 1 & 1 & x \end{vmatrix} + 1 \cdot -1 \begin{vmatrix} 1 & x & 1 \\ 0 & 1 & x \\ 1 & 0 & 1 \end{vmatrix} = 0 \tag{4.47}$$

Therefore,

$$x(x^3 - 2x) - (x^2 + 1 - 1) - (1 + x^2 - 1) = x^4 - 4x^2 = x^2(x^2 - 4) = 0 \tag{4.48}$$

Thus, $x = 0,\ 0,\ -2,\ +2$

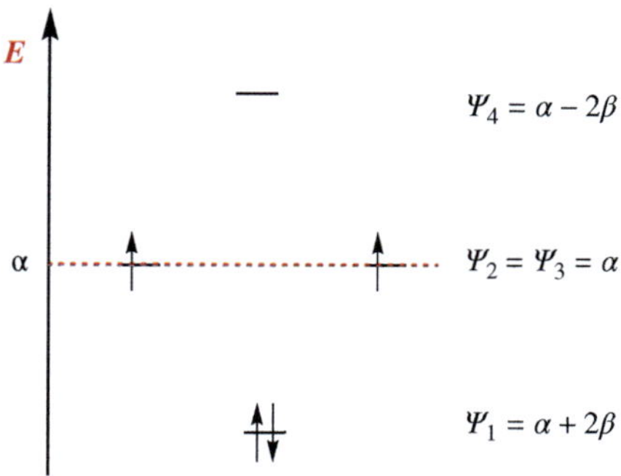

FIGURE 4.8   HMO diagram for neutral cyclobutadiene.

$$E = 2(\alpha + 2\beta) + 2\alpha$$

delocalised

$$= 4\alpha + 4\beta$$

vs

$$E = 2(2\alpha + 2\beta)$$

$$= 4\alpha + 4\beta$$

localised

$$DE = 0$$

FIGURE 4.9   Lack of delocalisation stabilisation in cyclobutadiene.

The corresponding values for x lead to the energy diagram shown in Figure 4.8. According to Hund's rule, neutral cyclobutadiene (four orbitals, four electrons) places two unpaired electrons in the isoenergetic Ψ_2 and Ψ_3 MOs. This open-shell, triplet electronic configuration confers cyclobutadiene its great chemical reactivity. In the same context, comparison with a localised analogue (two isolated ethylene units) indicates that the DE is equal to zero (Figure 4.9).

APPLICATION TO LARGER MOLECULES

For π-systems with more than four or five atoms, hand calculations become impractical unless the symmetry of the molecule permits some simplification. Thus, application of the Hückel method to larger molecules requires computer calculations. Nevertheless, available programmes can be run on microcomputers and require only a few seconds or minutes to calculate the results of medium-sized (10–40 atoms) molecules.

SCOPE AND LIMITATIONS OF THE HÜCKEL MOLECULAR ORBITAL METHOD

In spite of the drastic approximations introduced in the Hückel method, this simple procedure allows estimation of quite useful data, such as the energy associated to the MOs (eigenvalues), π-electron charge densities and bond orders. Indeed, these parameters may be applied to the estimation of bond strengths and bond lengths, as well as oxidation and reduction potentials, or to predict sites of electrophilic, nucleophilic and radical attack. Furthermore, HMO theory is applicable to monocyclic and fused-ring π-systems, as well as molecules incorporating heteroatoms.

The main weakness of the Hückel theory is that it does not take into account electron–electron repulsion; that is, it assumes that the potential of one electron is independent of the position of the other electrons present in the π-system. However, the energy of an electron in a given MO is strongly dependent on the manner in which the other electrons are distributed in the molecule. As a consequence, the Hückel MO method fails in certain cases, and in general, gives only qualitative or semi-quantitative information. Nevertheless, its results are a useful guide to more complete calculations and provide the basis for fruitful speculations in physical organic chemistry. Therefore, the main justification for the use of HMO theory is that it works!

HÜCKEL MOLECULAR ORBITAL METHOD IN CYCLIC π-SYSTEMS

Hückel (1931) introduced the so-called HMO theory to explain the remarkable stability of benzene. Indeed, studies of aromaticity have been rather fruitful via the Hückel MO method.

Indeed, the total π-electron energy of benzene is estimated to be $6\alpha + 8\beta$, corresponding to a DE of 2β. By contrast, cyclobutadiene is predicted to have zero DE, so in agreement with experimental observation, HMO theory indicates that there will be no stabilisation of cyclobutadiene as a result of the conjugation of the double bonds.

A useful mnemonic device for deriving energy diagrams for cyclic π-systems is *Frost's circle*, which consists of three simple steps:

1. A circle, whose centre is the α reference level of radius equal to 2β is drawn.
2. The appropriate regular polygon of n sides (equivalent to the n p-orbitals involved) is inscribed in the circle, with one corner at the lowest point.
3. The points at which the corners of the polygon touch the circle define the energy of the MOs (Figure 4.10).

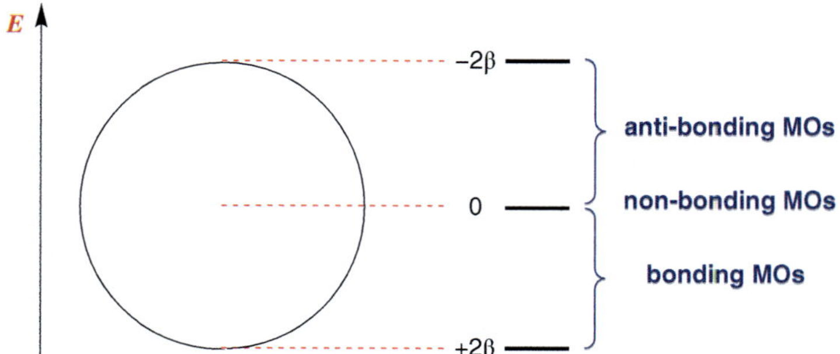

FIGURE 4.10   Frost's circle.

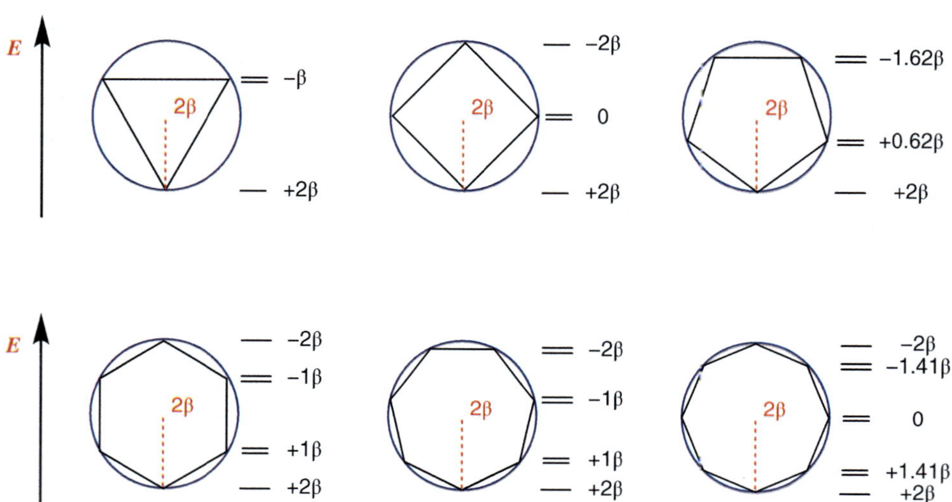

FIGURE 4.11   Energy level diagrams for the Hückel cyclic π-systems involving three to eight p-orbitals, illustrating Frost's circle.

Even a quick inspection of the illustrative examples collected in Figure 4.11 shows that stable (aromatic) systems are correctly predicted when the number of π-electrons is equal to $4n + 2$ (Hückel's aromaticity rule).

Indeed, stable closed-shell (aromatic) systems involve $(4n + 2)$ π electrons. Examples are the cyclopropenyl cation with 2 electrons, the cyclopentadienyl anion, neutral benzene, and the cycloheptatrienyl cation with 6 electrons, the cyclooctatetraenyl dianion with 10 electrons, and so on (Figure 4.12).

By contrast, π-systems with $4n$ electrons have highly reactive, open-shell electronic configurations. As a consequence, neutral cyclobutadiene, and the radical anion of cyclooctatetraene are planar species presenting bond-alternated geometries, that is, single and double bonds rather than delocalised electronic species.

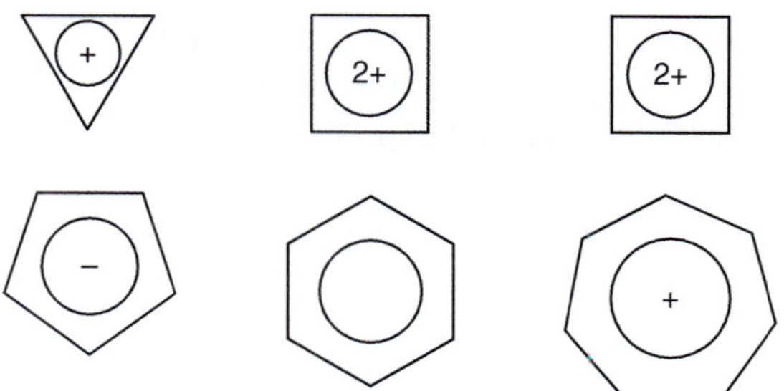

FIGURE 4.12 Examples of aromatic molecules presenting closed-shell electronic configurations, and no electrons in anti-bonding MOs.

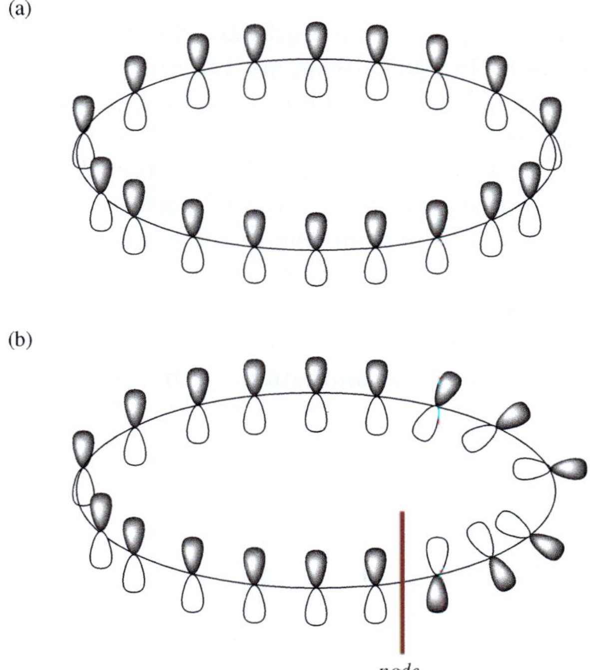

(a)

(b)

node

FIGURE 4.13 (a) Hückel π-system, stable with $(4n + 2)$ electrons. (b) Möbius π-system presenting one phase reversal, stable with $(4n)$ electrons.

For the application of the LCAO method to cyclic π-systems, the initial choice of signs (phases) of the *basis set* of AOs is quite arbitrary; that is, for p AOs one is free to choose which lobe is labelled positive and which one is negative. Nevertheless, it is convenient to assign the AO phases in such a way as to make the overlap between pairs of AOs positive (in-phase overlap) (Figure 4.13a).

However, when a conjugated chain of atoms in which the AOs all overlap in-phase is twisted through 180° and then united to form a ring (Figure 4.13b), the AOs at the point if union will *overlap out of phase.* This second type of topology (Möbius, by analogy with a Möbius strip) provoke that the rules of aromaticity are the opposite of those in Hückel systems:

Hückel orbital array	Möbius orbital
(4n + 2) aromatic	(4n) aromatic
(4n) anti-aromatic	(4n + 2) anti-aromatic

A modified Frost's circle mnemonics device inscribes now the polygon with one side horizontally situated in the lower part of the circle (Figure 4.14).

There is a fundamental difference between the orbital energies of cyclic Hückel and Möbius π-systems (Figures 4.11 and 4.14): Hückel systems present a single lowest-energy molecular orbital, whereas Möbius systems present two. Furthermore, in Hückel systems, the number of bonding orbitals is odd, accommodating 2, 6, 10, 14 or $4n + 2$ electrons when filled. In contrast, Möbius systems present an even number of bonding orbitals, accommodating 4, 8, 12 or $4n$ electrons. In addition, Hückel molecular orbitals are associated to zero or an even number of nodes, while Möbius molecular orbitals are known to present one or an odd number of nodes. The relevance of these observations becomes significant in the context of pericyclic reactions: in particular, alkenes, dienes, and olefinic compounds can approach one another to form Möbius transition states where $4n$ electrons are involved (see Chapter 7, examination of the nodal properties in the transition state).

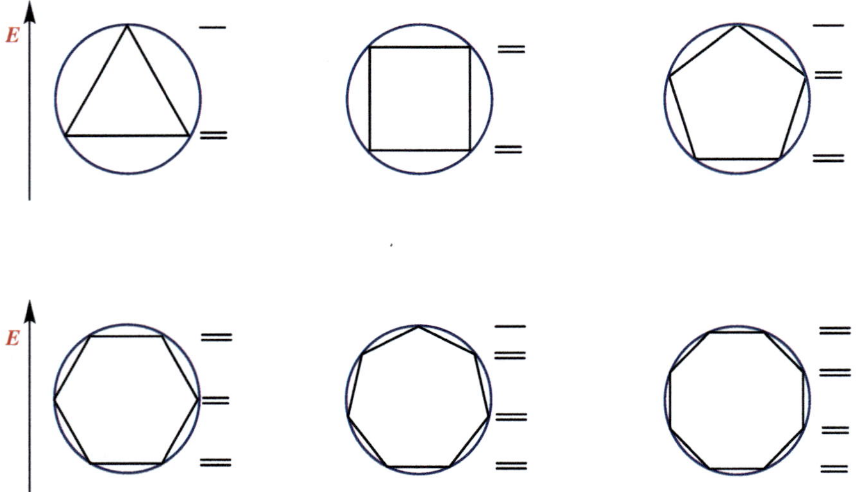

FIGURE 4.14   Energy level diagrams for Möbius cyclic π-systems. Stable, closed-shell configurations are obtained with $4n$ electrons.

ENERGY DIAGRAMS FOR ACYCLIC POLYENES

A mnemonic device applicable to acyclic polyenes, closely related to Frost's circle consists of the following steps:

1. A circle centred at $E = \alpha$, and with a radius equal to 2β is drawn.
2. The open-chain π-system is inscribed symmetrically disposed with respect to the α level, including two additional lines (of the same length than C—C bonds) touching the highest and lowest points of the circle.
3. The energy levels are gathered from the position of the points at which the molecule touches the circle (Figure 4.15).

In this regard, in order to obtain the appropriate eigenvalues in both cyclic and acyclic π-systems making use of Frost's circle, the following method may be used:

1. The n atoms of the inscribed molecules are labelled as $l = 0, 1, 2, \ldots$ starting at the position that touches the lowest point of the circle.
2. The energy value is obtained according to Eq. (4.49), where

$$E = \alpha + 2\beta\sin\theta \tag{4.49}$$

$$\theta(in\ degrees) = 90 - \frac{(360 \cdot l)}{(2n + 2)} \tag{4.50}$$

The application of this simple method affords the correct eigenvalues (in terms of parameters α and β) for Hückel cyclic π-systems (Figure 4.16a) as well as open-chain polyenes (Figure 4.16b).

A modification of this procedure, which allows for a rapid determination of the MO energy values in Möbius π-systems, is the following:

1. The right half of the properly inscribed polyene is labelled with odd integers, $l = 1, 3, 5, \ldots$ and so on, starting at the lower position.

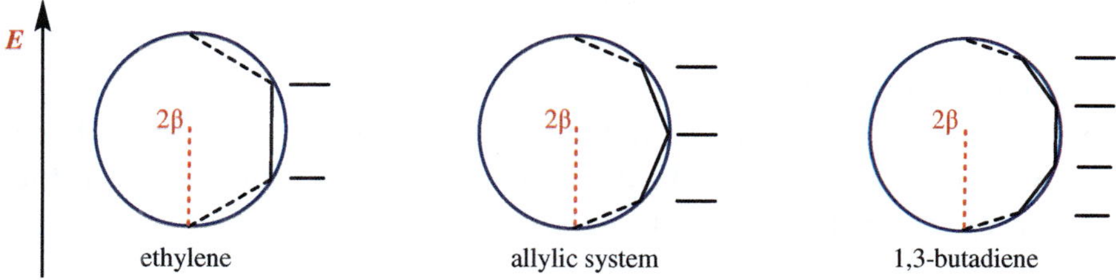

FIGURE 4.15 Energy diagram for open-chain π-systems (mnemonic device).

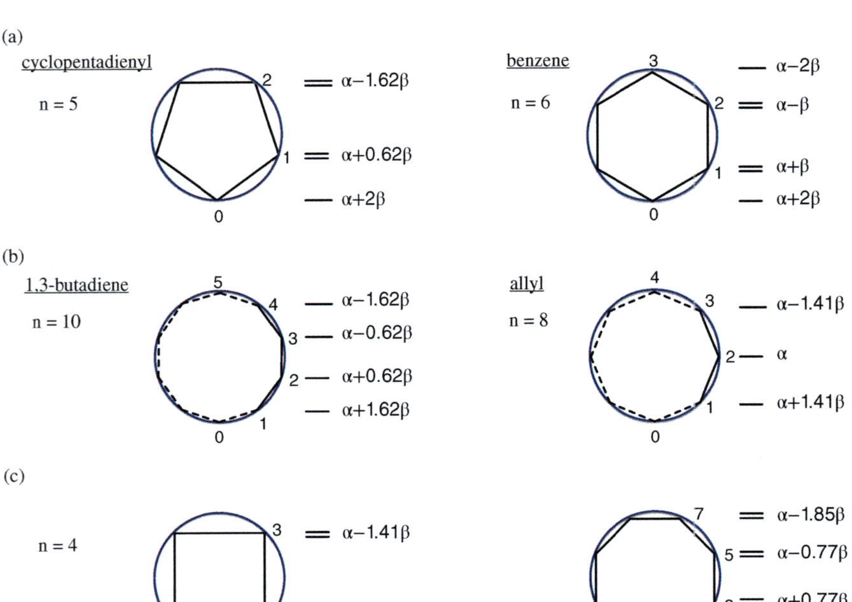

FIGURE 4.16   Energy values as a function of α and β for (a) Hückel cyclic π-systems, (b) open-chain polyenes and (c) Möbius cyclic π-systems.

2. The energy values are obtained by means of Eq. (4.49), where now

$$\theta(in\ degrees) = 90 - \frac{360l}{2m} \tag{4.51}$$

Two examples are included in Figure 4.16c.

π-SYSTEMS CONTAINING HETEROATOMS

Molecules such as pyridine, and functional groups like the carbonyl can be studied by the Hückel method as long as certain parameters are modified owing to the presence of the heteroatom (Figure 4.17).

1. *Coulombic Integral*
    Because this term refers to the energy of the electron in an isolated *p*-orbital, the substitution of carbon by a certain heteroatom X affects the energy of such electron; thus

$$\alpha \rightarrow \alpha_X \tag{4.52}$$

where

$$\alpha_X \rightarrow \alpha_C + h_X \beta_C \tag{4.53}$$

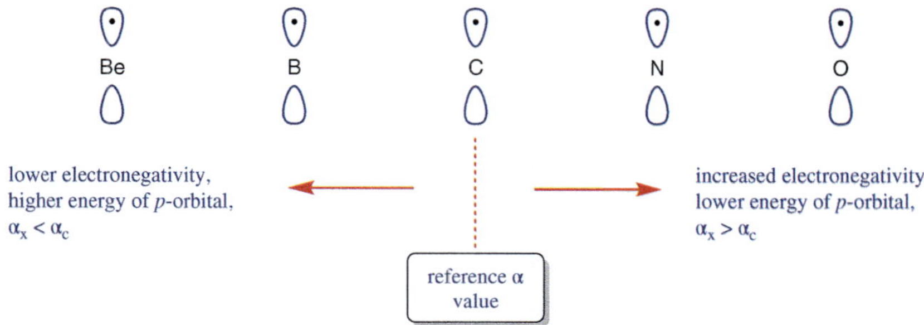

FIGURE 4.17   Incorporation of heteroatoms into π-systems.

FIGURE 4.18   Relative α values in heteroatoms.

Indeed, elements more electronegative than carbon will stabilise their electrons to a greater extent by comparison with the electron in its isolated orbital. This effect is expressed by the $h_X\beta_C$ term: for heteroatoms more electronegative than carbon $h_C > 0$ (that is, h_X for carbon is equal to zero). An opposite effect is expected for heteroatoms less electronegative than carbon so that h_X becomes negative, $h_X < 0$ (Figure 4.18).

2. *Resonance Integral*

This term corresponds to the stabilisation energy gained by the electron interacting with two nuclei. The replacement C=C → C=X modifies the value of this integral:

$$\beta_{CC} \to \beta_{CX} \tag{4.54}$$

where

$$\beta_{CX} = k_{CX} \cdot \beta_{CC} \tag{4.55}$$

Indeed, for heteroatoms with covalent radii longer than carbon, their overlap with vicinal p orbitals dwindles, and the magnitude of β_{CX} is reduced; that is, $k_{CX} < 1$.

Table 4.1 presents a set of values for parameters h_X and k_{CX}, in the most common heteroatoms.

TABLE 4.1   Recommended Parameters for the Evaluation of Coulombic and Resonance Integrals for Heteroatoms

Heteroatom (X)	h_X	k_X
B·	−0.45	0.73
N·	0.51	1.02
N:	1.37	0.89
O·	0.97	1.06
O:	2.09	0.66
F	2.71	0.52
Cl	1.48	0.62
Br	1.50	0.30
P·	0.19	0.77
P:	0.75	0.76
Si	0	0.75

From (a) F. A. Van-Catledge J. Org. Chem. **1980**, 45, 4801 and (b) A. Streitwieser, Molecular Orbital Theory for Organic Chemists, Wiley, New York **1961.**

EXAMPLE 1

Compare the energy values for the MOs in ethylene and a carbonyl group.

ethylene   carbonyl group

1. The corresponding determinant, as a function of the Coulombic and resonance integrals, is:

$$\begin{array}{cc} & \phi_C \qquad\qquad \phi_O \end{array}$$
$$\begin{array}{c} \phi_C \\ \phi_O \end{array} \begin{vmatrix} H_{11} - ES_{11} & H_{12} - ES_{12} \\ H_{21} - ES_{21} & H_{22} - ES_{22} \end{vmatrix} = 0 \tag{4.56}$$

$$\begin{vmatrix} \alpha - E & \beta_{CO} \\ \beta_{CO} & \alpha_O - E \end{vmatrix} = 0 \tag{4.57}$$

$$\begin{vmatrix} \alpha - E & k_{CO}\beta \\ k_{CO}\beta & (\alpha + h_O\beta) - E \end{vmatrix} = 0 \tag{4.58}$$

$$\begin{vmatrix} X & k_{CO} \\ k_{CO} & X + h_O \end{vmatrix} = 0 \tag{4.59}$$

From Table 4.1:

$$\begin{vmatrix} x & 1.06 \\ 1 & x + 0.97 \end{vmatrix} = x^2 + 0.97x - 1.06 = 0 \tag{4.60}$$

$$x = \frac{-0.97 \pm \sqrt{(0.97)^2 + 4.24}}{2} \tag{4.61}$$

so,

$$x = \frac{-0.97 + 2.276}{2} = 0.653 \tag{4.62}$$

and,

$$x = \frac{-0.97 - 2.76}{2} = -1.623 \tag{4.63}$$

Figure 4.19 compares the MO energy levels for ethylene and formaldehyde, $CH_2{=}O$. It can be noticed that both the bonding and anti-bonding MOs are of lower energy in the system containing the more electronegative oxygen atom.

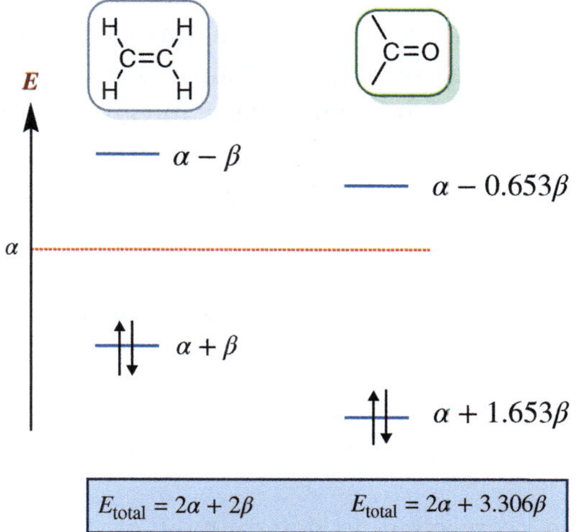

FIGURE 4.19 Energy diagram for the MOs and total energy in ethylene and formaldehyde.

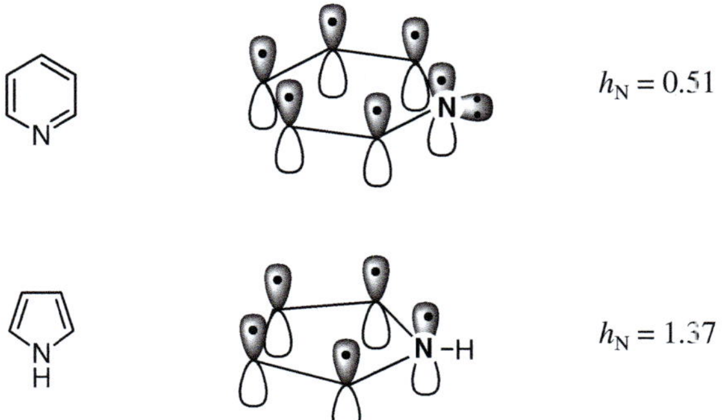

FIGURE 4.20　Parameter h_X for nitrogen in pyridine and pyrrole.

EXAMPLE 2

Regarding the h_X parameters for pyridine and pyrrole (Figure 4.20).

Parameter h_X for a particular heteroatom may vary depending on whether its participation in the π-system involves one or two electrons. In the pyridine molecule, the nitrogen atom contributes a single electron, whereas in pyrrole nitrogen provides both electrons to the aromatic system. In the latter case, the heteroatom acts effectively as a more electronegative atom, and this is reflected in a higher value for h_X.

INCLUSION OF OVERLAP BETWEEN VICINAL ATOMS

The Hückel MO method facilitates the solution of the necessary determinants by assuming $S_{rs} = 0$ whenever r ≠ s. In fact, for bonded atoms, $S_{rs} \approx 0.25$–0.29, and inclusion of this value is possible via incorporation of a proportionality constant, ε_{rs}. Indeed, assuming that all overlap integrals in vicinal atoms have the same magnitude, and that the overlap and resonance integrals are proportional:

$$\varepsilon_{rs} = \frac{H_{rs}}{\beta} = \frac{S_{rs}}{S} \tag{4.64}$$

where β and S are standard values of the resonance and overlap integrals in benzene. Thus,

$$H_{rs} - ES_{rs} = \varepsilon_{rs}(\beta - ES_{rs}) \tag{4.65}$$

Accordingly, the corresponding determinant of, for example, the allylic system takes the form

$$\begin{array}{cccc} & \phi_1 & \phi_2 & \phi_3 \\ \phi_1 & \begin{vmatrix} (\alpha - E) & \varepsilon_{12}(\beta - ES) & \varepsilon_{13}(\beta - ES) \\ \phi_2 & \varepsilon_{21}(\beta - ES) & (\alpha - E) & \varepsilon_{23}(\beta - ES) \\ \phi_3 & \varepsilon_{31}(\beta - ES) & \varepsilon_{32}(\beta - ES) & (\alpha - E) \end{vmatrix} = 0 \end{array} \qquad (4.66)$$

Nevertheless, the resonance integral for non-vicinal atoms is negligible, so that for this example $\varepsilon_{13} = \varepsilon_{31} = 0$. Dividing by $(\beta - ES)$:

$$\begin{vmatrix} (\alpha - E)/(\beta - ES) & \varepsilon_{12} & 0 \\ \varepsilon_{21} & (\alpha - E)/(\beta - ES) & \varepsilon_{23} \\ 0 & \varepsilon_{32} & (\alpha - E)/(\beta - ES) \end{vmatrix} = 0 \qquad (4.67)$$

If now, $x' = (\alpha - E)/(\beta - ES)$:

$$\begin{vmatrix} x' & \varepsilon_{12} & 0 \\ \varepsilon_{21} & x' & \varepsilon_{23} \\ 0 & \varepsilon_{32} & x' \end{vmatrix} = \begin{vmatrix} x' & 1 & 0 \\ 1 & x' & 1 \\ 0 & 1 & x' \end{vmatrix} = 0 \qquad (4.68)$$

Although the solution of the simultaneous equations requires of the usual types of $n \times n$ determinants, the meaning of x has changed:

$$x' = \frac{(\alpha - E)}{(\beta - ES)} \qquad \neq \qquad x = \frac{(\alpha - E)}{\beta} \qquad (4.69)$$

It can be demonstrated that inclusion of overlap S_{rs} affords

$$E = \alpha - x''\gamma \qquad (4.70)$$

where

$$x'' = \frac{x'}{1 - x'S} \qquad (4.71)$$

and

$$\gamma = (\beta - \alpha S) \qquad (4.72)$$

The similarity of Eq. (4.70) and $E = \alpha - x\beta$ is apparent.

From the experimental value for the DE in benzene (by comparison with 1,3,5-cyclohexatriene) equal to $36\,\text{kcal mol}^{-1}$, one obtains $\beta = -18\ \text{kcal mol}^{-1}$, and $\gamma = -36\ \text{kcal mol}^{-1}$. That is, roughly similar values for the DE in conjugated

TABLE 4.2   *DE*, Including and Neglecting Overlap Between Bonded Atoms, in Several π-Systems

Compound	DE (neglecting overlap)	DE (with overlap)
Benzene	$2.00\,\beta$	$1.00\,\gamma$
Naphthalene	$3.68\,\beta$	$1.86\,\gamma$
Styrene	$2.42\,\beta$	$1.21\,\gamma$
Azulene	$3.64\,\beta$	$1.60\,\gamma$
Butadiene	$0.47\,\beta$	$0.18\,\gamma$
Hexatriene	$0.99\,\beta$	$0.39\,\gamma$

π-systems are derived *with* or *without* inclusion of overlap between vicinal atoms (Table 4.2).

Nevertheless, an important consequence of the inclusion of overlap is that bonding MOs turn out nearer to α, whereas the anti-bonding MOs are now estimated to be higher in energy (farther from α). This result is closer to reality.

THE SHAPE OF THE MOLECULAR ORBITALS

Previous sections in this chapter have illustrated the use of HMO for finding the energy associated with π MOs. In this section, we focus on the three-dimensional shape that MOs display; that is, one seeks to gather a mental image of how they look like.

CONTRIBUTION OF THE AOs IN A MOLECULAR ORBITAL

Since Hückel's approach treats the MOs as a linear combination of the AOs (LCAO method, Eq. 4.73):

$$\Psi_i = C_{i1}\phi_1 + C_{i2}\phi_2 + C_{i3}\phi_3 \tag{4.73}$$

What is needed is to find the coefficients C_{i1}, C_{i2}, C_{i3}, ..., i.e. the contribution of AOs ϕ_1, ϕ_2, ϕ_3, etc. in the MO Ψ_i. As will be shown later in this chapter, the shape of a MO allows for the prediction of molecular properties such as charge density, bond order, reactivity index, and others.

In order to estimate the value of the coefficients in any MO, the same determinant developed to derive MO energies is employed. Let us consider an allylic π-system (Figure 4.21):

FIGURE 4.21   The allylic π-system.

and its corresponding determinant:

$$
\begin{array}{c}
\begin{array}{ccc} \phi_1 & \phi_2 & \phi_3 \end{array} \\
\begin{array}{c} \phi_1 \\ \phi_2 \\ \phi_3 \end{array}
\begin{vmatrix} x & 1 & 0 \\ 1 & x & 1 \\ 0 & 1 & x \end{vmatrix} = 0
\end{array}
\tag{4.74}
$$

which is equivalent to three simultaneous equations (Eqs. 4.75–4.77):

$$C_1 x + C_2 + 0 = 0 \tag{4.75}$$

$$C_1 + C_2 x + C_3 = 0 \tag{4.76}$$

$$0 + C_2 + C_3 x = 0 \tag{4.77}$$

where the values of x have already been calculated:

$$x = -\sqrt{2},\ 0, +\sqrt{2} \tag{4.78}$$

Therefore, for $x = -\sqrt{2}$ (the MO Ψ of lowest energy since $E = \alpha - x\beta$) Eqs. (4.75 and 4.76) become, respectively,

$$-\sqrt{2}\, C_1 + C_2 = 0 \tag{4.79}$$

$$C_1 - \sqrt{2}\, C_2 + C_3 = 0 \tag{4.80}$$

From Eq. (4.79),

$$C_2 = \sqrt{2}\, C_1 \tag{4.81}$$

Therefore, Eq. (4.80) may be written as

$$C_1 - 2C_1 + C_3 = 0 \tag{4.82}$$

or

$$C_3 = C_1 \tag{4.83}$$

For an appropriate MO, the normalisation condition must be met:

$$C_1^2 + C_2^2 + C_3^2 = 1 \tag{4.84}$$

since, $C_3 = C_1$ (Eq. 4.83) and $C_2 = \sqrt{2}\,C_1$ (Eq. 4.81), then

$$C_1^2 + 2C_1^2 + C_1^2 = 4C_1^2 = 1 \tag{4.85}$$

or

$$C_1 = 1/2 \tag{4.86}$$

Substitution of C_1 in Eqs. 4.81 and 4.83 provides the values for C_2 and C_3:

$$C_2 = \sqrt{2}/2 \tag{4.87}$$

$$C_3 = 1/2 \tag{4.88}$$

Therefore, the MO of lowest energy (Ψ_1) can now be written

$$\Psi_1 = \frac{1}{2}\phi_1 + \frac{\sqrt{2}}{2}\phi_2 + \frac{1}{2}\phi_3 \tag{4.89}$$

where all the coefficients are positive, but $C_2 > C_1 = C_3$. Pictorially, Ψ_1 can be represented as in Figure 4.22.

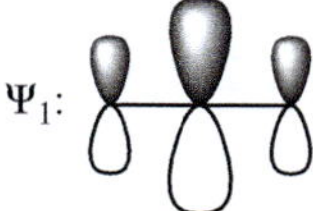

FIGURE 4.22   Graphical representation of Ψ_1 for the allylic π-system. This MO is bonding because the overlap among the AOs is positive ($S > 0$), and its energy ($E = \alpha + \sqrt{2}\,\beta$) is lower than that present in the original orbitals ($E = \alpha$).

The coefficients corresponding to Ψ_2 ($x = 0$, $E_2 = \alpha$) are then obtained following the same procedure.

$$C_1 = \frac{1}{\sqrt{2}} \tag{4.90}$$

$$C_2 = 0 \tag{4.91}$$

$$C_3 = -\frac{1}{\sqrt{2}} \tag{4.92}$$

Therefore,

$$\Psi_2 = \frac{1}{\sqrt{2}}\phi_1 - \frac{1}{\sqrt{2}}\phi_3 \tag{4.93}$$

Pictorially (Figure 4.23),

This MO is non-bonding because even though ϕ_1 and ϕ_3 present opposite phase (i.e. anti-bonding character), they are sufficiently far from each other so that their overlap is negligible ($S \approx 0$), which means that there is no difference in energy relative to the reference level, $E = \alpha$.

The MO of highest energy ($x = +\sqrt{2}$, $E_3 = \alpha - \sqrt{2}\,\beta$) is then estimated:

$$\Psi_3 = \frac{1}{2}\phi_1 - \frac{\sqrt{2}}{2}\phi_2 + \frac{1}{2}\phi_3 \tag{4.94}$$

and pictorially (Figure 4.24),

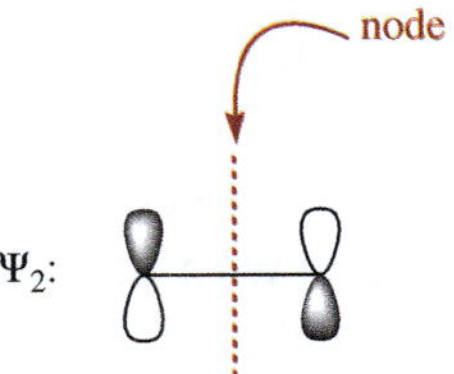

FIGURE 4.23 Graphical representation of Ψ_2 for the allylic π-system.

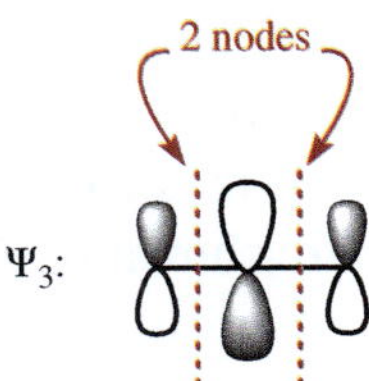

FIGURE 4.24 Graphical representation of Ψ_3 for the allylic π-system.

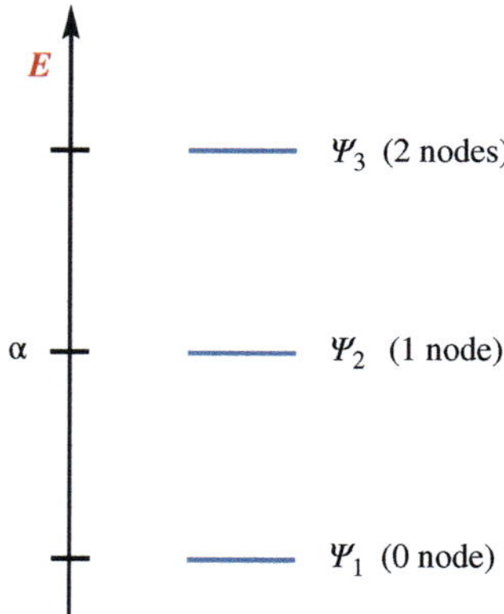

FIGURE 4.25   Energy diagram for the MOs in the allylic system, showing the direct relationship of energy with the number of nodes.

Ψ_3 is clearly an anti-bonding orbital since $S < 0$ and $E_{\Psi_3} > \alpha$. The relationship between the energy associated to each MO and their nodal properties should also be noted (Figure 4.25).

SYMMETRY SIMPLIFICATIONS IN ALTERNANT HYDROCARBONS

It has been observed that those π-systems that can be uniformly labelled on non-adjacent carbons (see for example, Figure 4.26) exhibit a symmetrical distribution of their MOs above and below the $E = \alpha$ level (Figure 4.27).

Furthermore, it has been found that an anti-bonding orbital is obtained for these systems by simply changing the sign of the coefficients of alternant AOs, i.e. those without asterisks. For example, in 1,3-butadiene the MOs Ψ_3 and Ψ_4 can be obtained from Ψ_2 and Ψ_1, respectively, by simply changing the sign of the unmarked carbons (Figure 4.28).

This procedure cannot be applied to π-systems where such alternant labelling is not possible. So, it is appropriate with naphthalene but not with azulene (and in general, with odd-number-membered rings).

For some systems, the labels may be distributed in an alternant fashion, but n* $\neq$ n°. This happens with non-Kekulé structures such as diradicals (Figure 4.30).

In contrast with benzene, 1,3-butadiene, naphthalene and other *balanced alternant hydrocarbons, non-balanced alternant hydrocarbons* present an even number of AOs. A second class of non-balanced alternant systems presents an odd number of AOs. Two examples are shown in Figure 4.31:

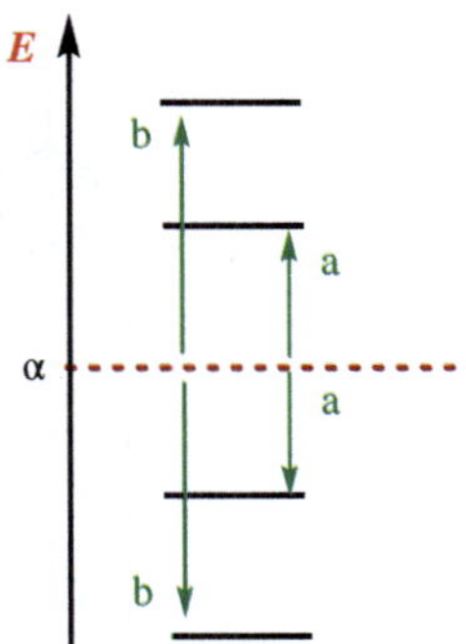

$$n^* = n^\circ$$

FIGURE 4.26   Example of a π-system with non-adjacent labelled carbons.

FIGURE 4.27   Symmetrical distribution of MOs in alternant hydrocarbons.

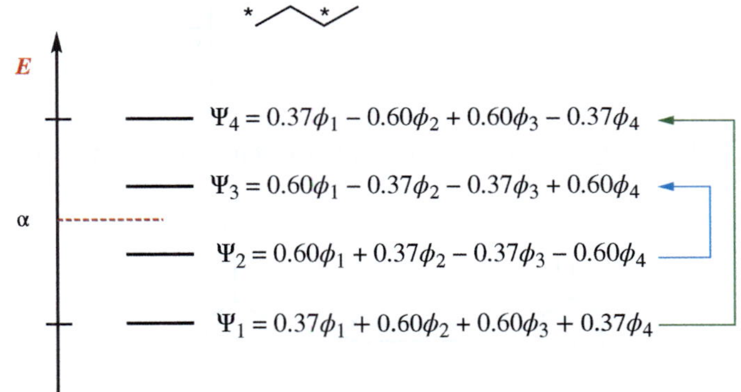

$$\Psi_4 = 0.37\phi_1 - 0.60\phi_2 + 0.60\phi_3 - 0.37\phi_4$$

$$\Psi_3 = 0.60\phi_1 - 0.37\phi_2 - 0.37\phi_3 + 0.60\phi_4$$

$$\Psi_2 = 0.60\phi_1 + 0.37\phi_2 - 0.37\phi_3 - 0.60\phi_4$$

$$\Psi_1 = 0.37\phi_1 + 0.60\phi_2 + 0.60\phi_3 + 0.37\phi_4$$

FIGURE 4.28   The anti-bonding MOs in 1,3-butadiene (alternant hydrocarbon) can be obtained by sign inversion of the coefficients of alternant AOs.

alternant

non-alternant

FIGURE 4.29   Examples of π-systems with alternant and non-alternant labelling.

etc.

$$4n^* \neq 2n^\circ$$

FIGURE 4.30   Example of a non-balanced alternant hydrocarbon.

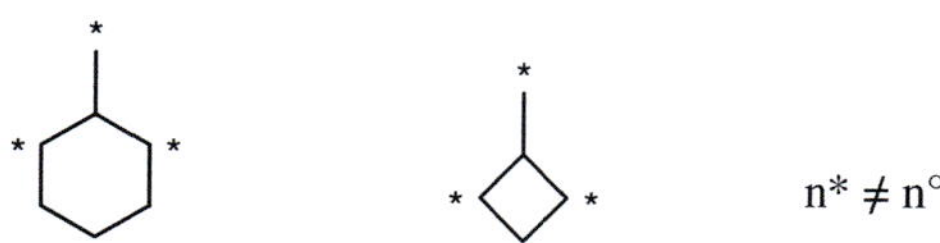

$$n^* \neq n^\circ$$

FIGURE 4.31   Non-equivalent alternant systems present an odd number of AOs.

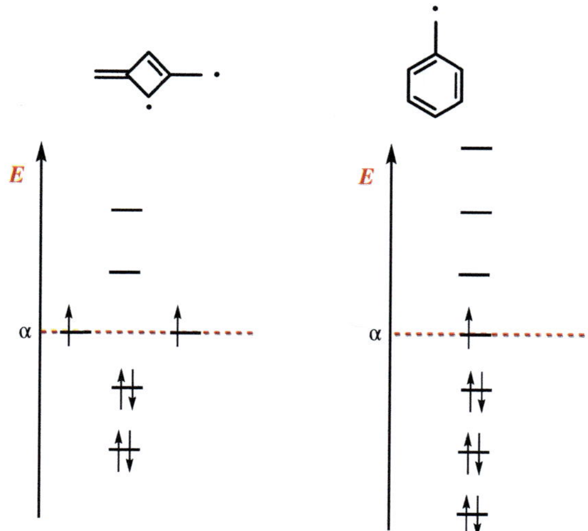

FIGURE 4.32   MO energy diagrams for representative non-balanced alternant hydrocarbons.

FIGURE 4.33   The benzylic π-system.

Non-balanced alternant systems present $n^* - n^\circ$ non-bonding molecular orbitals (NBMO). Because of Pauli's exclusion principle this implies the existence of radical species for the neutral molecules. Bonding and anti-bonding MOs are again symmetrically disposed around the reference α level (Figure 4.32).

An important characteristic of non-balanced alternant hydrocarbons is that, as was the case with balanced AH, the anti-bonding MOs may be obtained by changing the sign to alternant coefficients in the bonding MOs. Furthermore, unlabelled positions do not contribute to the non-bonding MO ($C = 0$), although the sum of the coefficients in the labelled positions around any specific unlabelled (non-active, $C=0$) atom is also zero. These properties permit the rapid generation of the coefficients in the NBMO. As an example, one can consider the benzylic π-system shown in Figure 4.33. The desired coefficients are obtained according to Eqs. 4.95–4.99.

$$\Psi_{\text{NBMO}} = -2C\phi_1 + C\phi_3 - C\phi_5 + C\phi_7 \tag{4.95}$$

from the normalisation condition,

$$C_1^2 + C_3^2 + C_5^2 + C_7^2 = 1 \tag{4.96}$$

thus,

$$(-2C)^2 + (C)^2 + (-C)^2 + (C)^2 = 1 \tag{4.97}$$

$$C = \frac{1}{\sqrt{7}} \tag{4.98}$$

Therefore,

$$\Psi_{\text{NBMO}} = -\frac{2}{\sqrt{7}}\phi_1 + \frac{1}{\sqrt{7}}\phi_3 - \frac{1}{\sqrt{7}}\phi_5 + \frac{1}{\sqrt{7}}\phi_7 \tag{4.99}$$

ESTIMATION OF MO ENERGIES AND COEFFICIENTS

The energy of any MO is defined as $E_i = \alpha - x_i\beta$. The values x_i can be easily calculated as follows:

(a) *For open chains with n atoms*

$$x_i = -2\cos\frac{i\pi}{n+1} \quad i = 1, 2, 3 \tag{4.100}$$

(b) *For rings with n atoms*

$$x_i = -2\cos\frac{2i\pi}{n} \quad i = 0, \pm 1, \pm 2\ldots \tag{4.101}$$

up to $\pm\frac{n-1}{2}$ (odd systems), and up to $\pm\frac{n}{2}$ (even systems).
For example, in 1,3-butadiene,

$$x_1 = -2\cos\frac{180}{5} = -1.618 \tag{4.102}$$

$$x_2 = -2\cos\frac{360}{5} = -0.618 \tag{4.103}$$

since x_3 and x_4 are symmetrically disposed (*vide supra*)

$$x_3 = +0.618 \tag{4.104}$$

$$x_4 = +1.618 \tag{4.105}$$

The appropriate coefficients may be determined according to the following equations:

(c) *For open chains with n atoms*

$$C_{rj} = \left(\frac{2}{n+1}\right)^{1/2}\left(sin\frac{rj\pi}{n+1}\right) \tag{4.106}$$

where r is the subscript distinguishing each AO in MO Ψ_j.

For example, Ψ_1, $x = -\sqrt{2}$, in the allylic system

$$C_{11} = \left(\frac{1}{2}\right)^{1/2} sin\left(\frac{\pi}{4}\right) = 0.5 \tag{4.107}$$

$$C_{12} = \left(\frac{1}{2}\right)^{1/2} sin\left(\frac{\pi}{2}\right) = 0.707 \tag{4.108}$$

$$C_{13} = \left(\frac{1}{2}\right)^{1/2} sin\left(\frac{3\pi}{4}\right) = 0.5 \tag{4.109}$$

(d) *For rings with n atoms*

$$C_{rj} = \frac{1}{\sqrt{n}}\, e^{\frac{2\pi ijr}{n}} \tag{4.110}$$

where r is defined as in (a) above, and $i = \sqrt{-1}$.

Note that Eq. (4.105) involves complex numbers, and consequently, the proper solutions are not obtained in a simple manner, especially in cases involving degenerate MOs.

BOND ORDERS (P_{ij})

From the discussion of the consequence of nodal properties on MO energy content, it is reasonable to anticipate that the amount of π bonding between adjacent bonds is related to the corresponding coefficients of the involved AOs. Qualitatively, one can examine the *occupied* MOs, and deduce the bonding characteristics of the species. For example, Ψ_1 in the 1,3-butadiene system shows medium-strength bonding between C(1) and C(2) (or between C[3] and C[4]) but strong bonding between C(2) and C(3) (Figure 4.34).

$$\Psi_1:$$

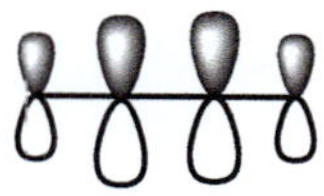

FIGURE 4.34 Graphical representation for Ψ_1 of 1,3-butadiene.

$$\Psi_2:$$

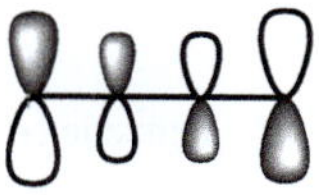

FIGURE 4.35 Graphical representation for Ψ_2 in 1,3-butadiene.

On the other hand, Ψ_2 presents medium-strength bonding among C(1)–C(2) [or C(3)–C(4)], but anti-bonding character among C(2)–C(3) (Figure 4.35).

Therefore, 1,3-butadiene exhibits strong C(1)–C(2) and C(3)–C(4) bonds, but a weaker C(2)–C(3) bond because the anti-bonding character in occupied Ψ_2 cancels some of the bonding character between C(2)–C(3) in Ψ_1.

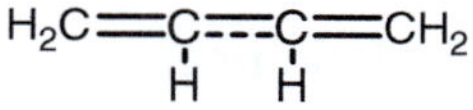

$$H_2C = C\cdots C = CH_2$$
$$\quad\;\; H \quad\; H$$

A more quantitative treatment is carried out by means of Coulson's Eq. (4.111):

$$P_{ij} = \sum^{\Psi_{occ}} N C_i C_j \tag{4.111}$$

where N is the number of electrons in each occupied MO, and C_i, C_j are the normalised coefficients for atoms i, j in the same MO. As expected, the product $C_i \cdot C_j > 0$ in bonding situations, and $C_i \cdot C_j < 0$ for anti-bonding interactions.

Let us consider once more 1,3-butadiene as a working example:

$$\begin{aligned} P_{12} = P_{34} &= (N_1 C_1 C_2)_{\Psi_1} + (N_2 C_1 C_2)_{\Psi_2} \\ &= (2 \times 0.37 \times 0.60) + (2 \times 0.60 \times 0.37) = 0.89 \end{aligned} \tag{4.112}$$

$$P_{23} = (2 \times 0.60 \times 0.60) + (2 \times 0.37 \times -0.37) = 0.45 \tag{4.113}$$

The order of π-bonding is then,

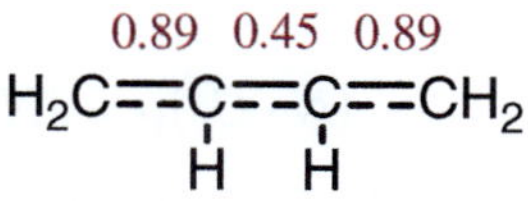

$$\begin{array}{ccc} 0.89 & 0.45 & 0.89 \end{array}$$
$$H_2C = C = C = CH_2$$
$$\quad\;\; H \quad\; H$$

Taking into account the bonding arising from σ bonds, the total bonding order is, therefore,

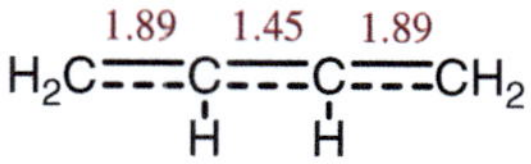

$$\begin{array}{ccc} 1.89 & 1.45 & 1.89 \end{array}$$
$$H_2C \cdots C \cdots C \cdots CH_2$$
$$\quad\;\; H \quad\; H$$

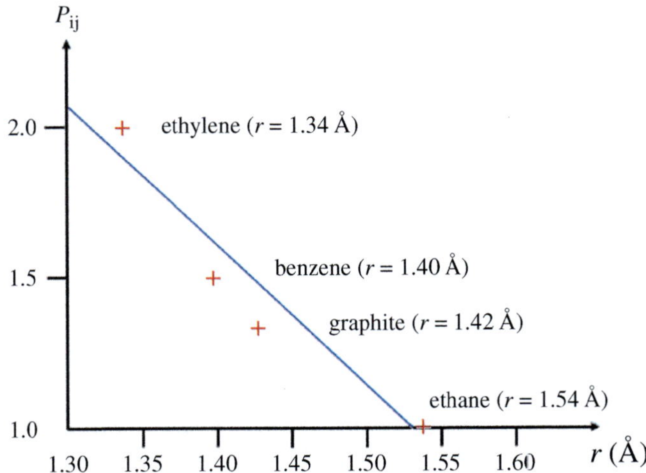

FIGURE 4.36   Linear correlation between calculated total bond order and experimental bond length.

There is usually a good correlation between calculated total bonding (the amount of π-bonding varies between 0 and 1) and experimental bond lengths (r) (Figure 4.36).

CHARGE DISTRIBUTION (q_i)

The electron density of atoms that participate in the construction of a π-system may be different than the normal ($q_i = 0$) for a neutral carbon. That is, a participating orbital may be initially assumed to contribute 1 electron to the π-system, but deviations from such 1-electron atomic density in the resulting molecule will manifest as charge (q_i). The magnitude of q_i is determined from the summation of the electronic contributions in each AO at the occupied MOs (Eq. 4.114).

$$q_i = 1.0 - \sum^{\psi_{OCC}} N \cdot C_1^2 \tag{4.114}$$

For 1,3-butadiene,

$$q_1 = q_4 = 1.0 - \left(N \cdot C_1^2\right)_{\psi_1} - \left(N \cdot C_1^2\right)_{\psi_2} \tag{4.115}$$
$$= 1.0 - \left[2(0.37)^2\right] - \left[2(0.60)^2\right] = 0$$

$$q_2 = q_3 = 1.0 - \left[2(0.60)^2\right] - \left[2(0.37)^2\right] = 0 \tag{4.116}$$

This result suggests that there is no charge associated with any carbon of 1,3-butadiene. Noting that the sum of charges for the atoms involved should be equal to the total charge in the molecule.

INDEX OF FREE VALENCE (F$_i$)

Although not as commonly used as P_{ij} or q_i, the concept of free valence is useful to predict relative chemical reactivities of the carbon atoms present in a molecular π-system. Free valence refers to the bonding degree of the atoms in a molecule as compared with a presumed 'maximum bonding potential'. That is, if a particular atom is not extensively bonded to its surrounding partners, it has significant free valance, and therefore, considerable relative reactivity. Coulson's Eq. (4.117) is employed to estimate the F_i.

$$F_i = \text{maximum bonding potential} - \sum^{\psi occ} P_{ij} \tag{4.117}$$

The amount of bonding at the central carbon in trimethylenemethane has been used to calculate the 'maximum bonding potential', since this carbon presents π bonding to the three methylene groups, which do not enter into any π bonding with any other atom. $\sum^{\psi occ} P_{ij}$ for the central carbon in this model was estimated as 4.73 (Figure 4.37).

Therefore, for 1,3-butadiene:

$$F_1 = F_4 = 4.73 - (2P_{\sigma_{C-H}} + P_{\sigma_{12}} + P\pi_{12} = 4.73 - (2 + 1 + 0.89) = 0.84 \tag{4.118}$$

$$F_2 = F_3 = 4.73 - [3P_{\sigma_{C-C}} + P_{\pi_{12}} + P_{\pi_{23}}]$$
$$= 4.73 - (3 + 0.89 + 0.45) = 0.39 \tag{4.119}$$

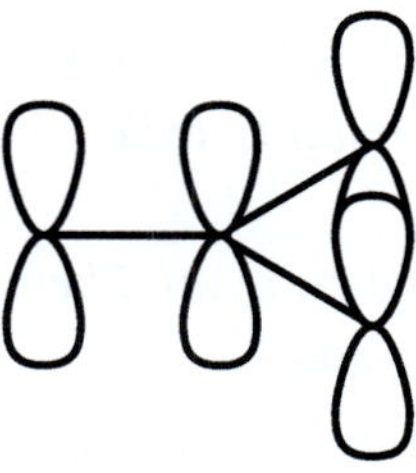

FIGURE 4.37   The trimethylenemethane π-system.

$$C(CH_2)_3 \qquad\qquad \qquad\qquad H_2C{=}CH_2$$

$$F = 0 \qquad\qquad F = 0.23 \qquad\qquad F = 0.73$$

$$H_2C{=}\!\!\bigcirc\!\!{=}CH_2 \qquad\qquad \bigcirc\!\!-CH_2\bullet \qquad\qquad CH_3\bullet$$

$$F = 0.92 \qquad\qquad F = 1.04 \qquad\qquad F = 1.73$$

FIGURE 4.38 F_i in illustrative examples.

These results (Eqs. 4.118 and 4.119) indicate that C(1,4) tend to be more reactive than C(2,3). Other illustrative examples, in order of increasing reactivity indexes, are collected in Figure 4.38.

FURTHER READING

A. A. Frost, B. Musulin, *J. Chem. Phys.* **1953**, *21*, 572.

A. D. Baker, M. D. Baker, *J. Chem. Educ.* **1984**, *61*, 770.

D. Feller, E. R. Davidson, W. T. Borden, *J. Am. Chem. Soc.* **1984**, *106*, 2513.

E. Heilbronner, H. Bock, *The HMO Model and its Application*, Wiley, New York and Verlag Chemie, Weinheim, **1976**.

E. Heilbronner, *Tetrahedron Lett.* **1964**, 1923.

E. Hückel, *Z. Physik.* **1931**, *70*, 204.

E. S. Huyser, D. Feller, W. T. Borden, E. R. Davidson, *J. Am. Chem. Soc.* **1982**, *104*, 2956.

F. A. Carey, R. J. Sundberg, *Advanced Organic Chemistry*, 3rd ed., Plenum, New York, **1990**.

F. A. Van-Catledge, *J. Org. Chem.* **1980**, *45*, 4801.

G. Ercolani, P. Mencarelli, *Educ. Chem.* **1987**, *24*, 178.

G. Taubmann, *J. Chem. Educ.* **1992**, *69*, 97.

H. E. Zimmerman, *Quantum Mechanics for Organic Chemists*, Academic Press, New York, **1975**.

H. Ichikawa, K. Sameshima, *Bull. Chem. Soc. Jpn.* **1990**, *63*, 3248.

J. D. Roberts, *Notes on Molecular Orbital Calculations*, Benjamin, New York, **1962**, Chapter 3.

J. H. Hammons, D. A. Hrovat, T. Borden, *J. Am. Chem. Soc.* **1991**, *113*, 4500.

J. H. Reeder, *J. Chem. Educ.* **1987**, *64*, 499.

J. J. Farrell, H. H. Haddon, *J. Chem. Educ.* **1978**, *66*, 839.

J. N. Murrell, S. F. A. Kettle, J. M. Tedder, *Valence Theory*, 2nd ed., Wiley, London, **1965**, Chapter 15.

J. P. Lowe, *Quantum Chemistry*, Academic Press, New York, **1978**.

J. Pramata, D. A. Dougherty, *J. Am. Chem. Soc.* **1987**, *109*, 1621.

A. Streitwieser, *Molecular Orbital Theory for Organic Chemists*, Wiley, New York, **1961**.

K. B. Wiberg, R. E. Rosenberg, P. R. Rablen, *J. Am. Chem. Soc.* **1991**, *113*, 2890.

L. Salem, *The Molecular Orbital Theory of Conjugated Systems*, Benjamin, New York, **1966**.

M. Isihara, *Bull. Chem. Soc. Jpn.* **1991**, *64*, 2284.

M. J. S. Dewar, *J. Am. Chem. Soc.* **1984**, *106*, 669.

M. J. S. Dewar, R. C. Dougherty, *The PMO Theory of Organic Chemistry*, Plenum, New York, **1975**.

P. J. Garratt, *Aromaticity*, Wiley, New York, **1986**.

P. v. R. Schleyer, D. Wilhelm, T. Clark, *J. Organomet. Chem.* **1984**, *273*, C1. R. Dagani, *Chem. Eng. News.* **1982**, *60*, 36.

EXERCISES

4.1 Apply the Hückel procedure, making use of the co-factors method for solving the determinant, in order to estimate the energy values associated with the four MOs in methylenecyclopropane.

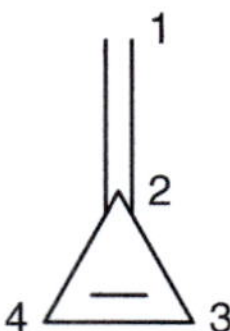

4.2 Solve the Hückel determinant for trimethylenemethane, and then (1) determine the delocalisation energy, (2) determine the appropriate coefficients in the MOs and (3) estimate bond orders and indexes of free valence.

4.3 Which of the following are alternant hydrocarbons?

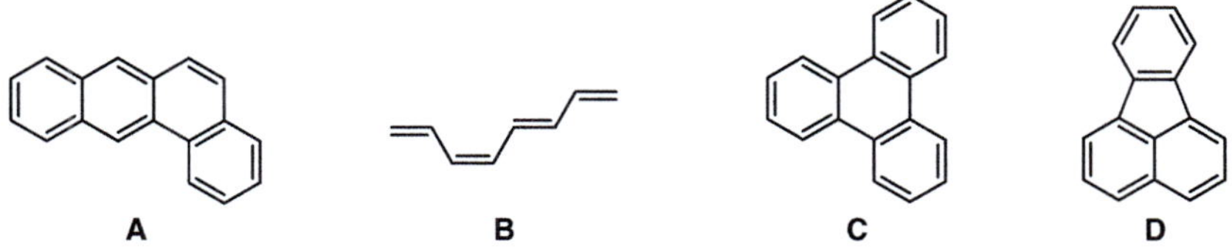

4.4 Obtain the coefficients and draw pictorial representations for all MOs in 1,3-butadiene. Comment on the bonding, non-bonding or anti-bonding properties of these orbitals.

4.5 How many NBMO possesses each of the following π-systems?

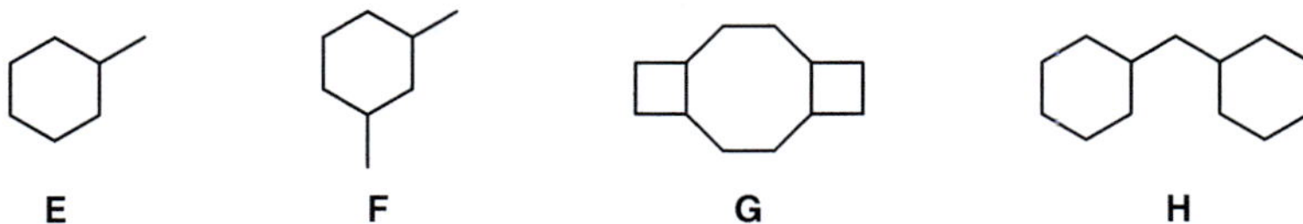

E   **F**   **G**   **H**

4.6 Solve the Hückel determinant for the following heterocyclic system. Determine the delocalisation energy.

4.7 Although annulene[10] ($C_{10}H_{10}$) fulfills Hückel's $4n + 2$ rule, it is not considered to be aromatic. Explain.

4.8 Making use of the Frost and Baker circles, compare the MOs for benzene and 1,3,5-hexatriene. Comment on the energy difference between the highest occupied MO (HOMO) and the lowest unoccupied MO (LUMO) in both π-systems.

4.9 Propose the most relevant resonance structures for azulene, as suggested by the indicated bond orders:

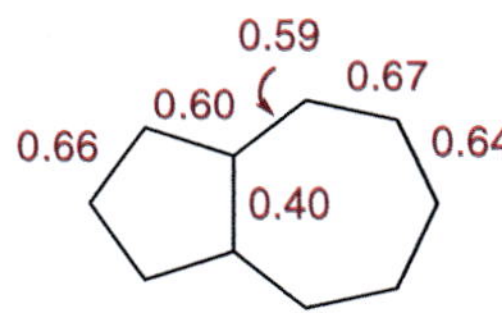

CHAPTER 5

Interactions Between Molecular Orbitals: Chemical Reactions

INTRODUCTION

In the course of a chemical reaction, as the original bonds are broken and the new ones are formed, the molecular orbitals (MOs) of the involved molecules interact with each other while their corresponding electrons are redistributed. This implies that the shapes and energies of the participating orbitals change during the process. In this regard, most reactions in chemistry are polar in nature, meaning that electrons flow from one molecule to another as the reaction proceeds. In terms of MO theory, electrons move from an occupied orbital in one reagent to an empty orbital in the other, resulting in the formation of a new covalent bond between atoms. It is important to recall that the MOs in the reactants are three-dimensional in shape, so they exert an important spatial control at the reaction's transition state. To begin with, for a new bond to be formed, the orbitals of the two reacting species must be correctly positioned in space. For example, not all interactions between ammonia (a Lewis base) and borane (a Lewis acid) are productive (Figure 5.1a). It is only when the occupied sp^3 orbital of nitrogen approaches the unoccupied $2p$ orbital of the boron atom in a properly aligned manner that the N—B bond will form (Figure 5.1b). As it was discussed above in the description of bond formation, when atomic or MOs interact in chemical reactions their energy levels are split giving rise to two new MOs, one energetically below relative to the original orbitals and the other one above. In the case of the reaction between ammonia and borane, a new bonding σ orbital and an empty anti-bonding σ^* orbital will be formed (Figure 5.1c). The initial nitrogen lone pair electrons at the filled sp^3 nitrogen orbital relocate at the

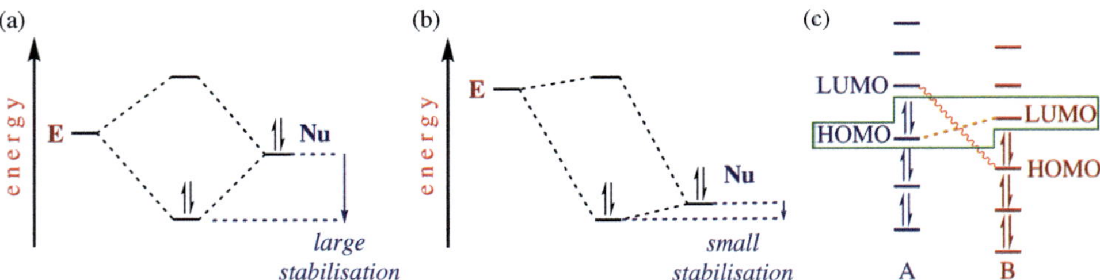

FIGURE 5.1   (a) Non-productive ammonia–borane interactions as a result of the incorrect orientation of the filled and empty orbitals at nitrogen and boron. (b) Productive orbital interaction. (c) MO energy level diagram of MOs for the bond-forming reaction between ammonia and borane.

FIGURE 5.2   Energy level MO diagram for (a) the reaction between a nucleophile (Nu:) and an electrophile (E) with rather similar energy. (b) Reaction between a nucleophile (Nu:) and an electrophile (E) with significantly different energy. (c) Effect of matching/mismatching orbital interactions in terms of energy difference. The wavy line suggests a weaker interaction relative to the broken line.

lower-energy sigma bond orbital, and this constitutes the driving force for the favourable (exergonic) N—B bond forming reaction.

Figure 5.2 shows the effect of relative energy matching between interacting orbitals in a qualitative manner: the closer the energy of the occupied and unoccupied interacting orbitals, the larger the energy difference among the newly formed orbitals. If the energies of the filled nucleophilic orbital (Nu:) and the empty orbital on the electrophile (E) are similar in magnitude, there will be a significant stabilisation of the system as a consequence of the sizeable lower energy of the new bonding orbital that is formed (Figure 5.2a). On the other hand, when the energy levels of the interacting orbitals are rather different, the energy lowering in the newly formed bonding orbital will be small and the process will be less favourable (Figure 5.2b). Therefore, reactions are more likely to occur when the two interacting orbitals are closer in energy.

Most commonly, reactions involve interaction between one occupied and one empty orbital (excluding radical reactions that implicate half-occupied MOs and present peculiar reactivity). The filled orbitals in most molecules are either bonding orbitals containing paired electrons or non-bonding orbitals containing lone electron pairs. The unoccupied orbitals are either anti-bonding MOs or empty atomic orbitals. An unoccupied orbital will always be associated with higher energy relative to an occupied orbital (Aufbau principle). Therefore, in a reaction between two molecules, the orbitals closest in energy most likely will be the **h**ighest **o**ccupied **m**olecular **o**rbital (HOMO) on one reactant and the **l**owest **u**noccupied **m**olecular **o**rbital (LUMO) on the other. The HOMO and LUMO orbitals are known as *frontier molecular orbitals* (FMO) and are widely used by chemists to understand the reactivity and regioselectivity of many chemical reactions.

In a reaction between two **A** and **B** reactants, there are two possible HOMO–LUMO interactions (Figure 5.2c). The HOMO–LUMO interaction presenting the smaller energy gap between the interacting MOs will be stronger and, therefore, more effective for chemical bonding. In fact, much of the reactivity of organic molecules can be predicted by knowing the shapes and relative energies of the frontier HOMO/LUMO orbitals involved in the reactions. Thus, a useful way to predict the reactivity between two molecules is to identify their respective FMOs and then proceed to anticipate the corresponding reactivity. In this regard, the approximate order in energy level among MOs is: $\sigma < \pi < n < \pi^* < \sigma^*$. As we have seen previously, a π orbital is normally higher in energy relative to a σ orbital, whereas a π^* orbital is usually lower in energy than a σ^* orbital in a particular molecular framework. On the other hand, the non-bonding n-orbitals (that is, lone electron pairs) lie energetically between the σ/σ^* and π/π^* orbital pairs. Figure 5.3 shows illustrative examples of HOMO and LUMO FMOs.

There are three general types of HOMO orbitals (Figure 5.3): an occupied σ orbital, an occupied π orbital, or an occupied non-bonding (n or n_π) orbital (for example, lone electron pairs in heteroatoms such as oxygen or nitrogen). The non-bonding orbitals (n or n_π) can also be occupied by a single electron (in free radicals), in which case they are called **s**ingly **o**ccupied **m**olecular **o**rbital (SOMO). The term n_π refers to occupied non-bonding orbitals in conjugated systems such as allyl or benzyl anions. There are also three general types of LUMO orbitals, empty σ^* or π^* orbitals and an empty non-bonding molecular or atomic

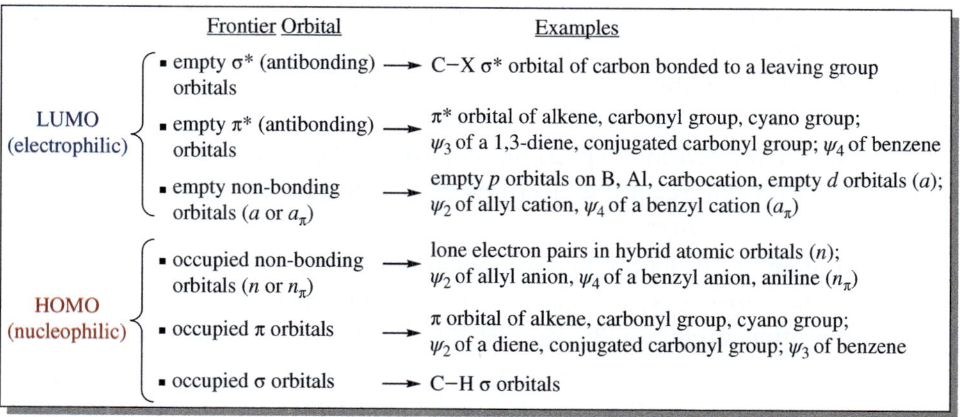

	Frontier Orbital	Examples
LUMO (electrophilic)	▪ empty σ^* (antibonding) orbitals	C–X σ^* orbital of carbon bonded to a leaving group
	▪ empty π^* (antibonding) orbitals	π^* orbital of alkene, carbonyl group, cyano group; ψ_3 of a 1,3-diene, conjugated carbonyl group; ψ_4 of benzene
	▪ empty non-bonding orbitals (a or a_π)	empty p orbitals on B, Al, carbocation, empty d orbitals (a); ψ_2 of allyl cation, ψ_4 of a benzyl cation (a_π)
HOMO (nucleophilic)	▪ occupied non-bonding orbitals (n or n_π)	lone electron pairs in hybrid atomic orbitals (n); ψ_2 of allyl anion, ψ_4 of a benzyl anion, aniline (n_π)
	▪ occupied π orbitals	π orbital of alkene, carbonyl group, cyano group; ψ_2 of a diene, conjugated carbonyl group; ψ_3 of benzene
	▪ occupied σ orbitals	C–H σ orbitals

FIGURE 5.3 Illustrative examples of FMOs.

orbitals (a or a_π). In this textbook, the symbol a is used to denote an unoccupied non-bonding MO (such as the empty $2p$ hybrid orbital in a carbocation). Whereas a_π will be used to denote unoccupied conjugated non-bonding orbitals, such as the allyl cation.

Commonly, a conjugated molecule will present different MOs capable of acting as the HOMO orbital, and accordingly, this will dictate its characteristic behaviour in terms of reactivity. Furthermore, the reaction's outcome may depend on the point at which the nucleophile attacks the LUMO orbital of the electrophilic reagent. Regarding an electrophilic reagent, for the proper selection of the LUMO among various anti-bonding MOs, it is useful to keep in mind that anti-bonding orbitals in C—X sigma* bonds (where X is an electronegative atom such as halogen or oxygen) usually present lower energy relative to $\sigma*$ orbitals in C—C or C—H bonds.

It is important to note that a molecule reacting through its filled HOMO orbital acts as an electron pair donor, that is, as a nucleophilic Lewis base. Conversely, a compound that reacts through its LUMO orbital is an electron pair acceptor, that is an electrophilic Lewis acid. Thus, much of organic chemistry can be treated as a Lewis acid reacting with a Lewis base or an electrophile reacting with a nucleophile. Therefore, an essential step for the proper understanding of organic reactions is to identify the correct HOMO and LUMO.

Using proper considerations of filled/empty orbital interactions, one can develop an appropriate MO interaction diagram for two reacting molecules. Furthermore, simple energy level diagrams can be convenient for the correct description of reactions in a qualitative way. However, the reader is warned that for more rigorous treatments of MO interactions, a higher level of computational calculations may be required.

In the following sections, we will examine some of the most important reactions in organic chemistry from the point of view of the interacting FMOs. This analysis will be qualitative and pictorial and will provide a guiding overview for the proper selection of the FMOs that determine the course and selectivity of the reactions of interest.

Before describing representative organic reactions in terms of MO theory (MOT), several additional considerations should be made. When two molecules approach each other, there are three main determining forces to consider:

1. Full shell electronic repulsion, that arises when the occupied orbitals of one molecule repel the electrons on the occupied orbitals of the other reacting molecule.

2. Coulombic attraction, that originates when polar or ionic regions in one reactant are exposed to opposite charges in the reacting partner, resulting in mutual attraction. By contrast, similar ionic charges repel each other.

3. Occupied orbitals (especially HOMO orbitals) of one molecule interact with the unoccupied orbitals (especially LUMO orbitals) of the other, providing the driving force for constructive bond forming reaction between the molecules.

In this regard, the Salem–Klopman equation is expressed as the sum of three terms that correspond to the general reaction controlling forces mentioned earlier. In its simplest expression, the Salem–Klopman equation assumes that the most significant interactions resulting in covalent bonding are those taking place between the HOMOs and LUMOs of the two reacting molecules. Accordingly, the simplified equation for chemical reactivity (Eq. 5.1) includes terms for quantifying the Coulombic and FMO interactions of two molecules:

$$\Delta E = -\frac{Q_{nuc}\,Q_{elec}}{\varepsilon R} + \frac{2\left(C_{HOMOa}\,C_{LUMOb}\,\beta_{ab}\right)^2}{E_{HOMO(nuc)} - E_{LUMO(elec)}} \tag{5.1}$$

where Q_{nuc} and Q_{elec} are the charges on the nucleophile and electrophile, respectively; ε is the local dielectric constant, C_{HOMOa} is the coefficient of atomic orbital **a** in the HOMO; C_{LUMOb} is the coefficient of atomic orbital **b** in the LUMO; β_{ab} is the resonance integral; E_{HOMO} and E_{LUMO} are the energies of the HOMO and the LUMO, respectively. Note that the FMO term depends on the coefficients of the overlapping orbitals, i.e. the degree of orbital overlap, as well as on the energy difference between the interacting HOMOs and LUMOs. Obviously, ΔE is negative as $E_{LUMO} > E_{HOMO}$. From this equation, one can deduce that the more efficient the overlap of the interacting orbitals, the stronger the resulting interaction and stabilisation will be. Furthermore, ΔE is greater when the term $2\left(C_{HOMOa}\,C_{LUMOb}\,\beta_{ab}\right)^2$ is larger. Qualitatively, the larger the size of the orbital lobes, the more efficient the overlap. Equation (5.1) also shows that ΔE is larger when the HOMO–LUMO energy difference is small, which is consistent with the discussion advanced above.

An energy level orbital correlation diagram is a visual tool used frequently in chemistry to understand the way MOs (or atomic orbitals) interact during a chemical reaction or during bond formation. Orbital correlation diagrams facilitate the visualisation of the way in which the orbitals of the reactant molecules or atoms combine to form the product molecules. Finally, energy level orbital correlation diagrams are particularly useful for understanding the feasibility of reactions and for predicting the relative stability of the reacting molecules and products.

MOLECULAR ORBITAL THEORY OF SELECTED ORGANIC REACTIONS

As discussed above, the overlap between a filled non-bonding orbital (n) and the LUMO associated to an empty non-bonding atomic orbital (a) leads to the formation of a bonding sigma orbital (σ) and an anti-bonding sigma* orbital (σ*) (Figure 5.1, right). The species providing the LUMO in this type of interaction is frequently a Lewis acid. The simplest example is a proton H^+ binding to water by interaction with its $1s$ AO (LUMO in the proton ion) with the lone electron pair on oxygen (n_O, HOMO in the water molecule).

In this context, aluminium chloride $AlCl_3$, borane BH_3 and boron trifluoride BF_3 are also classified as Lewis acids. The LUMO of BH_3 and BF_3 is the vacant $2p$ orbital at boron, whereas the LUMO for $AlCl_3$ is the $3p$ orbital on aluminium. In reality, the LUMO orbitals in $AlCl_3$ and BF_3 do have some contribution from the p orbitals of the halogens bonded to the central p orbital, but for simplicity, the LUMOs are treated here as simple vacant p orbitals.

An illustrative example of the interaction of a filled n orbital with an empty LUMO frontier orbital is the reaction between BH_3 and the hydride ion H^- (the simplest Lewis base, Figure 5.4). The highest energy electrons in this pair of reactants are the electrons in the $1s$ AO of the hydride. In fact, these electrons present higher energy than the $1s$ orbital of a neutral hydrogen atom. On the other hand, the lowest energy unoccupied MO in both species corresponds to the vacant $2p$ AO at the boron atom (Figure 5.4). As the hydride approaches the borane molecule, electrons present in the hydride ion are shared by the hydrogen and boron to afford the borohydride product BH_4^-. The angle of the hydride approach is determined by the direction in which the reactants come close to each other and provide the maximum overlap between the HOMO and LUMO orbitals. In the reaction between BH_3 and the hydride ion, the effective angle of attack is observed when the hydride ion approaches perpendicularly to the planar BH_3 molecule, resulting in a positive overlap of the $1s$ and $2p$ orbitals (Figure 5.4). The productive interaction is, therefore, on either side of the planar BH_3 molecule.

An additional example of a reaction involving the FMOs n and a is the bond-forming step in the S_N1 reaction, where the carbocation (the LUMO at the Lewis acid, a) reacts with a nucleophile (the HOMO in the Lewis base).

In this regard, carbocations such as the ethyl cation **1** are electron-deficient species with a formal positive charge at carbon, whose six valence electrons participate in three sigma C—H bonds (Figure 5.5a). The positively charged carbon atom adopts a planar trigonal shape (sp^2 hybridisation). The LUMO of the ion

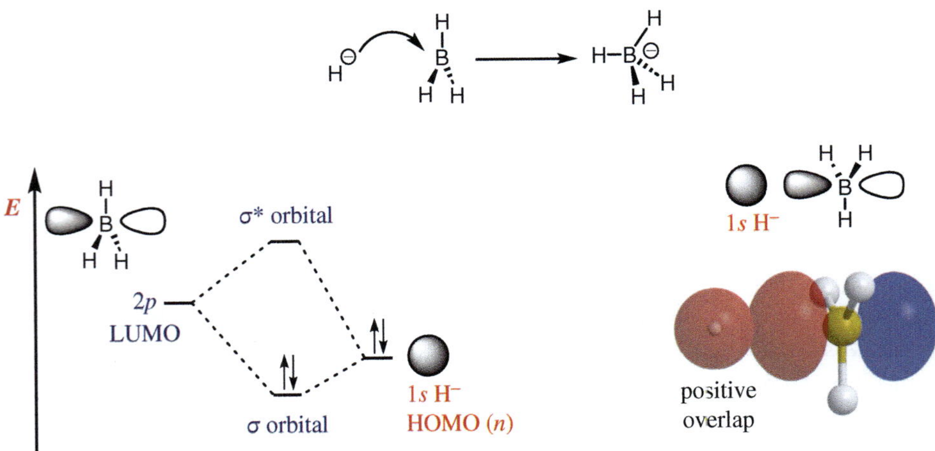

FIGURE 5.4 Interaction of the lone electron pair (n orbital) on the hydride ion H^- and the empty $2p$ orbital in Lewis acid BH_3.

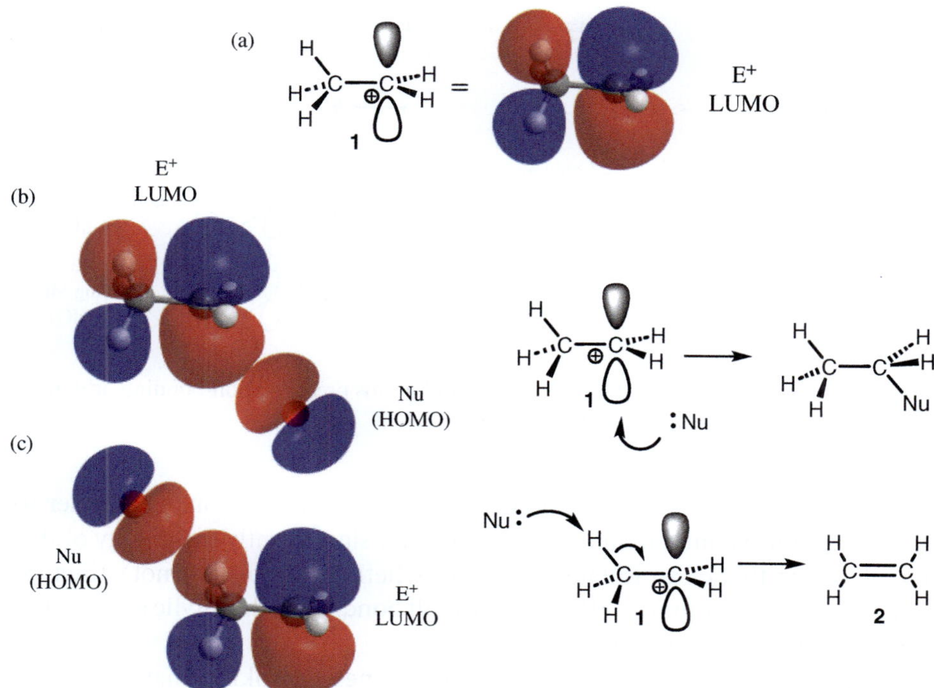

FIGURE 5.5   (a) Pictorial representation of the LUMO in ethyl cation **1**. (b) HOMO–LUMO overlap leading to S_N1 substitution reaction. (c) HOMO–LUMO overlap leading to E1 elimination reaction affording ethylene **2**.

corresponds essentially to the empty *p*-orbital centred on the carbon atom and dominates its reactivity. Nevertheless, it should be noted that there are two reactive sites in the LUMO of cation **1**, where the HOMO of a nucleophile can effectively overlap: the cationic carbon and two of the β-hydrogens (see Figure 5.5a). When the HOMO orbital of the nucleophile overlaps the positively charged carbon region of the LUMO, an S_N1 reaction occurs (Figure 5.5b). However, if overlap takes place with one of the β-hydrogen atoms, E1 elimination reaction takes place (Figure 5.5c). For this reason, S_N1 and E1 reactions are often in competition and chemoselectivity (substitution *versus* elimination) will depend on the specific reacting partners and reaction conditions.

As discussed in Chapter 4, π systems with an odd number of atoms (e.g. the allyl cation and anion), present a non-bonding orbital with energy content between the bonding orbital and the anti-bonding orbital (Figure 5.6a). The energy level MO diagram of the allyl cation is shown in Figure 5.6b. From the combination of the three atomic 2*p* orbitals on each carbon, three MOs at distinct energy levels are generated. The lowest energy MO combines the atomic orbitals all in phase (ψ_1). This is a bonding orbital because all the interactions between vicinal carbons are bonding (same phase overlap). The next MO (ψ_2) presents a nodal plane located through the central atom. This means that there will be no electron density in the central atom when this orbital is occupied.

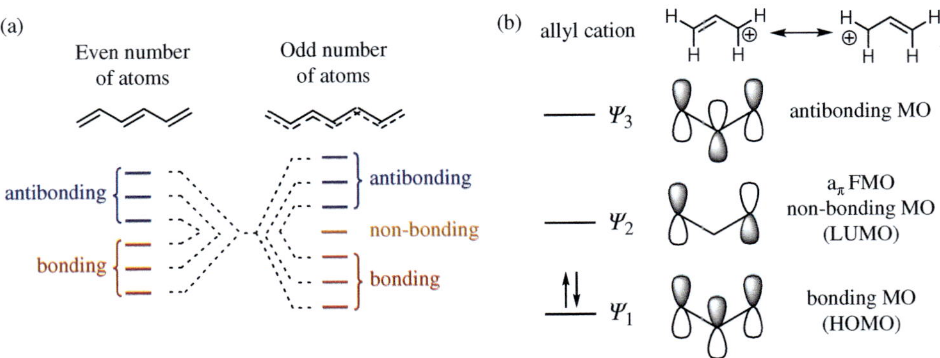

FIGURE 5.6   (a) π systems with an odd number of atoms present a non-bonding orbital. (b) Schematic representation of allyl cation MOs.

The atomic orbitals at C(1) and C(3) in ψ_2 are far apart from each other so that their repulsive interaction does not increase significantly the energy of this MO relative to that of an isolated p-orbital (whether occupied or not); thus, ψ_2 does not contribute significantly to the overall bonding in the allylic species and ψ_2 corresponds to a non-bonding orbital.

Finally, the ψ_3 MO presents two nodal planes provoking that all vicinal atomic orbital interactions are out of phase, affording an anti-bonding MO.

The LUMO orbital of the allyl cation corresponds to ψ_2 and it is consistent with a a_π type orbital. Unlike the n and a non-bonding orbitals, the n_π and a_π conjugated non-bonding orbitals are delocalised with orbital coefficients (lobes) in more than one position.

From the shape of the ψ_2 orbital in the allyl cation, one can notice that incoming nucleophiles will add on either one of the terminal carbons, leading in non-symmetric molecules; that is, to the formation of regioisomers when the terminal carbons are not equivalent. In terms of valence bond theory, these electrophilic points of attack are the result of electron delocalisation in species **3** and **4** (depicted by a delocalisation arrow $\leftrightarrow$) (Figure 5.7a). Figure 5.7a also shows how the positive charge is delocalised in the two terminal carbons.

It should be recalled here that the electrons in this system are not really 'moving around', they are simply distributed over all three carbons. One way to represent the delocalised cationic species is to draw the allyl cation as in **5**, showing the delocalisation of the p electrons with a broken line, placing the positive charge in the centre of the π system. Alternatively, the partial positive charges can be depicted at the terminal carbons (**6**). Representations **5** and **6** emphasise the similarity of the two terminal bonds and the distribution of the positive charge at both terminal carbons. The nucleophilic approach on these sites allows for efficient orbital overlap; indeed, the pictorial representation of the LUMO (Figure 5.7b) makes it clear that the terminal carbons are the preferred sites for nucleophilic addition. Solvolysis of the bromide Br^- atom in compound **7** leads to the formation of the allylic cation **8**, which is non-symmetrical ($R_1 \neq R_2$).

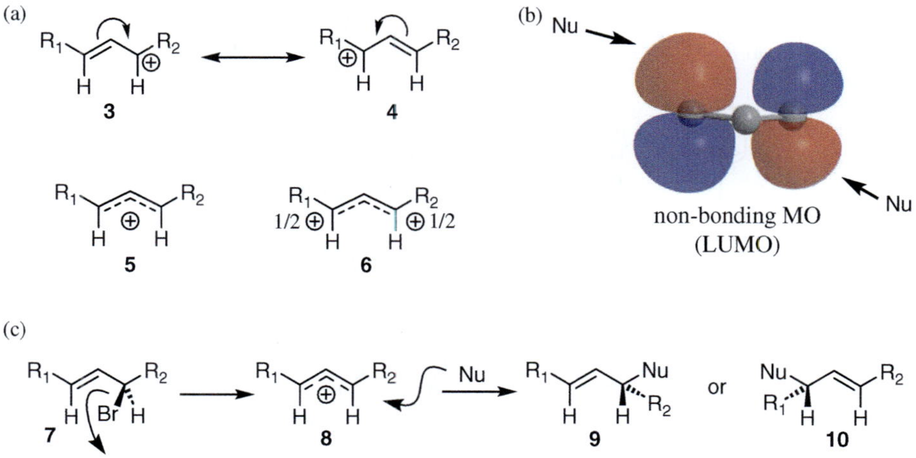

FIGURE 5.7 (a) Electron delocalisation in canonical cationic species **3** and **4** and charge distribution in delocalised representations **5** and **6**. (b) Nucleophilic attack at the terminal electrophilic carbons. (c) Bromide elimination with formation of cationic species **8** followed by nucleophilic attack to give isomeric products **9** and/or **10**.

Therefore, depending on where the nucleophile binds, either compound **9** or **10** is formed. This of course depends on the nature of R_1 and R_2 (Figure 5.7c).

Another type of reaction involving conjugated non-bonding orbitals as FMOs consists of the alkylation of enol ethers and their silylated derivatives (Figure 5.8). They can react, for example, with a $2p$ orbital of a carbocation (FMO type a, Figure 5.8a) or with an allyl cation (FMO type a_{π}, Figure 5.8b). The enol ether reacts through its ψ_2 HOMO (n_{π} type FMO).

Enol ethers are allyl anion-type systems (3 atoms, 4 electrons, Figure 5.8c). Enamines are another important class of allyl anion-type systems (Figure 5.8d). In particular, the HOMO energy of carbonyl-containing substrates is raised *via* the formation of their corresponding enamine derivative, that increases their nucleophilicity and facilitates their reaction with electrophiles relative to their enol counterparts. The HOMO energy of the enamine is higher than that of the corresponding enol because the lone pair electrons of the nitrogen atom are higher in energy relative to those at the oxygen atom. The activation of carbonyl compounds by means of enamines for α functionalisation has become a powerful tool in synthetic organic chemistry and asymmetric synthesis.

One of the most important reactions in organic chemistry consists in the nucleophilic addition to carbonyl compounds. The nucleophile may be the HOMO of an anion, such as cyanide or hydroxide ions, or it may be a neutral species, such as water or an amine. This process involves the anti-bonding LUMO (π*) of the carbonyl and the HOMO–LUMO overlap that initiates the reaction places electrons from the nucleophile in such anti-bonding π* orbital, resulting in the breaking of the π double bond. It is then appreciated that placing electrons in anti-bonding orbitals has a destabilising effect on the electron-accepting

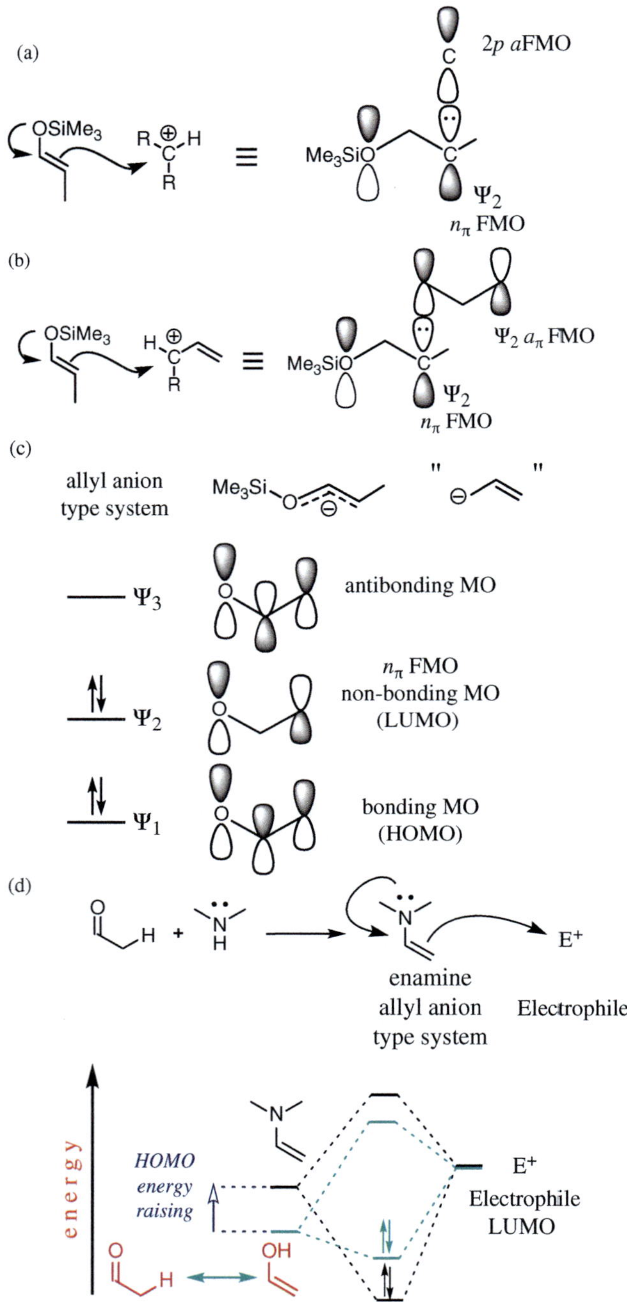

FIGURE 5.8   Non-bonding orbitals as FMOs in the alkylation of enol ethers and allyl anion type systems. (a) Enol ethers can react, for instance, with a $2p$ orbital of a carbocation. (b) Reaction of silyl enol ethers with carbocations highlighting the bonding interaction between ψ_2 HOMO (n_π type FMO) and the ψ_2 a_π allyl cation LUMO. (c) Enol ethers are allyl anion-type systems (3 atoms, 4 electrons). (d) Enamines are another important class of allyl anion-type systems. The HOMO of an enamine is higher in energy than that of the corresponding enol because the lone pair electrons of the nitrogen atom are higher in energy relative to those at the oxygen atom and this is reflected in increased nucleophilicity.

moiety. In the carbonyl group, the π C—O bond is formed by the overlap of the $2p$ orbital at carbon and the $2p$ orbital at oxygen (Figure 5.9a). Since oxygen is more electronegative than carbon, the carbonyl π orbital is closer in energy to the oxygen AO and the π^* orbital is closer in energy to the carbon AO. Therefore, the π orbital has a higher coefficient at oxygen and the π^* orbital has a higher coefficient at carbon (see also Chapter 3, Figure 3.36). The higher coefficient of the π^* orbital on carbon means that there is a better HOMO–LUMO overlap at carbon, so it is at carbon where the nucleophile will attack.

An understanding of the orbital interactions involved in the reaction of nucleophiles with the carbonyl group provides an insight into the most effective angle of attack by the nucleophile as it approaches the carbonyl (Figure 5.9b). If the nucleophile (HOMO) approaches from the back of the carbonyl towards the partially positive carbon atom at the planar carbonyl group, the in-phase and out-of-phase overlap with the π^* orbital (LUMO) cancel each other so that there will be no net interaction (Figure 5.9b). Experimental evidence and computational calculations suggest that the most favourable angle of attack by the nucleophile is sidewise, at approximately 105–109° relative to the carbonyl's plane (Figure 5.9c). This angle of approach is known as the Bürgi–Dunitz trajectory and corresponds to a balance between (i) the maximum orbital overlap of the nucleophile's HOMO with the carbonyl's π^* LUMO, (ii) the Coulombic attractive and repulsive forces between the molecules, and (iii) the minimum repulsion of the nucleophile's HOMO with the electron density at the bonding π-carbonyl bond. All these terms are included in the Salem–Klopman equation (see above). Potential trajectories of nucleophilic attack to the π-bond are considered in the Felkin–Anh model (Figure 5.9d) and the Baldwin's rules for nucleophilic ring closure (not shown).

As the nucleophile approaches the carbon atom in a carbonyl group, its HOMO electron pair interacts with the π^* LUMO to form a new Nu—C σ bond. When the nucleophile's electrons occupy the π^* orbital of the carbonyl group, the π bond is broken, leaving the σ C—O bond intact. The electrons that initially occupied the π bond move to the electronegative oxygen, which ends up with a negative charge. The HOMO of the nucleophile is frequently a sp^3 orbital lone electron pair or a carbon-metal σ orbital. The π^* orbital can also be part of a double bond in an imine or a triple C—N bond in a nitrile group, or the ψ_3 LUMO in an α,β-unsaturated carbonyl system. Furthermore, the π^* LUMO orbital can also be the lowest unoccupied π orbital of an electron-deficient aromatic ring system such as nitrobenzene or pyridine.

Competitive 1,2- or 1,4- nucleophilic addition to α,β-unsaturated carbonyl compounds can be explained by MOT. In this regard, the shape of the MOs in π-conjugated systems such as α,β-unsaturated carbonyl derivatives can be determined by means of the perturbation approximation of the orbital coefficients (Figure 5.10). For example, acrolein can be represented using traditional resonance models, with one major canonical form (**11**, neutral state) and two minor dipolar canonical forms **12** and **13** (Figure 5.10a). The π-orbital system of

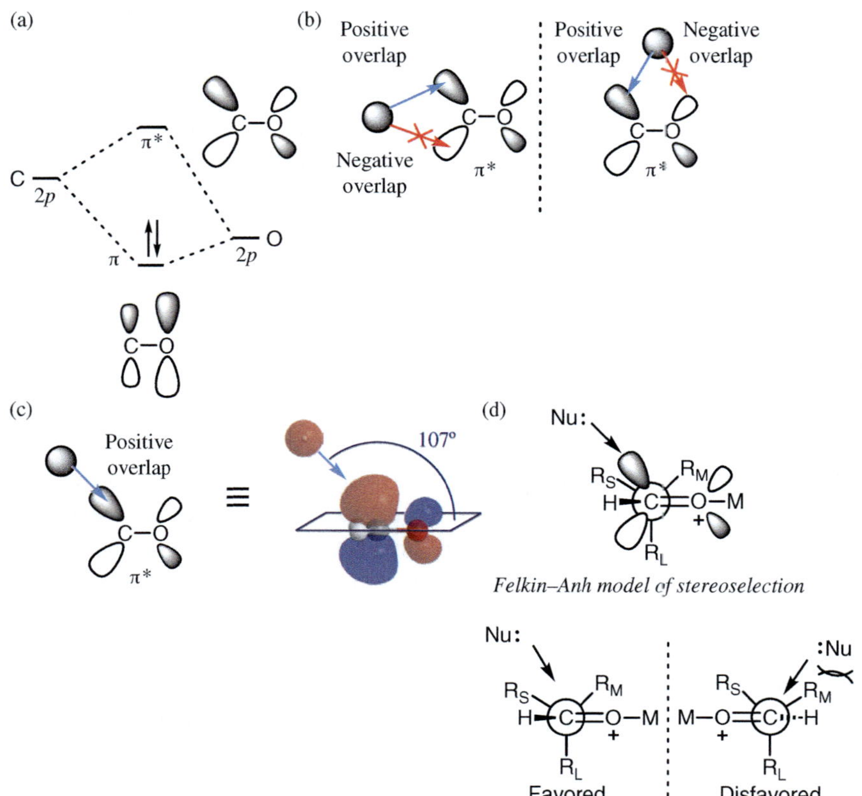

FIGURE 5.9   MOs involved in nucleophilic attack on carbonyl groups. (a) The carbonyl's bonding (π, HOMO) and anti-bonding (π^*, LUMO) MOs. (b) Coplanar nucleophile addition from the back of the carbonyl leads to in-phase and out-of-phase orbital overlap that cancel each other preventing bond formation. (c) Constructive in-phase overlap at the Bürgi–Dunitz angle leading to bond formation. (d) Felkin–Anh HOMO/LUMO model for addition of the nucleophile to the carbonyl.

acrolein consists of four atoms (the three carbon atoms plus the oxygen atom, all sp^2 hybridised) and four π electrons. Therefore, the system consists of two occupied and two unoccupied MOs. To define their FMOs the orbitals are split into a *base system* (Figure 5.10b), which corresponds to the hydrocarbon analogue that most closely resembles the neutral canonical form, in this case 1,3-butadiene, as well as *the perturbing influence*, which corresponds to the hydrocarbon ion that most closely approximates the bipolar canonical form, in this case the π orbitals of the allyl cation. Using the approximation method, the HOMO of acrolein is predicted to be similar to ψ_2 in 1,3-butadiene modified by ψ_1 of the allyl cation, and the LUMO is anticipated to be similar to ψ_3 in 1,3-butadiene modified by ψ_2 of the allyl cation. As it can be seen in a qualitative manner, the superposition of the ψ_1 of the allyl cation on the ψ_2 of 1,3-butadiene gives an orbital in which the largest orbital coefficient is located on the β-carbon, and it is therefore at this carbon that nucleophiles will attack. Furthermore, the second largest coefficient

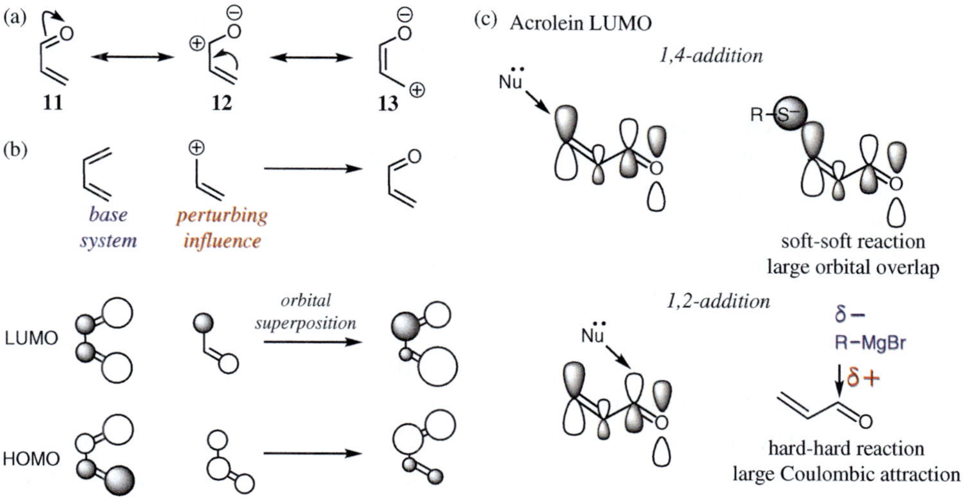

FIGURE 5.10 Perturbation approximation for the orbital coefficients in acrolein. Competitive nucleophilic 1,2- or 1,4- addition to α,β-unsaturated carbonyl compounds explained by MOT.

is located on the carbonyl's carbon atom, corresponding to the two electrophilic sites leading to different pathways, 1,2- or 1,4-addition.[1]

The 1,2-addition pathway will dominate when strongly basic nucleophiles are used so that the addition proceeds irreversibly at low temperatures under kinetic control. On the other hand, the 1,4-addition pathway tends to dominate when the nucleophile is a weak base, such as the conjugate base of alcohols or amines, or when the reaction is heated to reach equilibrium (under thermodynamic control). Nevertheless, the most important variable for attaining regioselectivity corresponds to the nature of the nucleophile. One helpful model used to explain these divergent reactivity patterns is the hard or soft acid–base (HSAB) theory (Figure 5.10c). The HSAB theory is a consequence of the Klopman–Salem equation (see above); in particular, 'hard' nucleophiles tend to react with 'hard' electrophiles and 'soft' nucleophiles tend to react with 'soft' electrophiles. The carbonyl carbon in α,β-unsaturated carbonyl substrates is more electropositive than the β carbon, making it a harder electrophile. As indicated above, hard nucleophiles with a low energy HOMO will then react in 1,2- manner. In Grignard and organolithium reagents, the carbon—metal bond has a highly ionic character owing to the very low electronegativity of Li and Mg. Therefore, the carbon in these reagents will present high electron density, associated with a hard nucleophilic site that will react preferentially at the carbonyl carbon (Figure 5.10c). A hard–hard reaction is fast because

---

[1] The perturbation approximation of the orbital coefficients is supported by a rigorous mathematical backing and allows for estimation of the relative energies of the FMOs of interest. This procedure can also be used to predict the regioselectivity of various processes such as Diels–Alder reactions. For more information, the reader is referred to the further readings cited at the end of this chapter.

of its large Coulombic attraction component. On the other hand, soft nucleophiles and soft electrophiles are usually relatively large, polarisable atoms with low charge density. The β-position of conjugated carbonyl compounds is 'soft' as it presents lower charge density.

Soft nucleophiles such as thiolate and iodide ion will preferentially react in 1,4-fashion (Figure 5.10c, top). In soft–soft reactions, it is the orbital overlap term that dominates their reactivity.

Many addition reactions to carbonyl compounds can be accelerated by the presence of acid or metal cations. This is mainly due to two effects:

1. The electrostatic interaction between the electron-rich nucleophile and the protonated or metal-coordinated carbonyl carbon becomes dominant.

2. The LUMO of the π-carbonyl system is lower in energy following protonation or metal coordination to the carbonyl, leading to a stronger interaction with the HOMO of the nucleophile (Figure 5.11a).

The addition of a proton or a metal cation to the oxygen atom increases its effective nuclear charge that lowers the energy of the corresponding atomic

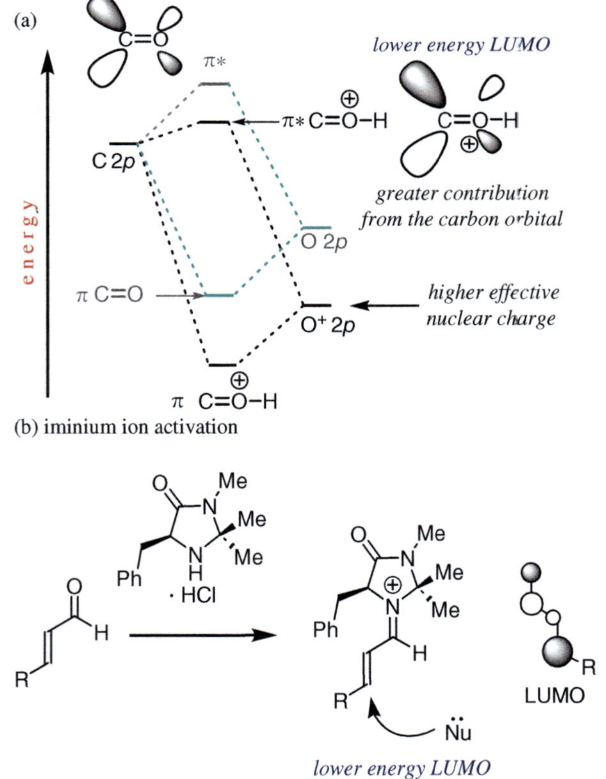

FIGURE 5.11   (a) Effect of protonation or metal coordination of carbonyl groups on MO energy levels. (b) Activation of conjugated carbonyl groups by iminium ion formation.

orbital relative to its original neutral state (Figure 5.11a), so that both the π and π^* orbitals are lower in energy relative to the neutral carbonyl group. By lowering its energy, the π^* orbital will present a larger contribution of the carbon coefficient (in position 2 for simple carbonyls and in position 4 for conjugated carbonyls), allowing a more efficient overlap with the incoming nucleophiles. This effect can be of practical use in asymmetric catalysis, where chiral Lewis acids can activate carbonyl compounds to increase their reactivity in a stereoselective reaction. In this regard, the working principle of iminium ion catalysis is associated to the lowering of the LUMO energy in an α,β-unsaturated aldehyde following condensation with a secondary amine catalyst (Figure 5.11b).

Bimolecular S_N2 nucleophilic substitutions are another relevant class of reactions in organic synthesis. These reactions involve the nucleophilic donation of electrons to the σ^* anti-bonding MO in a C–X bond, which results in the breaking of such bond. Figure 5.12a shows pictorially the LUMO of a general-type alkyl halide **14**. The S_N2 reaction takes place when the HOMO of the nucleophile donates electrons to the unoccupied σ^* orbital between carbon and the leaving group, for example, a halide. To achieve optimal orbital overlap, the nucleophile must attack at an 180° antiperiplanar angle relative to the leaving group, which induces the leaving group to move away. The product is formed with an inversion of the configuration at the central atom (Walden inversion). In the transition state, a planar sp^2-hybridised central carbon atom presents a p orbital containing a pair of electrons that are shared between the old and new bonds (Figure 5.12a). Therefore, the p orbital at the central carbon participates

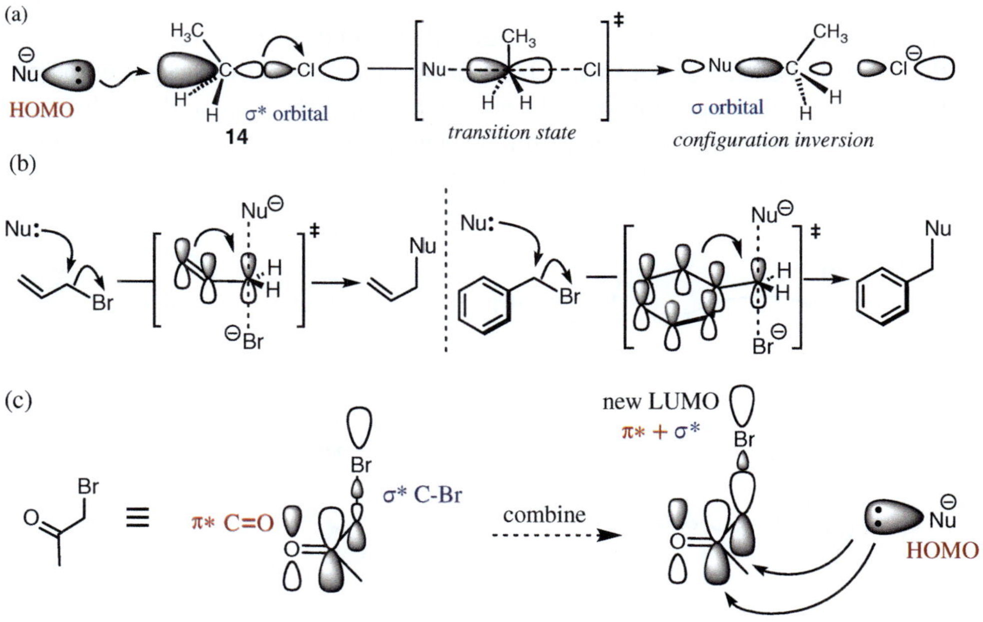

FIGURE 5.12 MOs involved in S_N2 reactions of alkyl halides, allyl bromide and α-halocarbonyl compounds.

in two partial bonds containing only two electrons, and that are therefore electron deficient. As a consequence of the nature of this transition state, adjacent C=C or C=O π systems increase the rate of S_N2 reactions by stabilisation of the electron-deficient carbon.

Allyl and benzyl groups adjacent to an electron-deficient carbon that is bonded to a leaving group accelerate the rate of nucleophilic substitution (Figure 5.12b). In the case of α-halocarbonyl compounds, two adjacent carbon atoms are strongly electrophilic sites, since each of them presents empty orbitals of low energy: the π^* orbital of the C=O group and the σ^* orbital of the C—Br bond. These two anti-bonding MOs can combine to form a new LUMO ($\pi^* + \sigma^*$) of lower energy than either of them (Figure 5.12c). Nucleophilic attack is then likely to occur at the position where this new orbital presents the largest coefficient. Although there will be competition between the two electrophilic sites, carbonyl attack may be reversible whereas bromide displacement is irreversible, a factor which constitutes a driving force in the direction of S_N2 reactions.

Electrophilic additions to alkenes and electrophilic aromatic substitutions are additional examples of reactions controlled by HOMO/LUMO interactions. In the bromination of alkenes, the filled π orbital of the alkene (HOMO) interacts with the empty σ^* orbital of bromine (Figure 5.13). The most efficient way for a π HOMO in an alkene to interact with the σ^* LUMO of Br_2 is to approach from the back side of the Br—Br bond (Figure 5.13). The initial product consists of a three-membered ring termed a bromonium ion **15**, which arises from the simultaneous cleavage of the π bond and cleavage of the σ Br—Br bond, releasing the Br^- ion as the leaving group. In the second step, the electrophilic bromonium ion **15** reacts with the bromide ion. This corresponds to an S_N2 substitution reaction, where the orbitals involved are the n filled (HOMO) of the bromide ion and the σ^* orbital of one of the two carbon—bromine bonds in the three-membered ring. This second step leads to the formation of one C—Br bond and the cleavage of one C—Br bond in the bromonium ion. The overall result is the *anti*-addition of the bromine molecule to the alkene double bond; that is, the Br^- ion attacks

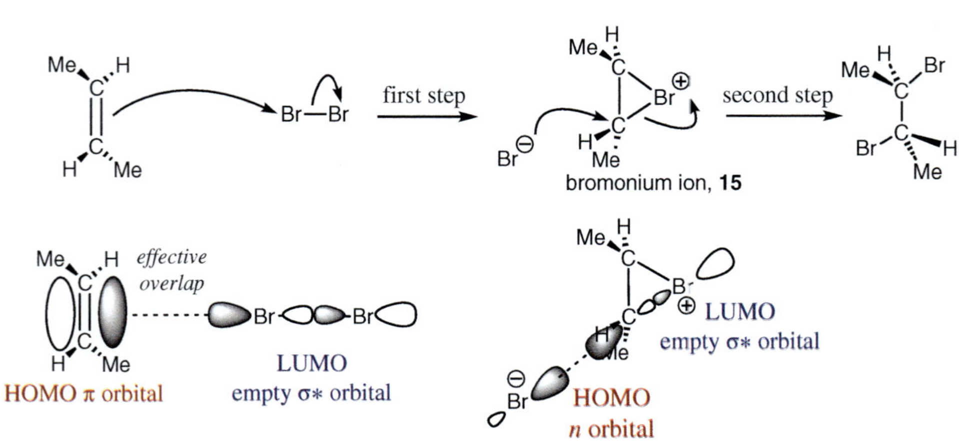

FIGURE 5.13   MOs involved in the electrophilic addition of bromine to alkenes.

the back of the carbon–halogen bond, which, as discussed above, allows for a better overlap of the HOMO–LUMO frontier MOs. Consequently, bromination of alkenes is stereospecific because the geometry of the starting alkene determines which diastereomeric product will be obtained.

In this context, the oxidation of alkenes with peracids to form epoxides involves the overlap of the filled π orbital of the alkene (HOMO) with the σ^* LUMO of the O–O bond in the corresponding peracid (Figure 5.14a). The reaction is also stereospecific as the two C–O bonds on the epoxide are positioned on the same side of the π bond of the alkene. Another example of this type of orbital interaction is the hydroboration reaction of alkenes, where the π-orbital of the alkene attacks the anti-bonding σ^* orbital of the B–O bond in its complex with THF (Figure 5.14b).

Electrophilic aromatic substitution (S_EAr) is a fundamental class of organic reactions in which an electrophile replaces a hydrogen atom in an aromatic ring. A classic example is the bromination of benzene, facilitated by Lewis acids such as $AlBr_3$. This reactivity can be explained using MOT. The reaction begins with the interaction between the activated electrophile (Figure 5.15) and the π-electrons of the aromatic ring, forming a π-complex (step 1). In this complex, the π-electrons of the aromatic ring interact with the σ^* antibonding orbital of the electrophile. For substituted aromatic rings, differences in the coefficients of

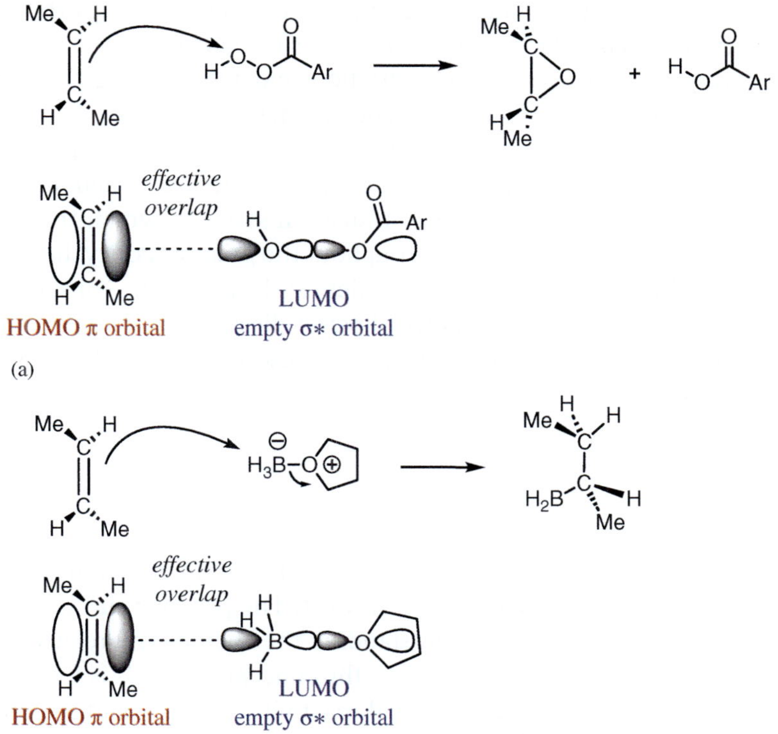

FIGURE 5.14 MOs involved in the oxidation and hydroboration of alkenes.

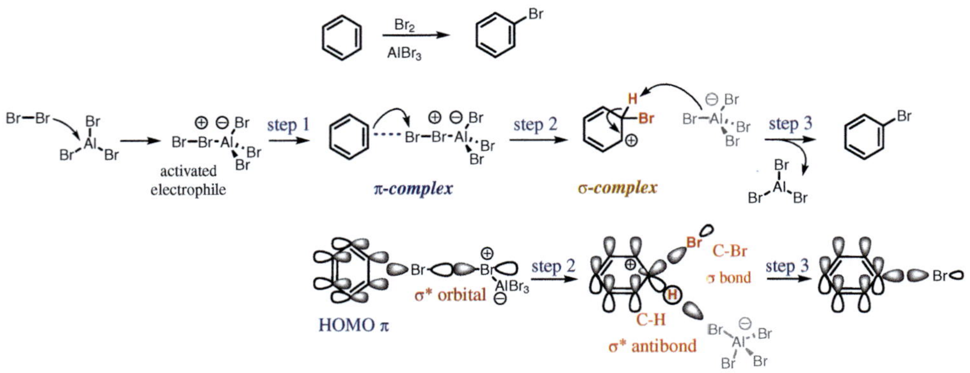

FIGURE 5.15 MOs involved in the electrophilic aromatic substitution of benzene.

the p orbitals result in variations in electron density across the ring. This causes the electrophile to associate more closely with certain atoms in the π-complex, influencing regiochemistry. Following this pre-organisation, the electrons on the HOMO of the aromatic ring attack the $\sigma*$ antibonding orbital of the activated electrophile, forming a new C—Br σ-bond (step 2). The involved carbon atom changes its hybridisation from sp^2 to sp^3, resulting in the loss of a p orbital at that position and the disruption of aromaticity. The resulting intermediate, a carbocation, is known as an arenium ion (or Wheland intermediate). In the final step, a base (e.g., $AlBr_4{}^-$) donates a pair of electrons to the $\sigma*$ antibonding orbital of the C—H bond (step 3). This electron transfer process results in cleavage of the C—H bond, restoring the carbon's sp^2 hybridisation and regenerating the aromatic π-system. The final product is the brominated aromatic derivative.

Pericyclic reactions are a special class of organic reactions of great importance in synthetic organic chemistry. This type of reaction is characterised by the fact that it proceeds through cyclic transition states in which the involved MOs maintain bonding interactions throughout the process. Pericyclic reactions follow the Woodward–Hoffmann rules, which are based on the conservation of orbital symmetry. Given the importance of pericyclic reactions and the extensive research based on MOT, these reactions will be described in detail in the following chapters.

SUMMARY

In this chapter, we have reviewed several types of reactions that are central to organic chemistry from the perspective of MOT. The primary aim was to highlight the importance of MOT in understanding the chemo- and regioselectivity of various processes. As we have explored, the concept of FMOs is fundamental to predicting and understanding chemical reactivity, stability, and the outcome of molecular interactions, which are very useful tools in the field of synthetic organic chemistry.

FURTHER READING

E. V. Anslyn, D. A. Dougherty, *Modern Physical Organic Chemistry*, University Science Books, Sausalito, California, **2005**.

J. Keeler, P. Wothers, *Chemical Structure and Reactivity: An Integrated Approach*, Oxford University Press, Oxford, UK, **2008**.

I. Fleming, Molecular *Orbitals and Organic Chemical Reactions*, John Wiley & Sons, NY, **2010**.

D. E. Lewis, *Advanced Organic Chemistry*, Oxford University Press, Oxford, UK, **2015**.

J. Clayden, N. Greeves, S. Warren, *Organic Chemistry* 2nd ed. Oxford University Press, Oxford, UK, **2012**.

I. V. Alabugin, *Stereoelectronic Effects: A Bridge Between Structure and Reactivity*, Wiley, Chichester, UK, **2016**.

EXERCISES

5.1 Which FMOs are involved in the following aldol reaction? Draw an energy level correlation diagram of the participating MOs. How would the addition of a secondary amine catalyst affect the FMOs?

5.2 In a qualitative way, construct the approximate FMOs of 1-alkoxy-1, 3-butadiene using the orbital coefficient perturbation method.

5.3 What is the expected regioselectivity of the reaction of carbonyl compound **16** with nucleophiles **a-d**?

a, CH$_3$MgBr
b, Me$_2$CuLi
c, EtNH$_2$
d, LiAlH$_4$

16

5.4 Which FMOs are involved in the electrophilic substitution reaction between benzene and *tert*-butyl chloride? What is the role of AlCl$_3$ in this reaction?

CHAPTER 6

Some Applications of Orbital Theory in Organic Chemistry

INTRODUCTION

Having discussed in previous chapters the qualitative and quantitative aspects of Hückel's method for describing molecular orbitals (MOs) and their interactions, it is now important to illustrate how these models can explain experimental techniques such as visible and ultraviolet spectroscopy, as well as photoelectron spectroscopy.

ULTRAVIOLET SPECTROSCOPY

In a spectroscopic measurement, electromagnetic radiation is applied to a sample of the compound of interest in order to promote the excitation of electrons at particular MOs to higher energy level orbitals, which usually go back to the original ground state electronic configuration via emission of light. Molecular orbital theory (MOT) provides insight into the electronic transitions that occur when molecules absorb or emit light. It is important to note that a given molecule will only absorb electromagnetic radiation whose wavelength corresponds to an amount of energy that is equivalent to the energy difference associated with the corresponding electronic transition. That is, the transition corresponds to an energy jump ΔE of electrons from the ground state **A** to the excited state **B** (Figure 6.1). Again, the molecule will only absorb radiation of a wavelength that corresponds to this particular energy gap, while other radiation of different wavelengths will pass through the sample without being absorbed.

Spectroscopic determination of the wavelength of the radiation that is absorbed by a molecule provides information on the feasible electronic transitions, and thus, on its molecular structure. In this regard, sample irradiation in the ultraviolet or near-visible range (200–400 nm) induces electronic transitions

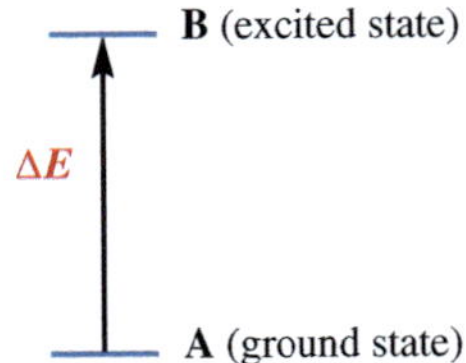

FIGURE 6.1   Energy difference for the electronic excitation from the ground state **A** to the excited state **B**.

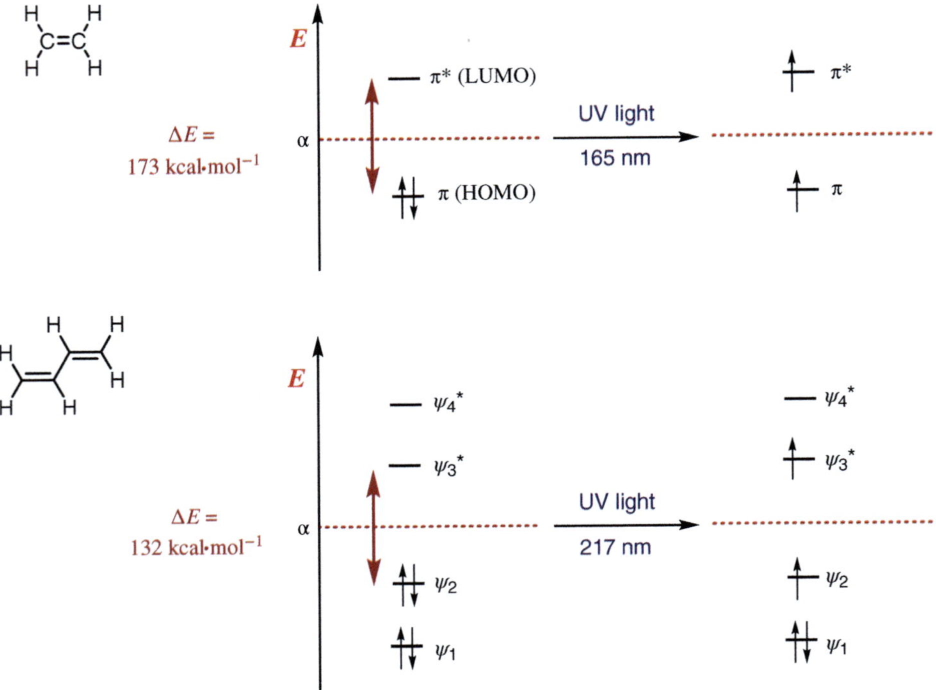

FIGURE 6.2   Electronic π-π^* transitions of ethylene and 1,3-butadiene.

in a large number of organic molecules. This indicates that when a molecule absorbs radiation in this energy range, one of its electrons is promoted from a low-energy MO to a higher-energy MO.

Let us consider the ethylene molecule (simplest π-system) as an illustrative example. When this organic compound absorbs ultraviolet light with a wavelength of 165 nm, it undergoes a π-π^* transition; in other words, electrons are promoted from a low-energy π molecular orbital (highest occupied molecular orbital, HOMO) to the higher-energy π^* level (lowest unoccupied molecular orbital, LUMO). The energy gap for these electronic transitions becomes narrower in conjugated systems compared to an isolated double bond (Figure 6.2). Indeed, the HOMO/LUMO energy difference in butadiene is smaller than that found in ethylene. In fact, the more conjugated a compound is, the smaller the HOMO/LUMO energy transition and the longer the wavelength of light that is absorbed. In this way, ultraviolet-visible (UV-vis) spectroscopy provides information about

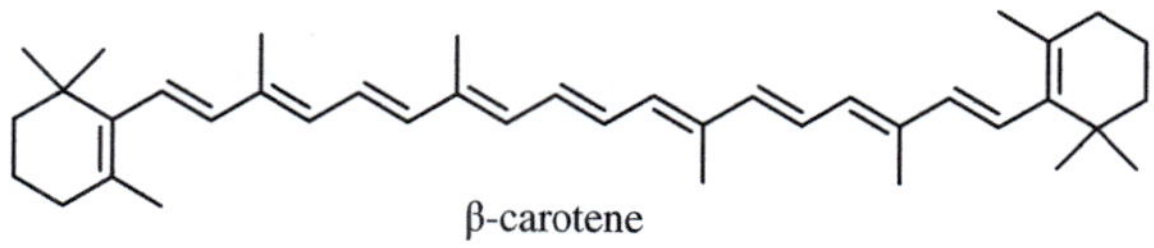

β-carotene

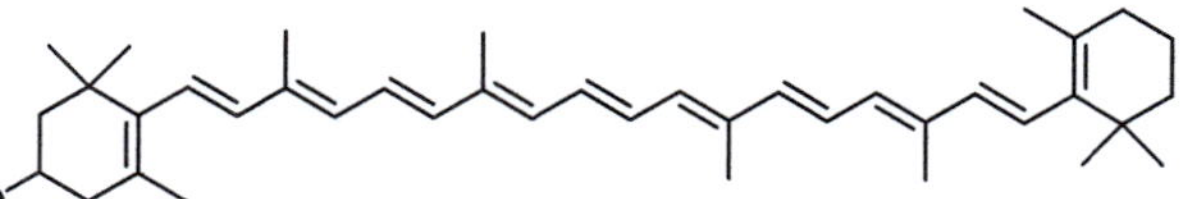

HO β-cryptoxanthin

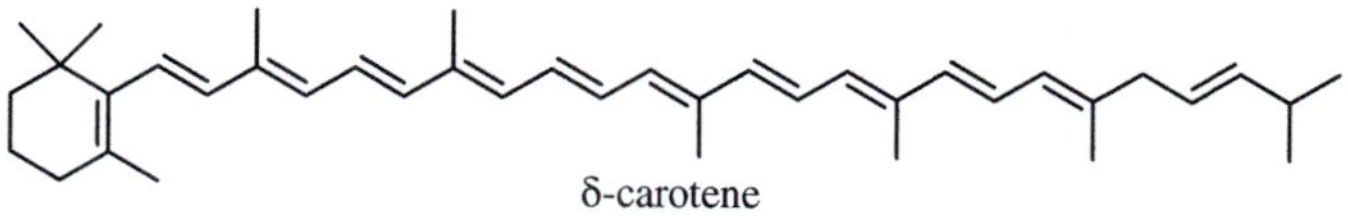

δ-carotene

FIGURE 6.3 Salient examples of molecules with extended conjugated π-systems.

the degree of conjugation that is present in a molecule. Molecules or molecular fragments containing the electrons involved in electronic transitions giving rise to absorption in the UV-visible region are called chromophores.

In molecules containing extended systems of conjugated double bonds, the energy difference between the HOMO and LUMO orbitals becomes so small that absorption takes place in the visible region of the electromagnetic spectrum. Examples of this are molecules of the carotene type (Figure 6.3). For example, β-carotene, the red pigment in carrots and other vegetables, presents 11 conjugated C=C double bonds, resulting in a very small energy difference between the π and π^* MOs. This gives rise to characteristic low-energy absorptions, up to 500 nm in wavelength.

Two characteristic phenomena are typically observed in UV-visible spectroscopy:

1. Bathochromic (red) shift that take place when the absorption wavelength of a substance shifts towards longer (lower energy) wavelengths. Accordingly, the modified substance will absorb light of lower energy relative to the absorption band present in the original system.
2. Hypsochromic (blue) shift, when the absorption wavelength shifts towards shorter wavelengths (higher frequencies), which induces absorption of higher energy light relative to the original absorption spectrum.

Bathochromic or hypsochromic shifts are indicative of interaction between two chromophores (usually π-systems or unshared electron pairs) in a molecule. These effects are observed when the overlap between the π orbitals of interest is non-zero. An emblematic example is found in the comparison of the UV spectra recorded for norbornene **1** and norbornadiene **2** (Figure 6.4).

The transannular (through space) overlap between the two bonding π orbitals and the two anti-bonding π^* orbitals in norbornadiene **2** gives rise to

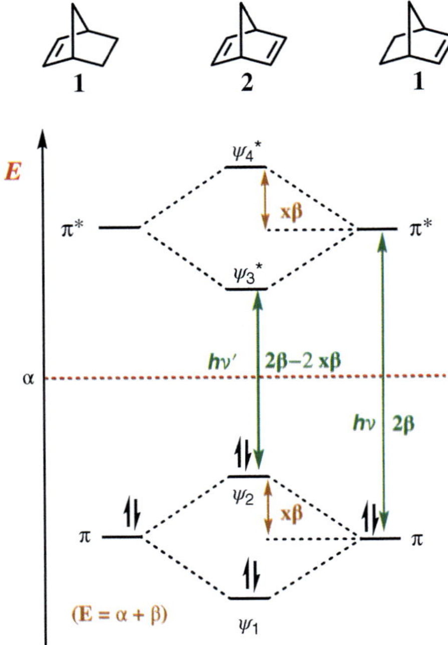

FIGURE 6.4 MO diagram of π-π^* transitions of norbornene and norbornadiene.

four new MOs ψ_1, ψ_2, ψ_3 and ψ_4; the first two of them (lower in energy) are occupied and the following two (higher in energy) are unoccupied. The magnitude of the through space interaction can be quantified as a fraction of the normal resonance integral: $x\beta$, since it is expected to be smaller than the resonance integral in the original π segments (β). The influence of the transannular interaction is reflected in a decrease of the energy $h\nu$ required for the $\pi \rightarrow \pi^*$ transition in norbornene **1** ($\Delta E = 2\beta$) compared with the transition energy in norbornadiene **2**, where the HOMO/LUMO difference ($\psi_2 \rightarrow \psi_3$) is $\Delta E = 2\beta - 2\,x\beta$. Experimentally it was found that $x \approx 0.3$.

IONISATION POTENTIALS

The ionisation potential (IP) corresponds to the amount of energy required to remove an electron from an isolated atom or molecule. Koopmans' theorem is a concept in quantum chemistry that provides a simple way to understand the relationship between electron energy levels in a molecule and its ionisation energy (*IE*). In general terms, Koopmans' theorem states that the energy required to remove an electron from an orbital is equal to the negative energy of the occupied molecular orbital (OMO).

$$-E_{OMO} = IP \tag{6.1}$$

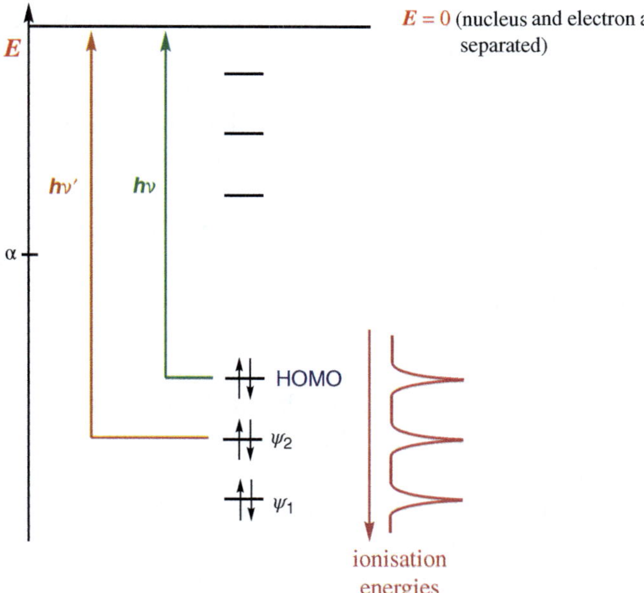

FIGURE 6.5   Correlation of ionisation energies with relative energy levels of occupied MOs.

As it can be appreciated in Figure 6.5, the lowest ionisation potential corresponds to electron removal from the HOMO, whereas the ionisation energies of the lower-energy occupied MOs will be higher. It is important to note that this is a simplified approximation since Koopmans' theorem does not take into account that the removal of an electron causes some changes in the energy level of the remaining electrons. Thus, the actual ionisation energy IE may be slightly different. Nevertheless, from photoelectron spectra, chemists can easily estimate approximate ionisation potentials without resource to complex calculations, making photoelectron spectroscopy a useful tool for the interpretation of molecular electronic structures.

PHOTOELECTRON SPECTROSCOPY (PES)

Actually, the ionisation process described in the previous section applies to both π and σ orbitals. When a high-energy photon hits an electron in an atom or molecule, the ionisation process takes place. The electron is ejected with an energy equal to the energy difference of the incident photon minus the IE (Eq. 6.2).

$$K = h\nu - IE \tag{6.2}$$

Normally, photoelectron spectra register the ionisation energies for the higher level atomic or MOs, typically occupied π or n (unshared electron pairs) orbitals. In addition, photoelectron spectroscopy can record the ionisation potential of lower energy orbitals, such as the σ orbitals and occupied orbitals closer to the atomic nuclei. This is achieved by irradiating the sample in the gaseous state with high-energy monochromatic light from a He lamp. The electrons ejected from the sample are quantified according to their kinetic energy (K). Thus, incorporation of the values for $h\nu$ and K in Eq. (6.2), allows the construction of a graph of ejected electrons and their IE. This graph provides essential information about the electronic structure of a molecule as well as relevant data on the composition of a material that is useful for the recognition of bonding characteristics by means of the determination of the energy distribution of the emitted electrons.

In this regard, the 58.4nm photon obtained from an He light source presents an associated energy of 21.2 eV, i.e. 488.9 kcalmol^{-1}, which is able to ionise electrons with IE lower in magnitude, in particular, valence electrons. This ionisation technique is called ultraviolet photoelectron spectroscopy (UV-PES). Furthermore, by using a higher energy source it is possible to achieve the ionisation of electrons that are closer, and therefore more attracted by the positive nucleus; for example, the X-ray Cr K_α emission line with an energy of 5414.7 eV is used in the method known as ESCA (electron spectroscopy for chemical analysis). For organic chemists, photoelectron spectroscopy is particularly informative as it provides information about the consequences of interactions between π and n orbitals.

INTERACTIONS BETWEEN π ORBITALS

EXAMPLE 1

Ethylene presents the lowest IE at 11.65 eV, which corresponds to the loss of an electron from the HOMO π orbital (Figure 6.6).

By comparison, less energy is required to remove an electron from the HOMO in 1,3-butadiene; i.e. 9.06 eV. This value indicates that Ψ_2 is about 2.5 eV higher in energy than the π orbital in ethylene and provides an estimate of the importance of the stabilising interaction of two conjugated ethylene units to 1,3-butadiene owing to increased charge delocalisation (Figure 6.7).

FIGURE 6.6 Ionisation of ethylene.

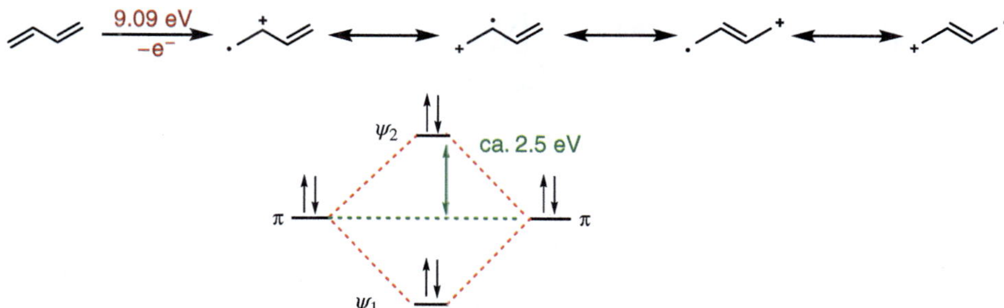

FIGURE 6.7   Energy required for the ionisation of 1,3-butadiene in comparison with the energy required for the ionisation of an isolated ethylene.

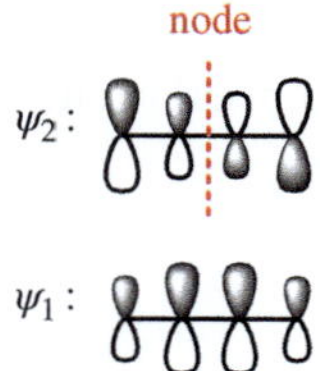

FIGURE 6.8   MOs ψ_1 and ψ_2 in 1,3-butadiene.

It should be noted that the ionisation of the HOMO (ψ_2) in 1,3-butadiene effectively reduces the contribution of this MO in the electronic nature of the derived cation. Thus, considering that ψ_2 presents a node, and therefore anti-bonding character between C(2) and C(3) (Figure 6.8), then a decrease in the participation of this orbital should result in greater double bond character between the central carbons C(2) and C(3) in the cation.

EXAMPLE 2

The magnitude of the interaction between the two π segments in norbornadiene is about one-fifth of that present in a true diene (0.43 eV vs. 2.5 eV, respectively) (Figure 6.9). This result is in qualitative agreement with the one obtained by UV/Vis spectroscopy (cf. Figure 6.4).

EXAMPLE 3

The ionisation energies for various radicals show the importance of the delocalisation of the positive charge on the derivatised radical cation: the greater the stability of the corresponding cation, the lower the *IE*. Similarly, it can be seen that the energy required to remove an electron from an anti-bonding level is much lower than that required to remove an electron from the bonding levels (Figure 6.10).

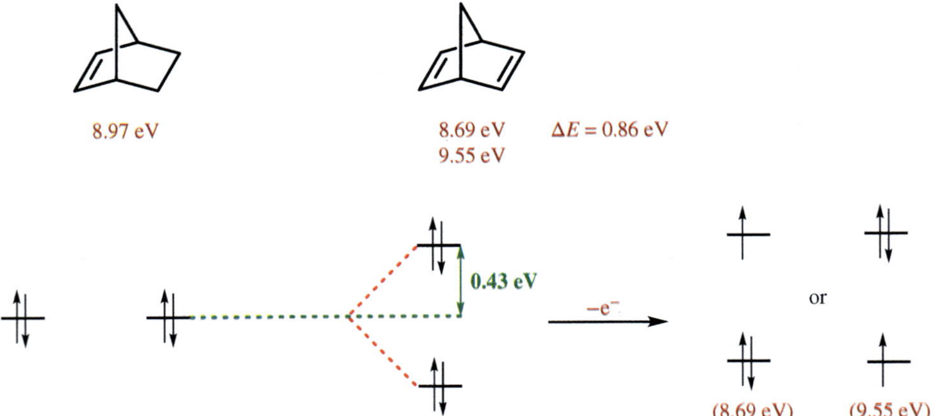

FIGURE 6.9　Difference between the ionisation energies of norbornene and norbornadiene.

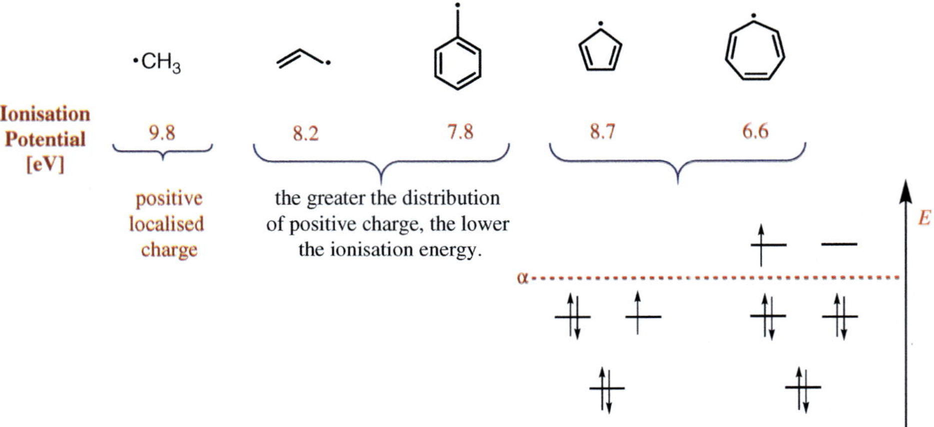

FIGURE 6.10　Ionisation potentials of various radicals.

The existence of spiroconjugation in 4,4-spirononatetraene was confirmed by photoelectron spectroscopy. By comparison with the photoelectron spectrum of cyclopentadiene, it can be seen that the ionisation potentials at 7.99 eV and 9.22 eV in the spiro system result from the through-space interaction of two ψ_2 orbitals with the appropriate symmetry (Figure 6.11).

Upon consideration of the symmetry of the orbitals in the 1,3-butadiene segments with respect to the σ plane (in the plane of the paper; i.e. the xz plane) and the σ' plane (perpendicular to the paper; i.e. the yz plane)) it is concluded that only the MOs ψ_2 and ψ_4 are of the proper AA symmetry to interact (Figure 6.12).

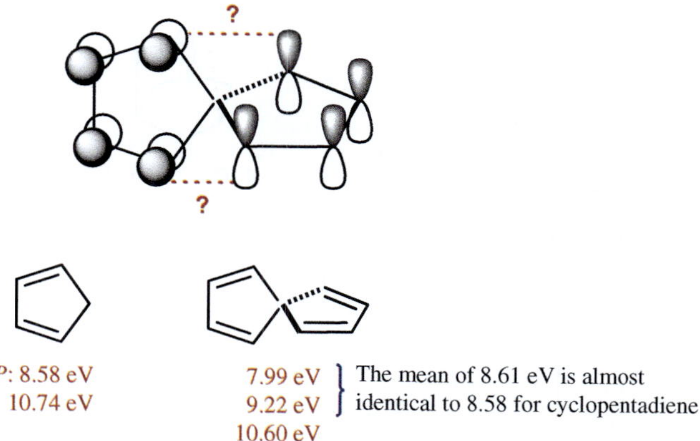

FIGURE 6.11 Ionisation potentials of 4,4-spirononatetraene.

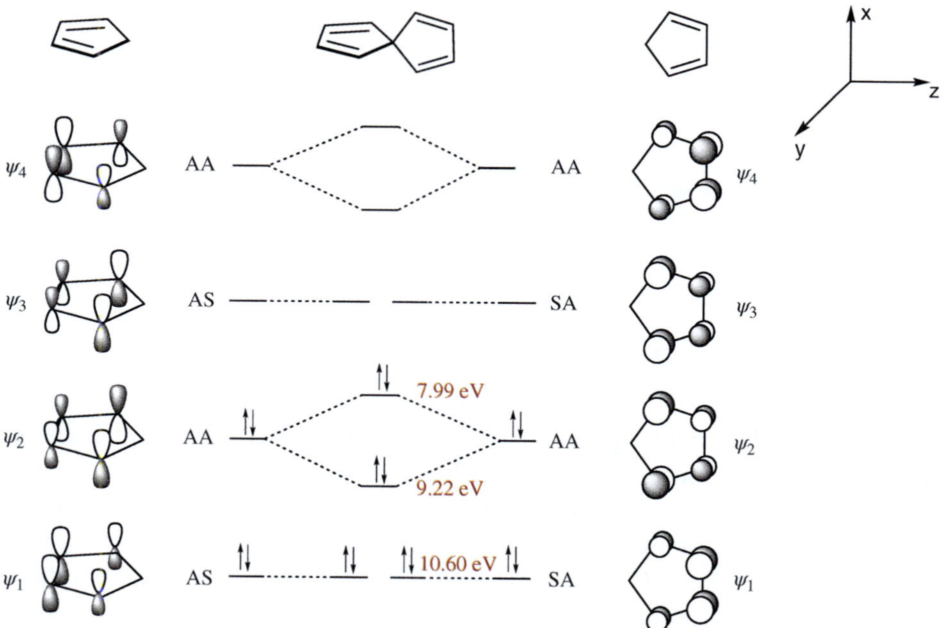

FIGURE 6.12 MO diagram for 1,3-butadiene and 4,4-spirononatetraene.

INTERACTIONS BETWEEN *n*-ORBITALS

EXAMPLE 1

1,3-*n*/*n* interactions are anticipated to exist in azaadamantane **3**. Let us begin the examination of this hypothesis taking into account monoamine **4**, where the *IE* of the electron pair on the bridge nitrogen is equal to 7.95 eV. In **3**, the second nitrogen can exert two effects: (a) an inductive effect associated with

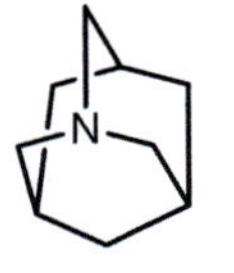

3 *IP* = 7.75 and 8.79 eV   **4** *IP* = 7.95 eV

FIGURE 6.13   Adamantane-derived amines.

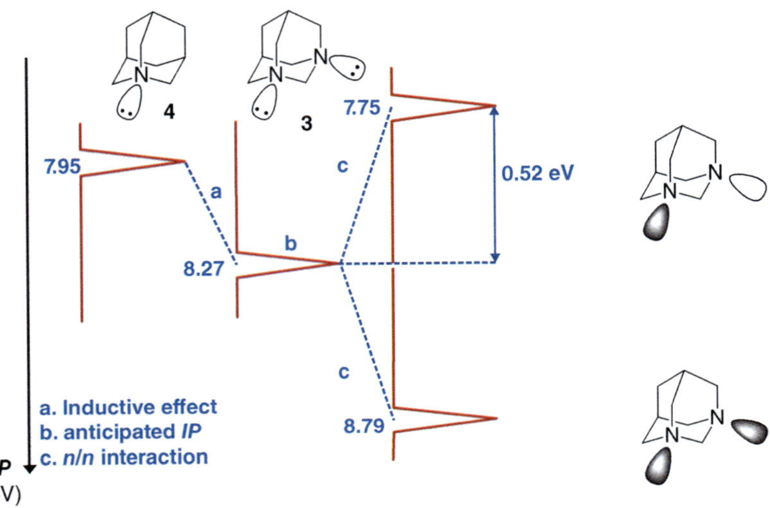

FIGURE 6.14   Ionisation energy differences for amines **3** and **4**.

the electronegativity of nitrogen, and (b) the effect produced by the through-space overlap between the occupied *n* orbitals (Figure 6.13).

Indeed, the study of open-chain diamines models has shown that the inductive effect provoked by the electronegative nitrogen atom lowers the energy of the n_N orbital by 0.32 eV (i.e. an increase in *IE* from 7.95 eV to 8.27 eV is anticipated in going from monoamine **4** to diamine **3**). Thus, the ionisation potentials observed at 7.75 eV and 8.79 eV in the photoelectron spectrum of **3** indicate that the magnitude of the through-space interaction between the *n* orbitals amounts to 0.52 eV (Figure 6.14).

EXAMPLE 2

The through-space 1,4-*n/n* interaction in coplanar α-dicarbonyl compounds results in a separation of approximately 2.1 eV for the HOMO in diketone **5** (Figure 6.15). Thus, the bonding combination across the space of the unshared

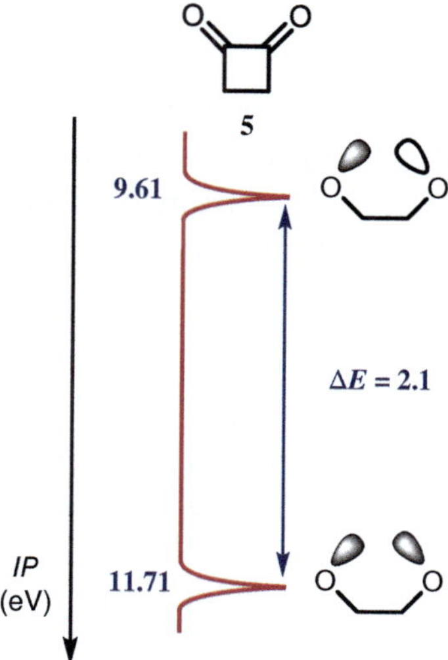

FIGURE 6.15   Through-space 1,4-*n/n* interaction in 1,2-cyclobutanedione **5**.

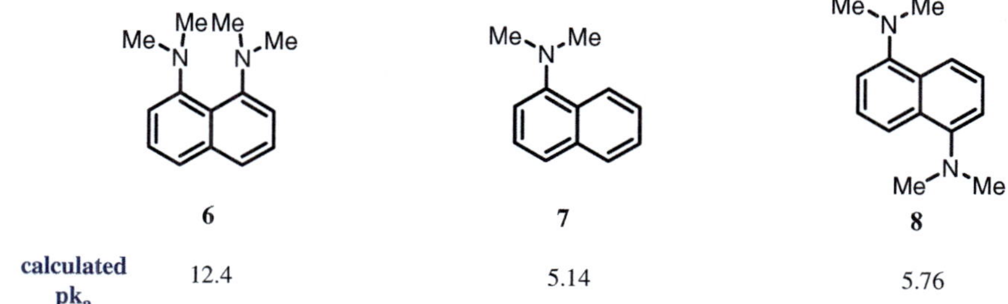

FIGURE 6.16   Consequence of through-space 1,5-*n/n* interaction on the pka value of aromatic amine **6**.

electron pairs is associated with an ionisation potential of 11.71 eV, whereas the anti-bonding combination presents $IE = 9.61$ eV.

EXAMPLE 3

An intramolecular 1,5-*n/n* interaction is clearly manifested in 1,8-bis(dimethyl-amino)naphthalene **6**, which is much more basic than amines **7** or **8** (Figure 6.16).

Indeed, *n/n* repulsion between the lone electron pairs in the amino groups present in **6** raises the energy of the HOMO, which results it a stronger basic character (Figure 6.17).

EXAMPLE 4

The topology adopted by diazophane **10** allows for through-space interaction across the 1,6-n/n gap. The magnitude of this interaction is manifested as a 0.8 eV *IP* separation in the photoelectron spectrum to be compared with the *IP* band recorded for pyridine derivative **9** (Figure 6.18).

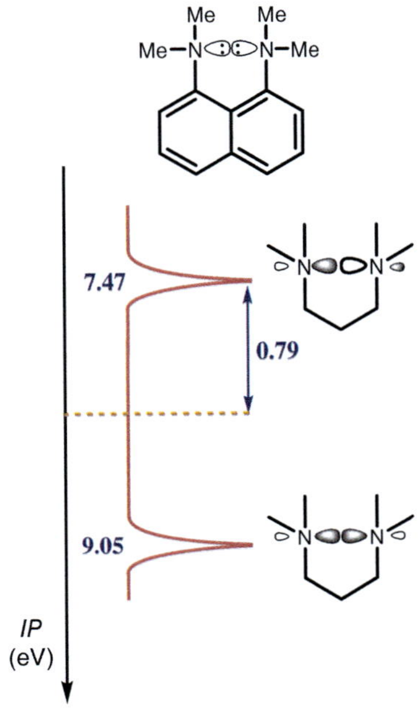

FIGURE 6.17 Ionisation energies for the lone electron pairs at nitrogen in diamine **6**.

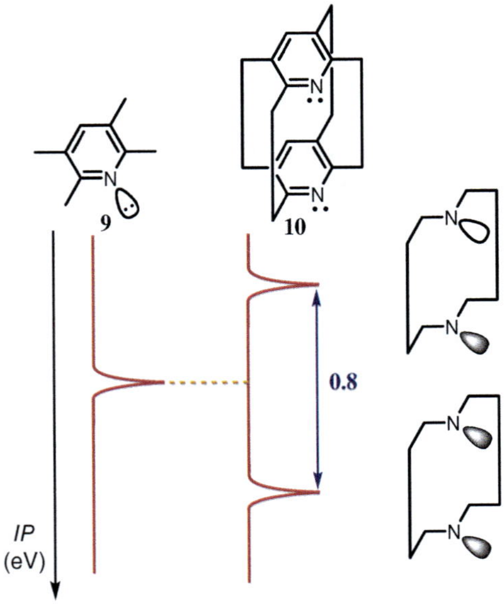

FIGURE 6.18 Through-space 1,6-n/n interaction in diazophane **10**.

ELECTRON SPECTROSCOPY FOR CHEMICAL ANALYSIS (ESCA) SPECTROSCOPY

The effect of the electronegativity of the substituent in C—X bonds is manifested in the *IE* for the C_{1s} orbitals in the series:

	C_{1s} (eV)
CH_4	285
CH_2O	288
CCl_4	293
CHF_3	299

This observation suggests that as the electronegativity of X increases then the $1s$ electrons attach themselves more strongly to the carbon atom nucleus.

CHARGE TRANSFER COMPLEXES (EDA COMPLEXES)

Molecules with high energy HOMO are good electron donors (D) that can interact effectively with other molecules whose LUMO is of low energy (electron acceptors A) when they present the appropriate symmetry (Figure 6.19). This new intermediate, known as the EDA complex, has physical properties that differ from those of the individual substrates. This is due to the formation of new MOs resulting from the electronic coupling of the frontier orbitals of D and A (HOMO/LUMO). This new intermediate is characterised by the appearance of a new absorption band, the charge transfer band ($h\nu_{CT}$), associated with an intra-complex electronic transfer $\Psi_{GS} \rightarrow \Psi_{ES}$ from the donor to the acceptor (Ψ_{GS} refers to the wave function associated with the ground state and Ψ_{ES} refers to the wave function associated with the excited state).

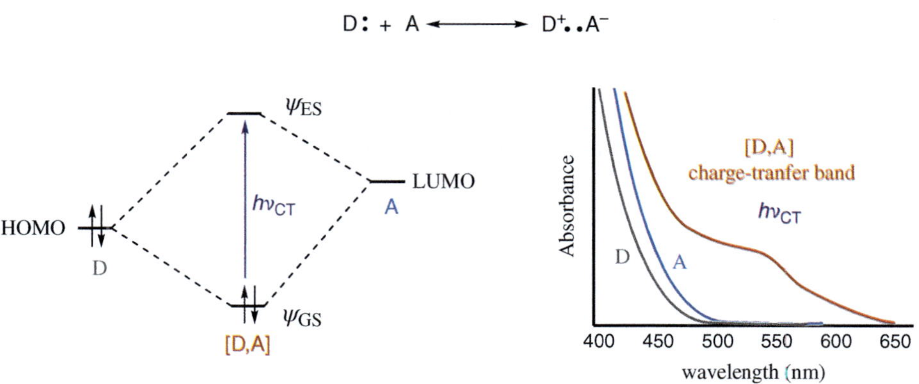

FIGURE 6.19 MO diagram for charge transfer complexes.

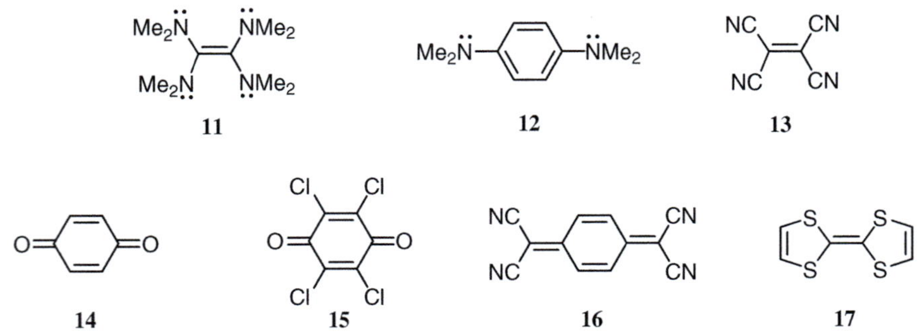

FIGURE 6.20   Some examples of molecules incorporating donor or electronegative substituents.

The charge transfer band typically appears in the visible region of the UV-visible spectrum and represents the colour change of a solution containing an electron donor (D) and an electron acceptor (A). While the individual components, A and D, may not absorb visible light, the resulting EDA complex does. Molecules **11** and **12** are examples of good donors, the electron-rich amino substituents increase the energy of the HOMO; on the other hand, molecules incorporating electronegative substituents such as halogens and cyano or carbonyl groups lower the energy of the LUMO as in compounds **13–16** (Figure 6.20).

Among other properties, charge transfer complexes impart semiconductor characteristics to certain organic materials. For example, adducts of tetrathiofulvalene **17** with tetracyanoethylene **13** constitutes the semiconductor material in electronic transistors.

In recent years, EDA complexes have also been used in organic synthesis, leading to many novel radical transformations under photochemical conditions.

FURTHER READING

A. K. Wortman, C. R. J. Stephenson. *Chem.* **2023**, *9*, 2390.

C. F. Wilcox, S. Winstein, W. G. McMillan, *J. Am. Chem. Soc.* **1960**, *82*, 5450.

D. C. Harris, M. D. Bertolucci, *Symmetry and Spectroscopy. An Introduction to Vibrational and Electronic Spectroscopy*, Dover Publications, New York, **1978**, Chapter 4.

H.-D. Martin, B. Mayer, *Angew. Chem. Int. Ed.* **1983**, *22*, 283.

V. K. Yadav, *Steric and Stereoelectronic Effects in Organic Chemistry*, 2nd ed., Springer, New York, **2021**.

EXERCISES

6.1 Draw the molecular orbital diagram for the π-π^* transition of 1,3,5-hexatriene when irradiated with ultraviolet light at 258 nm. Indicate the energy associated with this electronic transition.

6.2 Draw the molecular orbital diagram for the n-π^* transition of 4-methyl-3-penten-2-one knowing that its HOMO–LUMO gap is 91 kcal mol^{-1}. Indicate the wavelength associated with this electronic transition.

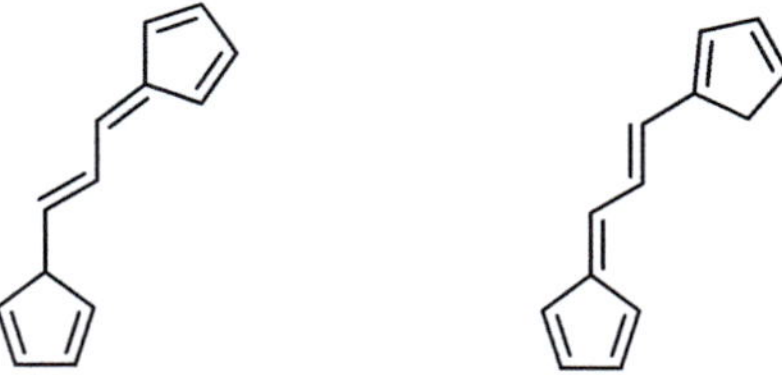

6.3 Which of the following molecules is expected to absorb at the longest wavelength in the UV region of the electromagnetic spectrum? Justify your answer

6.4 The photoelectron spectrum for norbornadiene shows two ionisation energies at 8.69 and 9.55 eV (cf. Figure 6.6 above) while the exocyclic π bond in **18** presents $IP = 8.49$ eV. Explain the ionisation energies at 7.97, 9.25, and 9.55 eV for triene **19**.

18   **19**

CHAPTER 7

Conservation of Molecular Orbital Symmetry: Introduction to Pericyclic Reactions – Cycloaddition Reactions

INTRODUCTION

As discussed in Chapter 4 in this book, the development of Hückel's method in the year 1930 consisted of the treatment of molecular orbitals (MO) in π-delocalised molecular systems as linear combinations of p-atomic orbitals, which allowed the prediction of the shape and energy of the corresponding MOs. This development paved the road for the analysis of a wide range of chemical and spectroscopic observations that had been recorded over the years in organic molecules. Nevertheless, it was not until Woodward and Hoffmann (1965) demonstrated that the chemical behaviour of a large number of *concerted reactions*; that is, reactions that take place in a single step, with bond breaking and bond forming happening simultaneously (specifically, concerted reactions do not involve the formation of intermediates) can be explained by the application of molecular orbital theory (MOT). For example, Woodward and Hoffmann were able to understand why certain conjugated polyenes close in a thermal process to afford a cyclic ring presenting a specific stereochemistry, whereas an analogous polyene with one more (or less) double bond gives rise to the ring with opposite configuration. As we will see below, Woodward and Hoffmann's interpretation was based on the symmetry characteristics of the MOs participating in the substrate to product transformation. Kenichi Fukui came to similar conclusions regarding the relevance of orbital

159

symmetry in pericyclic reactions. The concept was so innovative and useful that in 1981 Hoffmann and Fukui were awarded the Nobel Prize in Chemistry. Because this prize is not awarded posthumously, Woodward was unable to receive it (he had passed away in 1979).

Rarely in the history of organic chemistry has a discovery been greeted with such enthusiasm as that caused by the Woodward and Hoffmann orbital symmetry conservation rules.

The next chapters of this book are intended to familiarise the reader with the methods developed by Woodward and Hoffmann, and almost simultaneously by Dewar and Zimmerman, for the analysis of *pericyclic reactions*.

CONCERTED REACTIONS

Concerted reactions are those in which a reactant becomes a product without the intervention of an intermediate; that is, bond formation and bond breaking occur simultaneously, and the reaction takes place in a single kinetic step. Although in practice, there is no simple way to determine whether a reaction is concerted or not, in general, concerted reactions are: (i) stereospecific, (ii) insensitive to changes in solvent and (iii) not subject to interception of intermediates by special reagents during the reaction's course.

A concerted reaction can further be classified as *synchronised* when the various chemical changes associated with the bonds being formed or broken have proceeded to the same extent in the transition state. Figure 7.1 shows the possible transition states in a) a synchronised concerted [4 + 2] cycloaddition reaction, b) a *non-synchronised* concerted reaction, and c) a non-concerted reaction (proceeding in two stages).

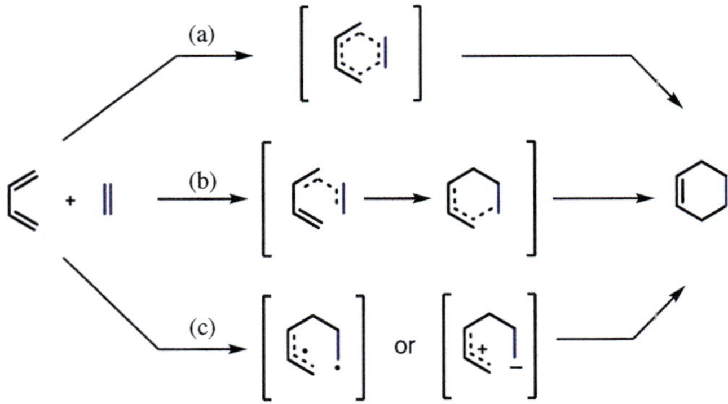

a) concerted, synchronised reaction
b) concerted, non-synchronised reaction
c) non-concerted reaction (via an intermediary)

FIGURE 7.1   Plausible transition states for concerted reactions.

PERICYCLIC REACTIONS

A pericyclic reaction is a type of organic chemical reaction that occurs through a concerted process involving a cyclic redistribution of bonding electrons. The key feature of pericyclic reactions is the involvement of π and/or σ orbitals in a cyclic array, leading to a highly ordered and predictable outcome.

There are four types of pericyclic reactions: 1) Cycloadditions, 2) Cheletropic reactions, 3) Electrocyclic reactions and 4) Sigmatropic reactions.

Cycloaddition reactions are those in which the concerted combination of two π-systems gives rise to a ring with the concomitant formation of two new σ-bonds. The iconic Diels–Alder reaction is an example of this class of reactions, and this chapter presents arguments based on orbital symmetry that help explain why [4 + 2] cycloadditions are *allowed* (Figure 7.2a), while [2 + 2] cycloaddition reactions (Figure 7.2b) or [4 + 4] cycloadditions (Figure 7.2c) are *forbidden*. It is important to note that pericyclic reactions are often represented using curly arrows, but any impression of directional electron pairs movement is incorrect. In a pericyclic reaction, the MOs of the starting materials evolve smoothly towards the MOs of the products. Figure 7.2d shows an example of a [6 + 4] cycloaddition reaction being *allowed*. It should be noticed how the formation of two σ bonds at the expense of two π bonds is thermodynamically favourable for all examples given here since sigma bonds are stronger than π bonds. Importantly, the reactions that are prohibited by the rules of conservation of orbital symmetry have a much higher activation energy for a *thermally concerted process*, and when they do react, they usually do so *via* a non-concerted mechanism induced typically by photochemical activation.

Chapter 8 in this book deals with *cheletropic reactions*, which are special cases of cycloaddition reactions in which the bonds being formed or broken converge on (or originate from) the same atom. Here, analysis in terms of orbital theory helps us understand why cyclopropanone does not dissociate under thermal activation to ethylene and carbon monoxide (Figure 7.3a), just as cyclopropenone does not dissociate to acetylene and carbon monoxide (Figure 7.3b). Similarly, the Woodward and Hoffmann rules help to understand why 3-cyclopentenone and 3-thiolene sulphone decompose when heated (thermal activation) to generate 1,3-butadiene and either carbon monoxide or sulphur dioxide, respectively (Figure 7.3c).

Electrocyclic reactions are discussed in Chapter 9; these reactions involve the concerted cyclisation or opening reaction of a π-system. Analysis of the symmetry properties of the MOs involved in the transformation helps explain why *cis-* rather than *trans*-tetramethylcyclobutene is formed during the cyclisation reaction of (Z,E)-3,4-dimethyl-2,4-hexadiene under thermal activation (Figure 7.4). Conversely, the (E,Z,E) diastereoisomer of 2,4,6-octatriene undergoes cyclisation to produce *cis*-dimethylcyclohexadiene.

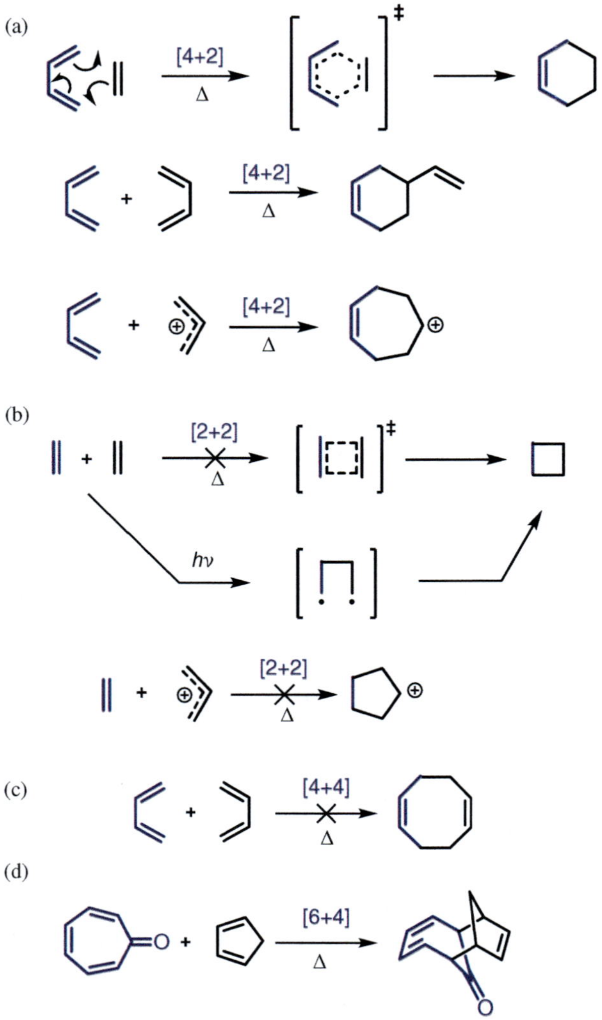

FIGURE 7.2   Various types of cycloaddition reactions.

On the other hand, *sigmatropic reactions* are molecular rearrangements in which a σ-bond flanked by one or more π-systems moves to a new position within those π-systems. Chapter 10 presents the MO analysis of the allowed *Claisen* reaction (Figure 7.5a), the allowed *Cope* reaction (Figure 7.5b) and the allowed (Figure 7.5c) or forbidden (Figure 7.5d) *ene* reactions.

It should be noted that, by the principle of microscopic reversibility, those reactions that are *allowed* in the direction indicated in Figures 7.1–7.5 are also allowed in the opposite direction, so that such reactions can be written as chemical equilibria, controlled thermodynamically by the relative energy of the reactants and products.

Furthermore, it should also be kept in mind that the reactions that are *thermically forbidden* are generally *allowed photochemically*. However, uncertainty

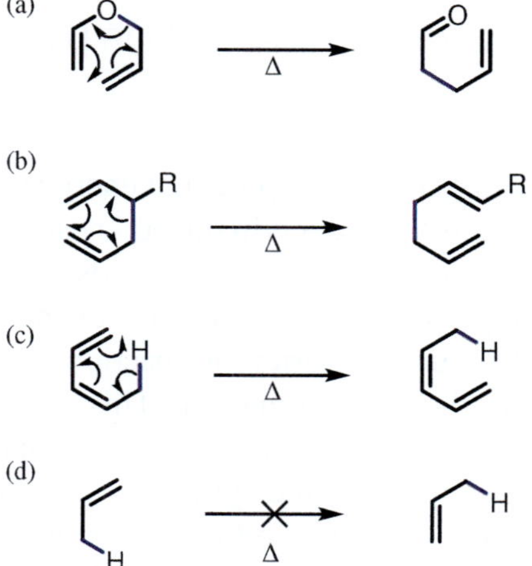

FIGURE 7.3 Various types of cheletropic reactions.

FIGURE 7.4 Two examples of stereospecific electrocyclic reactions.

FIGURE 7.5 Examples of *allowed* and *forbidden* sigmatropic reactions.

often exists regarding whether photochemical reactions are concerted. Therefore, this book focuses more on thermal processes where the evidence clearly supports concerted mechanisms.

PRINCIPLES OF THE CONSERVATION OF ORBITAL SYMMETRY

As mentioned above, concerted reactions that proceed through a cyclic arrangement of π-conjugated atoms involving a redistribution of electron density in each atom *via* a single transition state, are known as pericyclic reactions. The transition state of pericyclic reactions contains either 4n + 2 electrons, which are aromatic Hückel-type π-systems, or 4n electrons, which correspond to anti-aromatic Möbius-type π-systems. Such Hückel *vis-a-vis* Möbius stability will determine the stereochemistry of the product in pericyclic reactions.

An important characteristic of pericyclic reactions is that the energy barrier in the transition state is influenced by the symmetry of the participating MOs. Reactions with low-energy transition-state barriers are considered 'allowed,' while those with high-energy transition states are deemed 'forbidden'. The transition state energy of a symmetry-allowed process is inherently lower than that of a symmetry-forbidden alternative. The analysis of the orbital symmetry of a pericyclic reaction can be carried out by any of the three following procedures:

1. Symmetry correlation diagram analysis (advanced by Woodward and Hoffmann in 1969 and 1970).
2. Analysis of the symmetry of the HOMO/LUMO frontier orbitals (proposed by Fukui in 1982).
3. Examination of the nodal properties of the transition state (recommended by Dewar [1992] and Zimmerman [1966]).

SYMMETRY CORRELATION DIAGRAMS OF MOLECULAR ORBITALS

The idea in this strategy is to examine the potential energy surface that is generated when going from reactants to products in the pericyclic reaction. In this way, the surface with the lowest energy barrier corresponds to the allowed process, and the surface with the highest energy is indicative of the forbidden pathway. The key observation in the study of pericyclic reactions is *the conservation of MO symmetry* throughout the transformation. This principle ensures that symmetric orbitals remain symmetric whereas anti-symmetric orbitals remain anti-symmetric. Using this approach, the symmetry properties of the molecular orbitals involved in bond-breaking and bond-forming processes are analysed with respect to two symmetry elements, a *two fold (C_2) symmetry axis* and a *plane of symmetry σ*. A correlation diagram is then constructed to illustrate the corresponding symmetry properties. The MOs are then arranged according to their relative energies and a connection is established between MOs of similar symmetry in reactants and products. When the symmetry of the MOs in the reactants matches the symmetry of the MOs in the product's ground state, it is concluded that the reaction is thermally allowed. Conversely, if the symmetry of the reactant's MOs coincides with the symmetry of the MO in the product's first excited state, the reaction is photochemically allowed. If the symmetries of the MOs in reactants and product do not match, the reaction will not take place.

In all pericyclic reactions involving conjugated polyenes, the energy of MOs can be estimated by means of the Hückel method described in Chapter 4. On the other hand, the shape of the orbitals can be depicted by consideration of the phases of the atomic orbitals involved, so that the lowest energy orbital contains no nodes between atoms; the next lowest energy orbital contains one node, the next higher-energy MO contains two nodes, and so on.

For example, when examining the conversion of 1,3-butadiene to cyclobutene, two processes must be considered: the *conrotatory* and the *disrotatory* ring-forming cyclisation modes (Figure 7.6).

Upon examination of Figure 7.6, it can be appreciated that the *conrotatory* process leads to a transition state with no conflicting out-of-phase orbital interactions between the *p* orbitals participating in bond formation. By contrast, the *disrotatory* process results in an out of phase sign reversal between the two bond-forming *p* orbitals. Furthermore, upon operation of a C_2 symmetry axis reactant and product maintain their symmetry when the process is *conrotatory* but not when the process is *disrotatory*.

In conclusion, in the 1,3-butadiene to cyclobutene concerted cyclisation process, the *disrotatory* transition state is a *forbidden* (high energy) Hückel type with 4n electrons (Figure 7.7); by contrast, the *conrotatory* cyclisation

FIGURE 7.6 *Conrotatory vs. disrotatory* ring-forming processes for the conversion of 1,3-butadiene to cyclobutene.

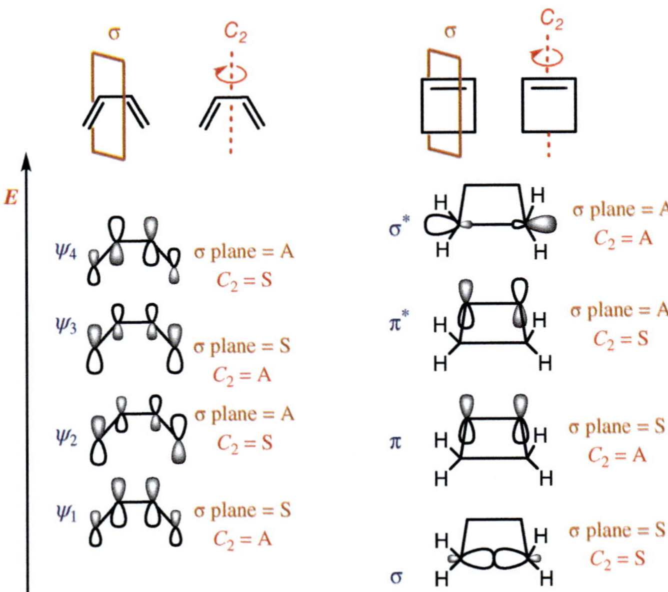

FIGURE 7.7 Symmetry properties of σ and π molecular orbitals of cyclobutene and 1.3-butadiene.

mode leads to a transition state of the Möbius type (with a change of sign along the π-system), which is *allowed* (low energy) for π-systems with 4n electrons (Figure 7.7).

The symmetry properties of the thermally-activated 1,3-butadiene to cyclobutene concerted cyclisation process with respect to the plane (σ) and two fold (C_2) of MOs of cyclobutene and 1,3-butadiene are also depicted in Figure 7.7.

The MO correlation diagram shown Figure 7.8 presents a symmetry correlation diagram between the ground state orbitals in cyclobutene and 1,3-butadiene. It is appreciated that a thermal *conrotatory* interconversion process is a symmetry-*allowed* process. In particular, Figure 7.8 shows that the σ ground state orbital of cyclobutene correlates with the Ψ_2 ground state orbital of 1,3-butadiene and π with Ψ_1. By contrast, in the disrotatory process the π orbital of cyclobutene correlates with Ψ_3 in 1,3-butadiene, that is an excited state MO. Consequently, the thermal cyclobutene $\leftrightarrows$ 1,3-butadiene interconversion *via* a *disrotatory* process is symmetry *forbidden*.

In addition, the MO correlation diagram involved in the *disrotatory* mechanism affords an open (diradicaloid, highly reactive and unstable) configured MO, whereas the corresponding route involved in the *conrotatory* entails a closed (stable) configuration (Figure 7.9).

Figure 7.10 shows the result of a similar MO correlation analysis for the interconversion 1,3,5-hexatriene $\leftrightarrows$ 1,3-cyclohexadiene. It can be appreciated that in this case the *conrotatory* process involves formation an unstable diradical species, whereas the *disrotatory* mechanism is the favourable one (*allowed*).

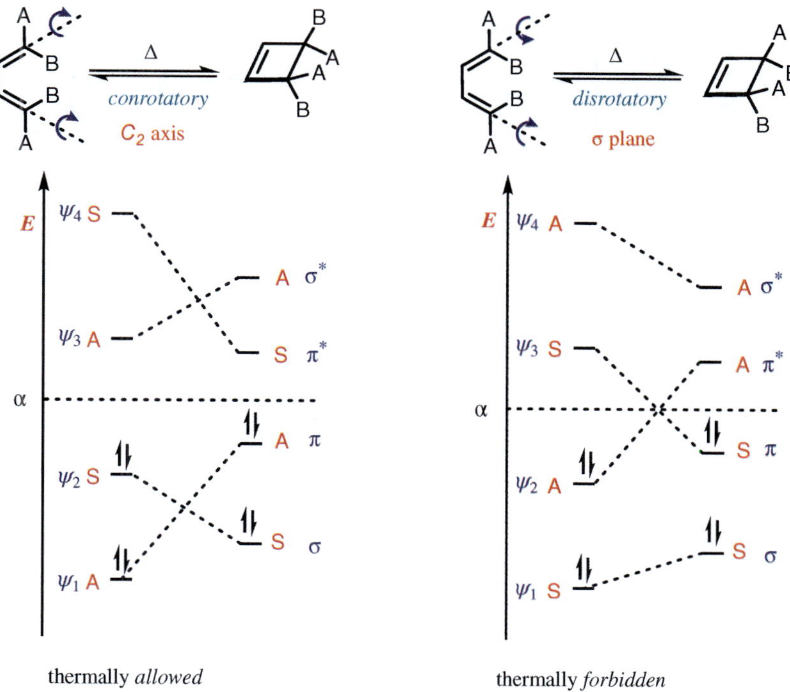

FIGURE 7.8 Correlation orbital diagrams for both *conrotatory* and *disrotatory* 1,3-butadiene ⇆ cyclobutene interconversion.

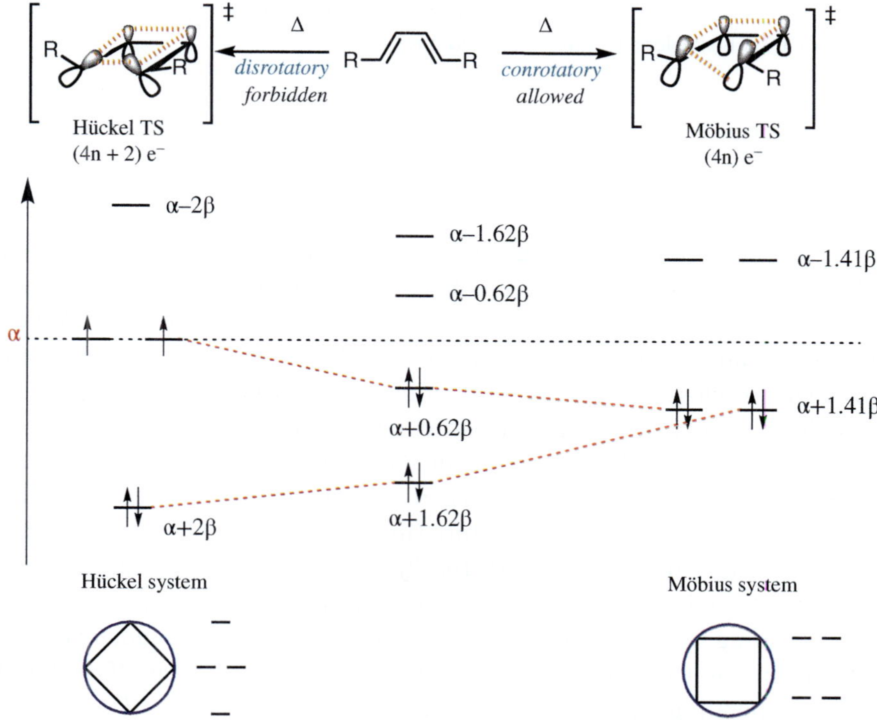

FIGURE 7.9 Correlation diagram for both *disrotatory* and *conrotatory* processes for the conversion of 1,3-butadiene to cyclobutene.

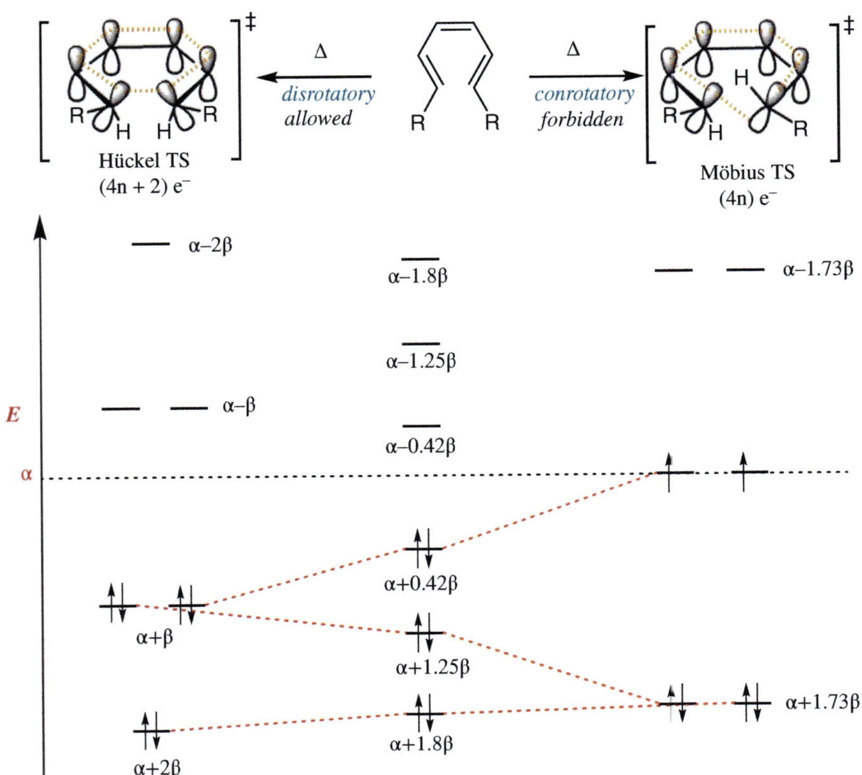

FIGURE 7.10 Correlation diagram for both *disrotatory* and *conrotatory* processes for the conversion of 1,3,5-hexatriene to 1,3-cyclohexadiene.

ANALYSIS OF THE SYMMETRY OF THE HOMO/LUMO FRONTIER ORBITALS (FMO)

Pericyclic reactions can be effectively examined by means of frontier molecular orbital (FMO) analysis. The FMO method focuses on the interactions of key frontier orbitals, rather than examining all molecular orbitals in detail. When two molecules approach each other, their MOs begin to interact, with those orbitals that are closer in energy interacting more strongly (see Chapter 5). When a filled molecular orbital (in particular the HOMO) interacts with an empty orbital (especially the LUMO) in the second reactant, the interaction can lead to a stabilisation of the system. When the interaction between two MOs in the ground state is of a bonding nature (in phase overlap of wave functions with the same sign), the reaction is *thermally allowed* (Figure 7.11a,b). By contrast, if the interaction is of an anti-bonding nature (out-of-phase overlap of the corresponding wave functions), the reaction is *thermally forbidden* (Figure 7.11c,d). In the excited state, however, if the interaction between HOMO and LUMO is of a bonding type, the reaction becomes *photochemically allowed*.

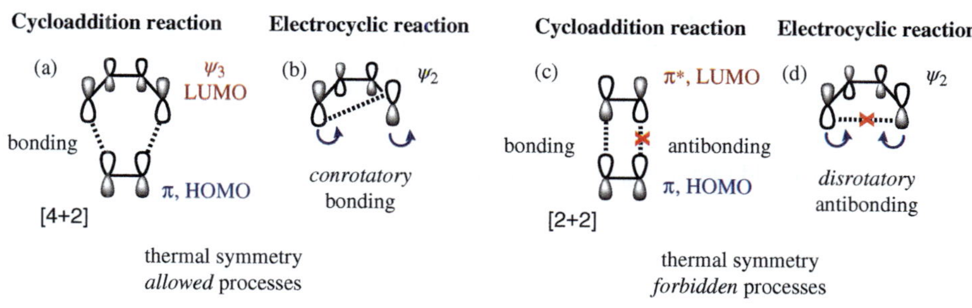

FIGURE 7.11   Analysis of FMOs in cycloaddition and electrocyclic reactions.

ANALYSIS OF THE NODAL PROPERTIES AT THE TRANSITION STATE OF A CYCLISATION REACTION

In the analysis of the nodal properties in the transition state, the aromatic or antiaromatic nature of a cyclic transition state is explained by the Hückel-Möbius concept of aromaticity (see Chapter 4). A Hückel-type π system is aromatic (stabilised by cyclic delocalisation) when it contains $4n + 2$ π-electrons, and antiaromatic (destabilised) if it contains $4n$ π-electrons. Conversely, in a Möbius-type system aromatic stabilisation is attained with $4n$ π-electrons, whereas $4n + 2$ π-electrons result in antiaromaticity.

This can be summarised as

> (a) $(4n + 2)$ electrons is therefore stable in Hückel systems.
>
> (b) $(4n)$ electrons are stable in Möbius systems.

In order to establish the stability of a cyclic transition state, the number of nodes in the π-system array and the number of involved electrons are counted. If the arrangement of π-orbitals presents no phase inversions or an even number of inversions, the process will be thermally allowed when it contains $4n + 2$ π-electrons. In contrast, if the transition state presents an odd number of orbital phase inversions, it must contain $4n$ π-electrons to be *thermally allowed*. When these conditions are not fulfilled, the energy content of the transition state will be high and the reaction will be *forbidden*.

EXAMPLE 1

In organometallic adducts it is common to observe the displacement of a ligand bound to a transition metal that is coordinated to a double bond towards the vicinal olefinic carbon affording the product of *cis*-addition (Figure 7.12):

Figure 7.13 depicts the corresponding cyclic transition state involving the ethylene π^* orbital, the transition metal d-orbital and the σ orbital at the metal–ligand bond.

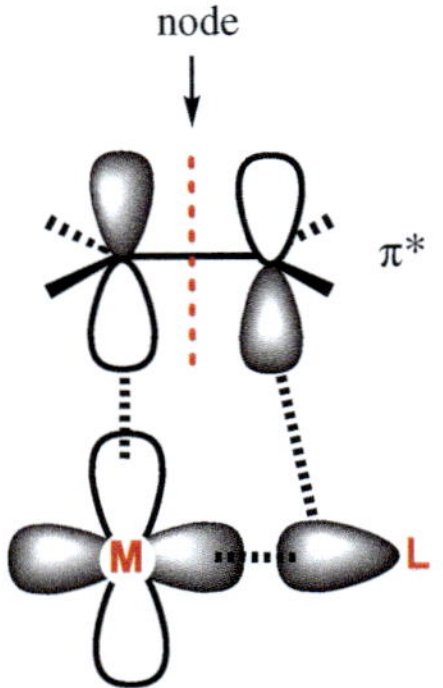

FIGURE 7.12   *Cis* addition of organometallic species to double bonds.

FIGURE 7.13   Cyclic transition state involved during the displacement of a ligand in a transition metal coordinated to a double bond towards the vicinal olefinic carbon.

Although the transition state depicted in Figure 7.13 shows constructive interactions between: (1) one olefinic carbon and the transition metal, (2) between the second olefinic carbon and the ligand and (3) between the metal and the ligand (L), the involvement of the π^* orbital of the olefin implies that one node is present between the olefinic carbons. Therefore, the transition state corresponds to a Möbius system (odd number of orbital phase sign reversals) containing four electrons (two olefinic electrons and two electrons present in the metal–ligand bond); thus, the reaction is *thermally allowed.*

EXAMPLE 2

By contrast with example 1, the 1,4-sigmatropic shift operative in a 1,3-butadienyl complex (Figure 7.14) corresponds to a forbidden process since the transition state is Möbius type: one sign change when Ψ_2 is HOMO (Figure 7.14a) and three sign changes when Ψ_3 is LUMO (Figure 7.14b). Thus, the 1,4-sigmatropic shift involves $4\pi + 2\sigma = 6$ electrons ($4n+2$) and the process is *thermally forbidden.*

EXAMPLE 3

[2+2] Cycloaddition reactions. In a *supra-supra* process, the new σ bonds form on the same side of each π reactant, whereas in a *supra-antara* process, bonds form on opposite sides of one of the π reactants. For a [2 + 2] cycloaddition (a $4n$ π-electron system), the *supra-supra cyclo*addition mode corresponds to a Hückel topology (with 0 nodes), which is antiaromatic and thus *thermally forbidden*

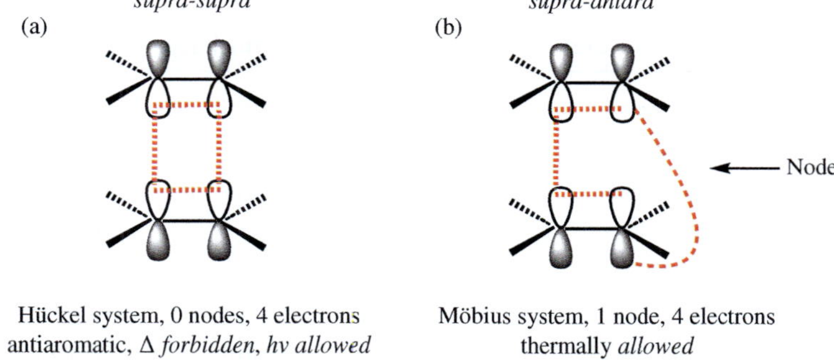

FIGURE 7.14 Forbidden [4n+2] Möbius type transition states between the 1,3-butadienyl π-system and the metal–ligand σ-bond.

supra-supra   *supra-antara*

Hückel system, 0 nodes, 4 electrons   Möbius system, 1 node, 4 electrons
antiaromatic, Δ *forbidden*, hv *allowed*   thermally *allowed*

FIGURE 7.15 Analysis of the nodal properties of the transition state in [2 + 2] cycloaddition reactions.

(Figure 7.15a). By contrast, the *supra-antara* cycloaddition mode corresponds to a Möbius topology (with 1 node), which is aromatic with $4n$ π-electrons, making the reaction *thermally allowed* (Figure 7.15b). In general, thermally activated pericyclic reactions proceed via Hückel aromatic transition states, while photochemical reactions occur through Hückel antiaromatic (Möbius aromatic) transition states. Therefore, the [2 + 2] cycloaddition is thermally forbidden unless the reaction proceeds in *supra-antara* fashion.

A Möbius-Hückel correlation diagram illustrates the contrasting transition states for the conversion of 1,3-butadiene to cyclobutene, a classic electrocyclic reaction (Figure 7.16). The *disrotatory* process maintains the same sign phase (0 nodes) between adjacent p orbitals, forming a Hückel-type transition state that is thermally forbidden for $4n$ electrons (Figure 7.16b). By contrast, the *conrotatory* process involves a phase sign inversion (1 node) between adjacent p orbitals, resulting in a Möbius-type transition state that is thermally allowed for $4n$ electrons (Figure 7.16a). Table 7.1 summarises the rules for nodal properties at the transition state of electrocyclic cyclisation reactions and their consequences.

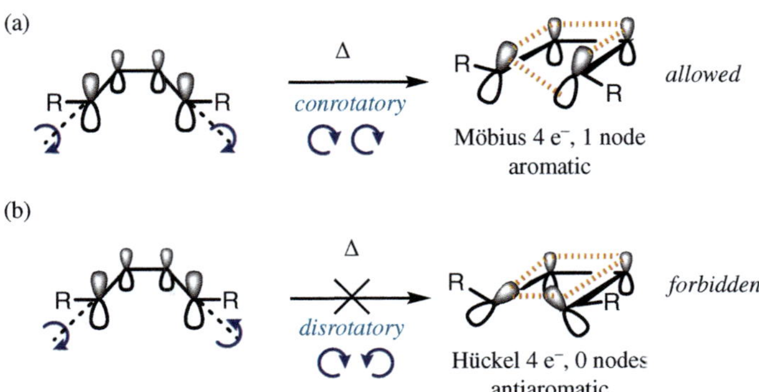

FIGURE 7.16 Analysis of the nodal properties in the transition state of electrocyclic pericyclic reactions.

TABLE 7.1 Rules for Nodal Properties at the Transition State of Electrocyclic Cyclisation Reactions and Their Consequences

No. of electrons	No. of nodes	TS type	Aromaticity	Feasibility
$4n + 2$	0 or even number	Hückel	aromatic	Δ allowed hν forbidden
$4n$	0 or even number	Hückel	antiaromatic	Δ forbidden hν allowed
$4n + 2$	Odd number	Möbius	antiaromatic	Δ forbidden hν allowed
$4n$	Odd number	Möbius	aromatic	Δ allowed hν forbidden

In the following Chapters, the application of orbital correlation diagrams and the analysis of nodal properties in transition states for the determination of the permissibility of pericyclic reactions will be discussed in a systematic way. The importance of the symmetry properties in the HOMO and LUMO orbitals of the participating reactants will also be emphasised.

CYCLOADDITION REACTIONS

As mentioned earlier, a cycloaddition reaction is a type of chemical reaction in which two or more unsaturated molecules combine to form a cyclic product. Cycloaddition reactions are often classified according to the number of π-electrons involved in the reacting systems. The most common types of cycloaddition reactions include:

- [2+2] cycloaddition: involves the combination of two unsaturated reactants, each contributing two π-electrons, to form a four-membered ring.

- [4+2] cycloaddition (Diels–Alder reaction): involves a conjugated *diene* (four π-electrons) and a *dienophile* (two π-electrons) to form a six-membered ring.
- [6+4] or higher-order cycloadditions: involve larger conjugated systems and are less common.

Cycloaddition reactions are widely used in organic synthesis for the construction of complex cyclic molecules, including natural products and pharmaceuticals. They are highly atom-economic and exhibit regio- and stereoselectivity. In particular, the Diels–Alder reaction is a cornerstone of synthetic organic chemistry. In the following section, we will first analyse several cycloaddition reactions using the methods introduced earlier in this Chapter: 1) analysis of starting materials ⇌ product correlation diagrams, 2) analysis of the symmetry properties of FMOs and 3) examination of the nodal properties of the transition state.

1,3-BUTADIENE + ETHYLENE ⇌ CYCLOHEXENE

Starting Material ⇌ Product Correlation Diagram

In the 1,3-butadiene+ethylene cycloaddition reaction affording cyclohexene there is an element of symmetry common to both starting materials and product: a σ-plane bisecting both 1,3-butadiene and ethylene (starting materials), and the π-system of the C=C double bond, as well as the MO combination of two sigma C—C bonds in the cyclohexene product (Figure 7.17).

Figure 7.18 presents the correlation diagram of the participant orbitals: it can be appreciated that the starting materials ⇌ product energy change maintains a bonding character (below the α level) throughout the reaction; this feature suggests that the reaction will be *allowed*. In addition, the correlation of the symmetry of the orbitals in starting materials and product is congruent: $[S_2A] \rightleftharpoons [S_2A]$, which again leads to the conclusion that the reaction is *allowed*.

Figure 7.19 presents the two through-space MO bonding ('constructive/positive') combinations of the C— sigma bonds being formed, as well as the two anti-bonding ('destructive/negative') combinations for the same σ^* orbitals.

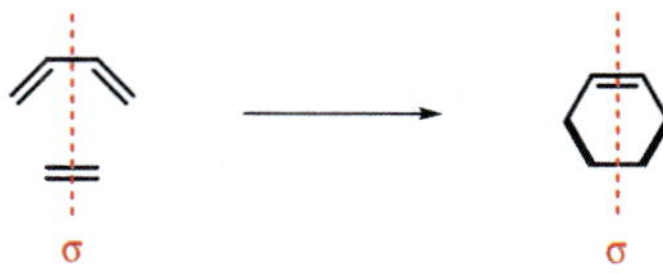

FIGURE 7.17   Elements of symmetry in the 1,3-butadiene+ethylene cycloaddition reaction.

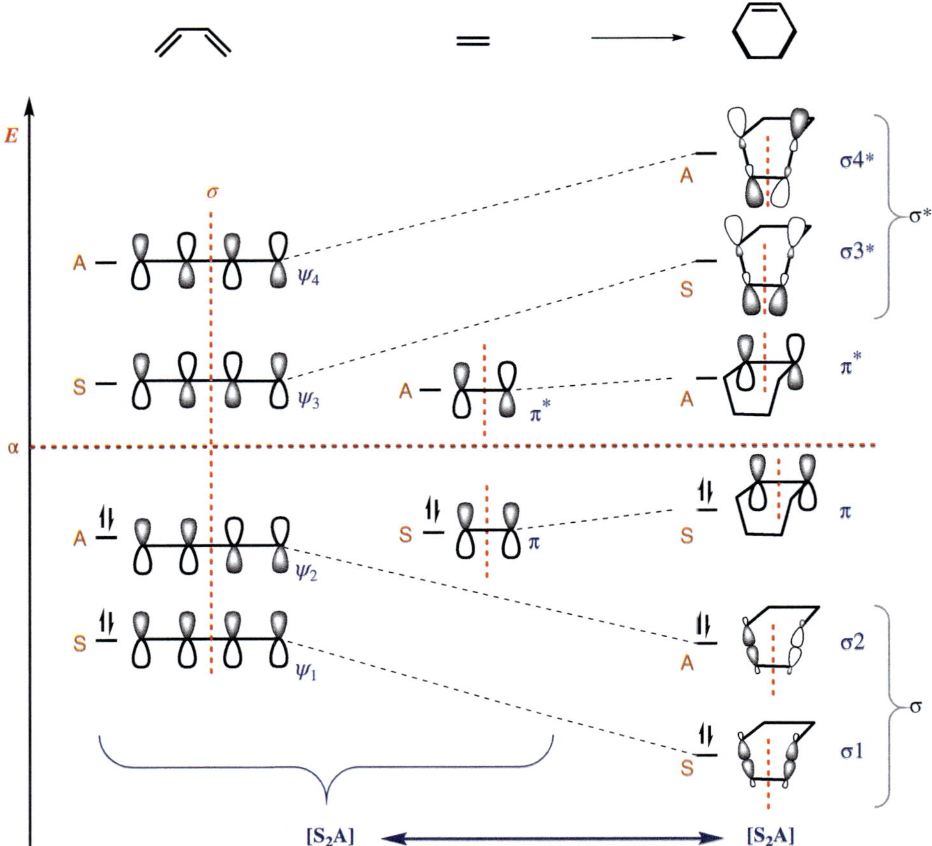

FIGURE 7.18   Correlation diagram for the Diels–Alder reaction between 1,3-butadiene and ethylene.

HOMO/LUMO Frontier MO Interactions

For analysis of the participating HOMO and LUMO frontier molecular orbitals (FMO), either the interaction of the HOMO in 1,3-butadiene (Ψ_2) with the LUMO in ethylene (π^*), or the interaction between the HOMO in ethylene (π) with the LUMO (Ψ_3) in 1,3-butadiene can be examined. In both cases, the symmetry of the interacting orbitals is appropriate (*vide infra*), so that a stabilising interaction through such combinations is anticipated to result in a favourable (*allowed*) process. Figure 7.20 presents the orbital interaction between the HOMO in 1,3-butadiene (Ψ_2, 4π electrons) and the LUMO in ethylene (π^*, 2π electrons). It can be appreciated that a) the symmetry properties of these orbitals are correctly 'matched', and that b) the participating FMO interact in phase.

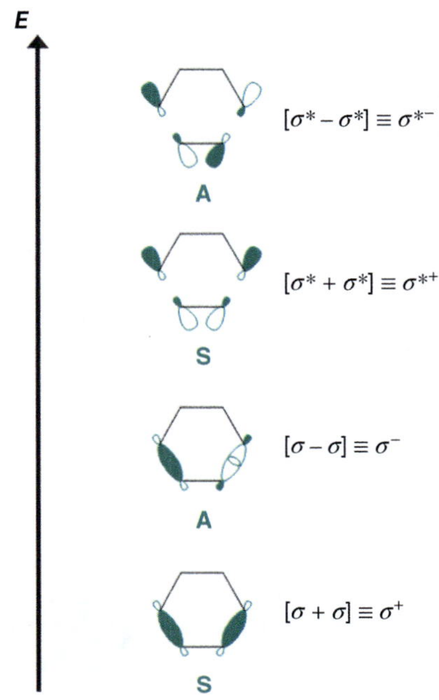

FIGURE 7.19   Correlation diagram for the σ-bond MOs combinations in the Diels–Alder product, cyclohexene.

Nodal Properties of the Transition State (TS)

One can also analyse the [4 + 2] cycloaddition reaction in terms of the nodal properties of the transition state (Figure 7.21). Here, the *supra-supra*-cycloaddition mode leads to a Hückel-type arrangement (0 nodes), which is aromatic with $(4n + 2)\,\pi$ electrons. The process is thus thermally allowed and photochemically forbidden. Alternatively, one can analyse the transition state between the HOMO of 1,3-butadiene (Ψ_2) and the LUMO of ethylene (π^*) MOs. Here, the transition state presents two out-of-phase interactions (nodes) between adjacent p orbitals; therefore, the transition state π-system is of the Hückel type (even number of phase reversals). Thus, since it contains six electrons $(4n + 2)$, it should then be *allowed* (Figure 7.21).

TWO ETHYLENE MOLECULES ⇌ CYCLOBUTANE

As it can be appreciated, the three different types of symmetry analysis performed on the [4 + 2] cycloaddition reaction indicate that this is an *allowed* cycloaddition reaction. Let us now examine [2 + 2] cycloaddition reactions.

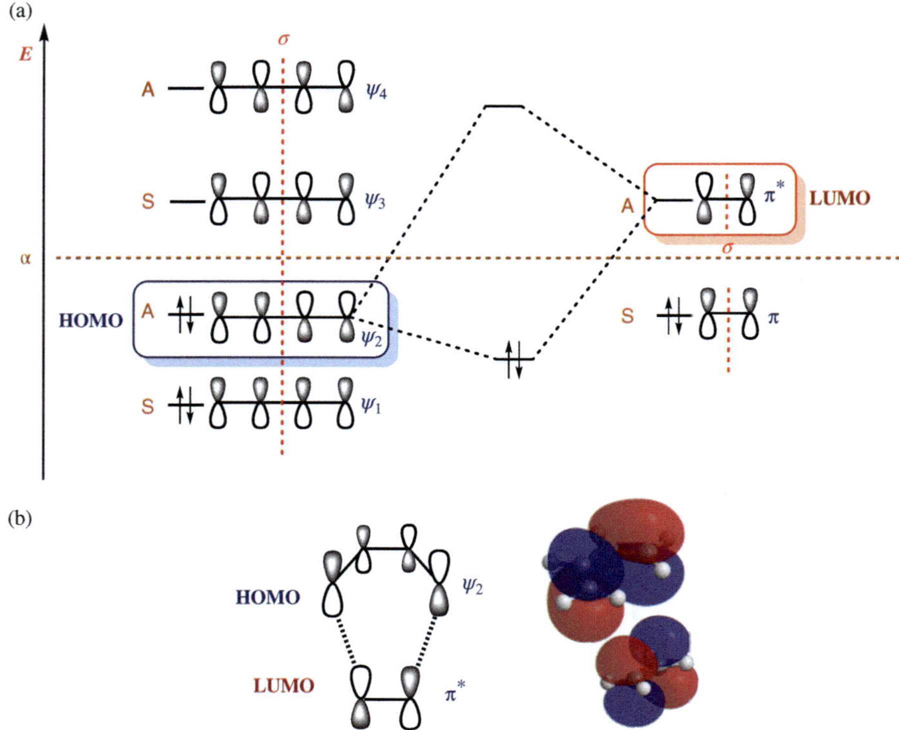

FIGURE 7.20   HOMO–LUMO symmetry analysis for the Diels–Alder reaction between 1,3-butadiene and ethylene.

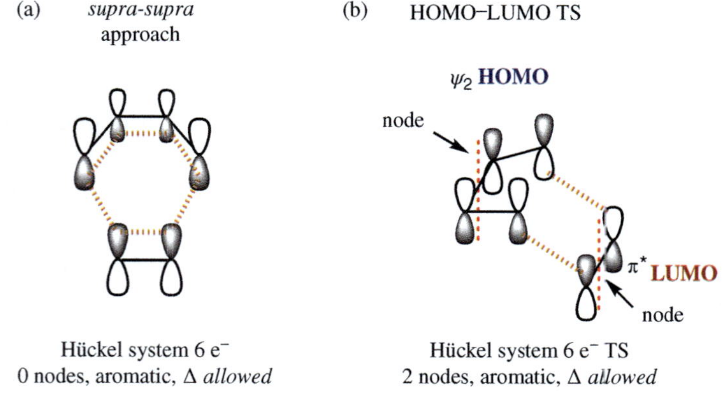

FIGURE 7.21   Examination of the nodal properties in the transition state of the Diels–Alder reaction between 1,3-butadiene and ethylene.

Correlation Diagram Starting Materials ⇌ Product

In this cycloaddition reaction, two ethylene π orbitals give rise to two sigma C—C bonds in the cyclobutane product. Again, a σ-plane of symmetry bisecting the ethylene groups, as well as the sigma combinations in cyclobutane

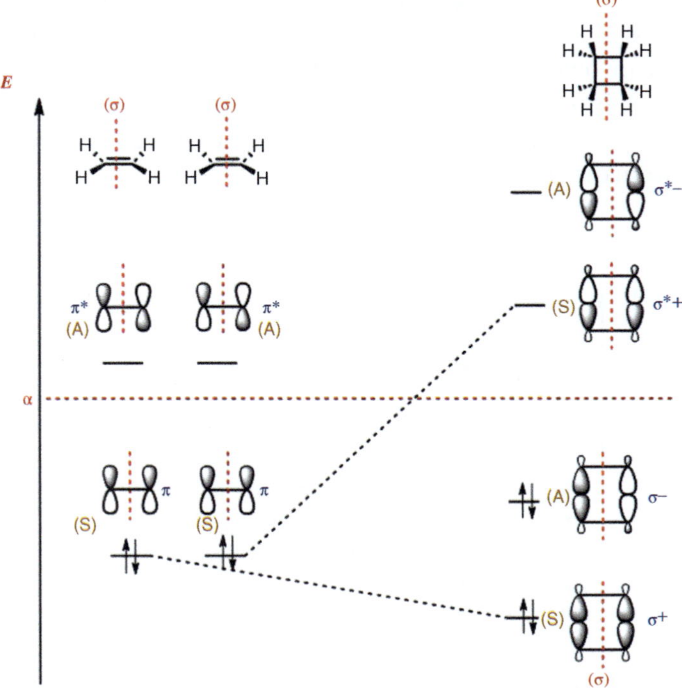

FIGURE 7.22 Correlation MO diagram for the [2 + 2] cycloaddition reaction between two ethylene molecules.

are convenient for assigning the corresponding symmetries in each MO (Figure 7.22).

The conversion of the two π MOs into two sigma C—C bonds is accompanied by a significant energy cost, since it leads from a bonding π MO in the reactants to an anti-bonding σ^* MO in the product (Figure 7.22). This situation implies that the reaction is *forbidden*. Furthermore, the symmetry associated with starting materials and product is not congruent: $[S_2] \rightleftharpoons [SA]$; so it is concluded again that the reaction is *forbidden*.

HOMO/LUMO Interaction

The interaction between FMOs in two ethylene molecules is not adequate [S] ↮ [A] (Figure 7.23a). Furthermore, it can be seen that the interaction between the HOMO of one ethylene molecule (π MO) and the LUMO of the second ethylene (π^* MO) involves out-of-phase overlap and therefore the reaction is thermally *forbidden* (Figure 7.23b).

Nodal Properties of the Transition State

In the case of a [2 + 2] cycloaddition reaction (a $4n$ π-electron system), the *supra-supra* mode of reaction corresponds to a Hückel arrangement, which being antiaromatic is thermally *forbidden*. In terms of FMO theory, when considering the reaction between the π-orbital HOMO of one ethylene and the π^* orbital (LUMO) in the second ethylene moiety, it can be appreciated that the transition state involves two sign changes between neighbouring p orbitals. This even number of inversions corresponds to an unfavourable π Hückel system for the [2 + 2] cycloaddition reaction under analysis, because its four electrons make it antiaromatic (Figure 7.24).

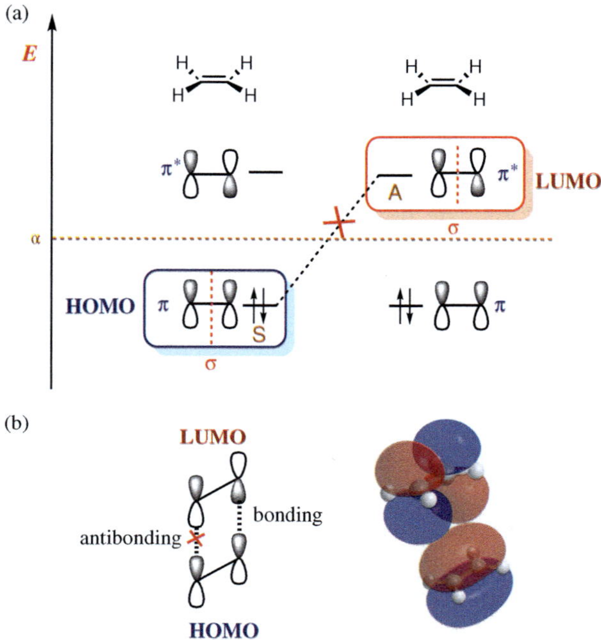

FIGURE 7.23   Analysis of the symmetry properties in the HOMO–LUMO frontier orbitals for the [2 + 2] cycloaddition reaction between two ethylene molecules.

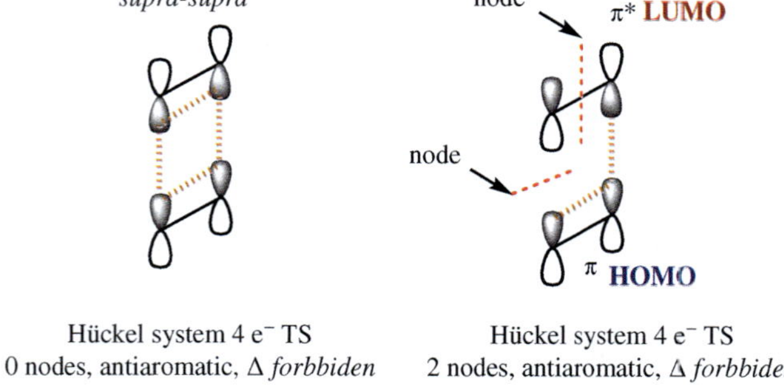

FIGURE 7.24   Examination of the nodal properties in the transition state of the [2 + 2] cycloaddition reaction between two ethylene molecules.

SUPRA- OR ANTARAFACIAL TOPICITY IN CYCLOADDITION REACTIONS

[4 + 2] Diels–Alder reactions, which are allowed according to the analysis described above, proceed by addition of the two π-systems from one side (Figure 7.25). These reactions are then classified as [4s + 2s] *suprafacial* cycloadditions. This type of approach is usually labelled by placing the subscript *s* (for *supra*) in each π component.

Antarafacial (or *supra-antara* approach) cycloaddition reactions are those pericyclic cyclisation reactions that proceed in such a way that opposite faces in the π-system are involved in the cycloaddition reaction. *Antarafacial* topicity is relatively rare because the participating π-system must be twisted for the concerted σ-bond formation to take place from opposite sides of the π nodal plane. An example of a reaction proceeding with *antarafacial* topicity is the cycloaddition of tetracyanoethylene to heptafulvalene (Figure 7.26a). The *trans*

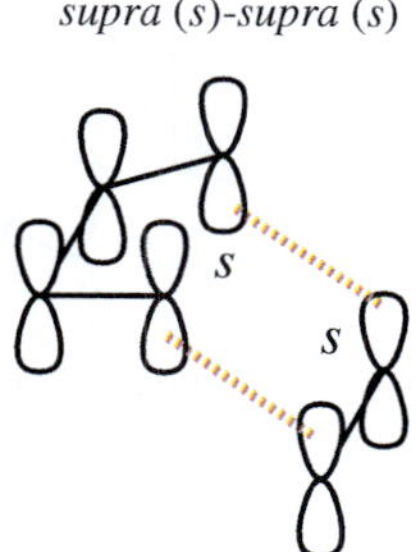

FIGURE 7.25   *Suprafacial* orbital interaction in [4 + 2] cycloadditions.

FIGURE 7.26   Transition state in the *antarafacial* addition of tetracyanoethylene to heptafulvalene.

configuration of the product shows that the addition is *anti* in heptafulvalene, a system large enough to allow the necessary deformation without appreciable loss in conjugation. In terms of FMO theory, for concerted bonding the orbital symmetry requirement is for one of the π–systems to react antarafacially and the other to react suprafacially (Figure 7.26b). This type of approach is usually labelled with the appropriate subscript s (*supra*) or a (*antara*) for each corresponding π partner.

EFFECT OF SECONDARY INTERACTIONS BETWEEN MOLECULAR ORBITALS

From a steric hindrance point of view, it can be anticipated that the addition of maleic anhydride to cyclopentadiene should proceed with *exo* stereochemistry; however, experimentally it has been found that the predominant product is the *endo* adduct.

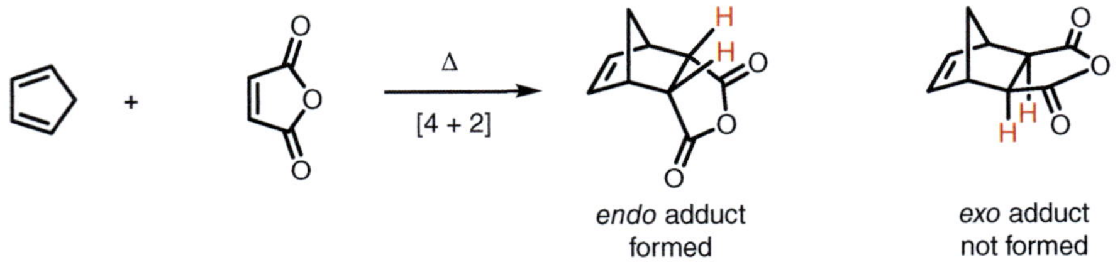

endo adduct
formed

exo adduct
not formed

Woodward and Hoffman explained this result in terms of a secondary bonding interaction between orbital segments separated from the principal active orbitals, leading to bond formation, that nevertheless stabilise the transition state of the pericyclic reaction.

Figure 7.27 shows the interaction between π orbitals leading to the [4s + 2s] dimerisation of 1,3-butadiene. The primary interactions leading to bond formation are those taking place between C-1/C-5 and C-4/C-6, while the secondary interaction, inducing the *endo* transition state, corresponds to the interaction between C-3/C-7.

By contrast, the *exo* transition state does not permit the stabilising secondary interactions observed in the *endo* transition state (Figure 7.28). As a result, the *endo* adduct, though less thermodynamically stable than the *exo* adduct, is often the predominant product due to kinetic control in the cycloaddition reaction. In the *endo* addition, secondary molecular orbital interactions stabilise the transition state, enhancing the reaction rate. These interactions are absent in the *exo* addition because the relevant centers are too distant for effective interaction.

To conclude this chapter, Table 7.2 summarises the selection rules for cycloaddition reactions based on FMO theory.

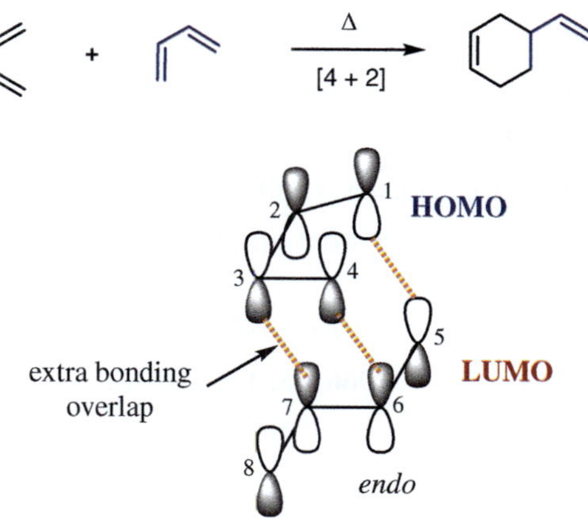

FIGURE 7.27 *Endo* transition state in the [4 + 2] dimerisation of 1,3-butadiene.

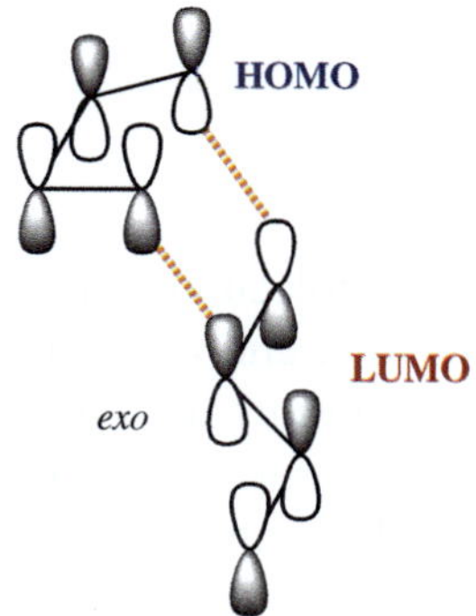

FIGURE 7.28 *Exo* transition state of the [4 + 2] dimerisation of 1,3-butadiene.

TABLE 7.2 Selection Rules for Cycloaddition Reactions as Deduced by FMO Analysis

No. of electrons	Stereochemical mode of reaction	Activation
4n	supra-supra antara-antara	$h\nu$
4n	supra-antara antara-supra	Δ
4n + 2	supra-supra antara-antara	Δ
4n + 2	supra-antara antara-supra	$h\nu$

FURTHER READING

A. Streitwieser, Jr., C. H. Heathcock, *Introduction to Organic Chemistry*, 2nd ed., Macmillan, New York, **1981**.

D. E. Lewis, *Advanced Organic Chemistry*, Oxford University Press, New York, **2016**.

E. Heilbronner, P. A. Straub, *Hückel Molecular Orbitals HMO*, Springer-Verlag, New York, **1966**.

F.-C. He, G. V. Pfeiffer, *J. Chem Ed.* **1984**, *61*, 948.

H. E. Zimmerman, *J. Am. Chem. Soc.* **1966**, *88*, 1564, 1566.

M. A. Fox, R. Cardona, N. J. Kiwiet, *J. Org. Chem.* **1987**, *52*, 1469.

M. J. S. Dewar, *Tetrahedron* **1966**, *8*, 75.

R. B. Woodward, R. Hoffman, *J. Am. Chem. Soc.* **1965**, *87*, 4388.

R. B. Woodward, R. Hoffmann, *The Conservation of Orbital Symmetry*, Academic Press, New York, **1970**.

S. Sugiyama, H. Takeshita, *Bull. Chem. Soc. Japan* **1987**, *60*, 977.

EXERCISES

7.1 Use correlation diagrams of molecular orbitals, analysis of the nodal properties of the transition state, and an examination of symmetry consequences in the HOMO/LUMO interaction to decide whether the reaction between the pentadienyl π-system (pentadienyl cation or anion) and ethylene to give cycloheptenyl cation or anion are allowed or forbidden cycloadditions.

7.2 Determine whether the following reaction is thermally allowed.

7.3 Use correlation diagrams of molecular orbitals, analysis of the nodal properties of the transition state, and an examination of symmetry consequences in the HOMO/LUMO interaction to determine whether the following cycloaddition reactions are allowed or forbidden.

(a) ... + ... $\xrightarrow{\Delta}$ ...

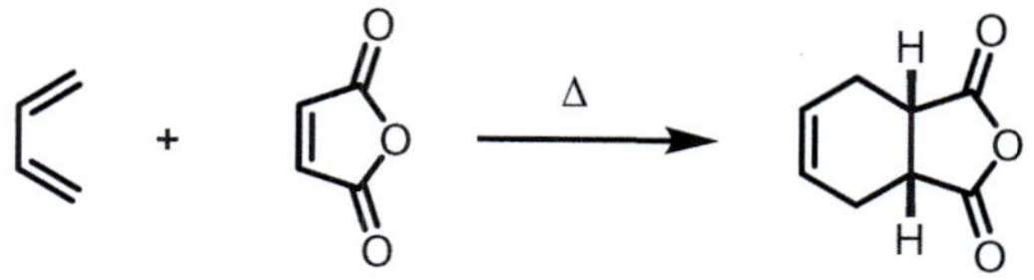

(b) ... + ... $\xrightarrow{\Delta}$ ...

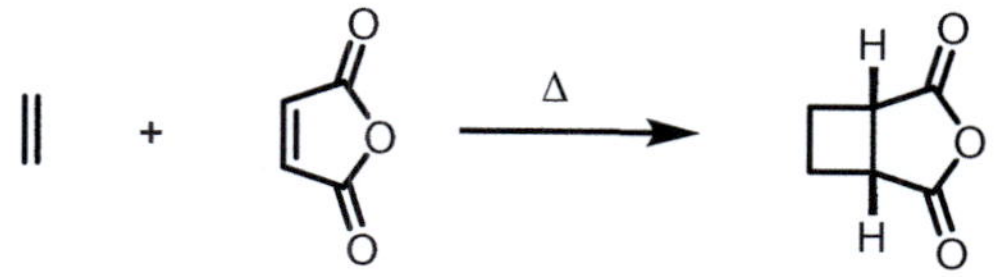

(c) 2 ... $\xrightarrow{\Delta}$ ...

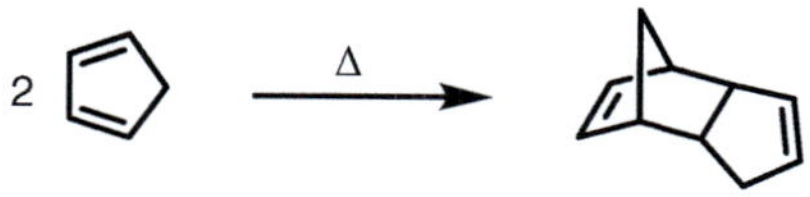

(d) ... + ... $\xrightarrow{\Delta}$ ...

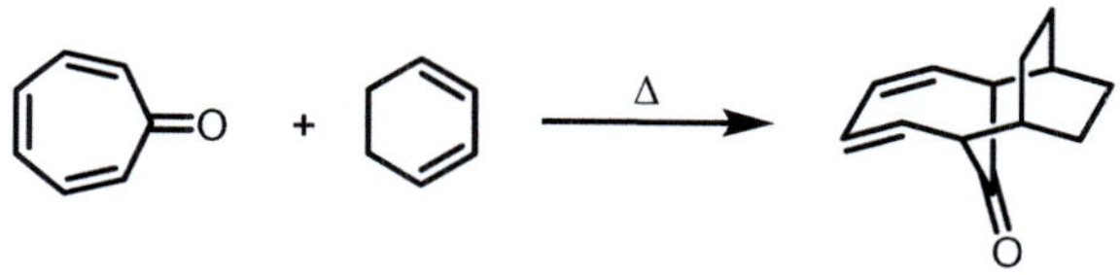

7.4 Following analysis of the correlation diagrams of molecular orbitals, the nodal properties of the transition state and symmetry consequences in the HOMO/LUMO interaction, predict the product of the following reactions:

(a) 2 ... $\xrightarrow{\Delta}$

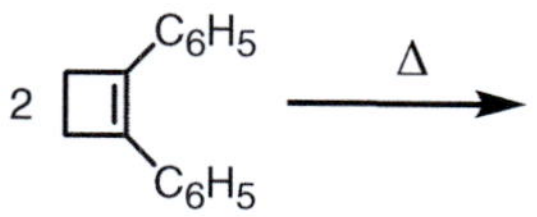

(b) ... + ... $\xrightarrow{\Delta}$

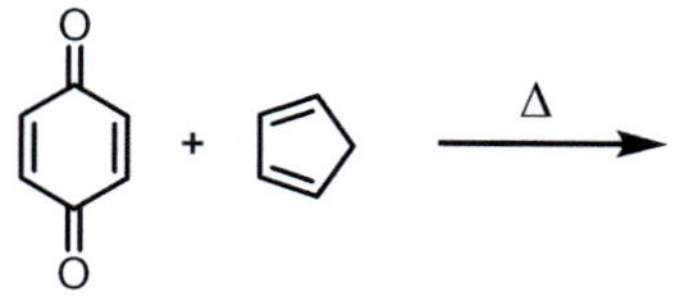

CHAPTER 8

Cheletropic Reactions

INTRODUCTION

Cheletropic reactions are a special group of concerted cycloadditions or cyclo-reversions in which two σ-bonds are simultaneously formed or broken on the same atom of the 'monocentric' species. Examples of this type of reactivity, involving bond formation or cleavage, are shown in Figure 8.1. In the cleavage direction (cycloreversion), where a small molecular fragment is eliminated, these reactions are often called *extrusions*. Cheletropic reactions usually involve the addition or extrusion of a small stable molecular fragment such as CO, N_2 or SO_2.

The analysis of the symmetry of the molecular orbitals (MOs) participating in cheletropic reactions is commonly carried out in the direction of addition, but symmetry analysis of the extrusion reaction is equally informative. The reaction will be *allowed* for certain specific molecular configurations, depending on whether the approach of component **1** to the π-system is linear or non-linear, and whether the π-system rotates in a *conrotatory* or *disrotatory* manner during cyclisation (Figure 8.2). In the linear approach, the *p*-orbital in *p*-type fragment **1** points towards the centre of the π-system to which it is incorpo-rating (Figure 8.2a), whereas in a nonlinear approach, the *p*-type orbital approaches sideways (Figure 8.2b). Analysis of the HOMO–LUMO frontier orbital interactions in the transition state for the linear approach of **1** (*p*-type highest occupied molecular orbital [HOMO] orbital) to the butadiene π-system (ψ_3 lowest unoccupied molecular orbital [LUMO] orbital) indicates a bonding, in-phase orbital interaction, as long as the terminal carbons of the diene rotate in *disrotatory* fashion (Figure 8.2a). On the other hand, the nonlinear sideway approach requires a *conrotatory* motion of the terminal carbons of the diene π-system to achieve an effective in-phase overlap with the HOMO of **1**. Below are some illustrative examples where this frontier orbital HOMO/LUMO sym-metry analysis is applied.

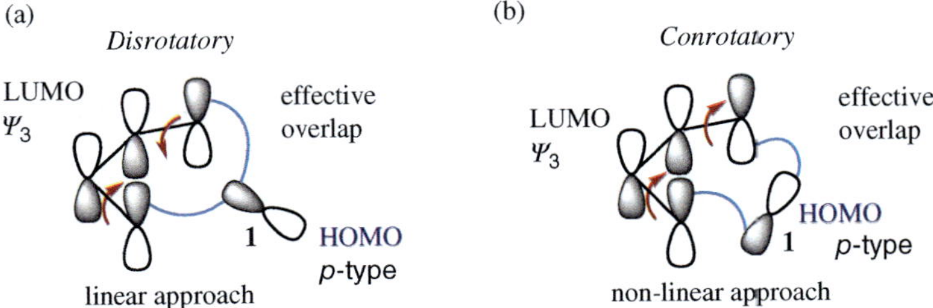

FIGURE 8.1   Examples of cheletropic reactions.

(a) *Disrotatory* (b) *Conrotatory*

FIGURE 8.2   HOMO/LUMO symmetry analysis in cheletropic reactions.

[2 + 2] CHELETROPIC REACTIONS

A most representative [2+2] cheletropic cycloaddition reaction corresponds to the addition of singlet carbenes to olefins to form cyclopropanes. Only singlet carbenes are considered in the analysis of the conservation of symmetry since the pericyclic selection rules cannot be applied to triplet states. Singlet carbenes are added to olefins in a concerted, and therefore *stereospecific* manner. As shown in Figure 8.3, the stereochemistry of the olefin (E or Z) determines the configuration of the cyclic product (*trans* or *cis*).

To understand the reactivity of singlet carbenes, it is necessary to recall their structure and associated orbitals (Figure 8.4). Singlet carbenes are bent, with bond angles between 100° and 110°, indicative of a trigonal sp^2 hybridisation state. Two of the three sp^2 hybrid orbitals are used to form two single bonds with substituents 'R' attached to the carbon atom. The third sp^2 hybrid orbital contains the non-bonded electron pair that acts as the HOMO orbital, while the non-hybridised p-orbital is empty and corresponds to the LUMO orbital.

In cheletropic cycloadditions, the incorporation of singlet carbene to an alkene can be conveniently analysed by the rules of orbital symmetry. In

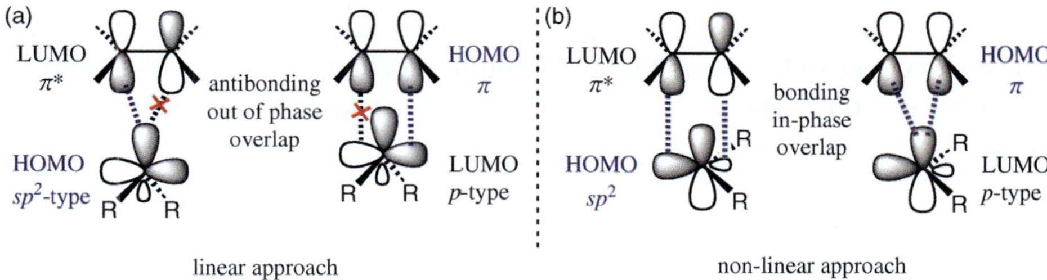

FIGURE 8.3 Stereospecificity of the cheletropic addition of singlet carbene to olefins.

FIGURE 8.4 Structure and orbitals in a general singlet carbene.

FIGURE 8.5 HOMO–LUMO interactions in the addition of carbene to an olefin. (a) Linear approach, out-of-phase orbital interaction, *forbidden*. (b) In the nonlinear approach, in-phase orbital interaction, *allowed* cheletropic cycloaddition.

particular, the carbene can approach the olefin in a linear or nonlinear mode (Figure 8.5). In the linear approach, the bond plane of two substituents on carbene is perpendicular to the bond plane of the olefinic C—C bond. As it can be appreciated in Figure 8.5a, for a linear approach of carbene, both the HOMO and LUMO orbitals of carbene involve out-of-phase anti-bonding interactions; that is, the cheletropic cycloaddition is *forbidden* by symmetry. By contrast, during a nonlinear approach of the carbene moiety, the bonding plane of two substituents on the carbene moiety will be parallel to that of the C—C bond of the π-system (Figure 8.5b). The reader will notice that during nonlinear carbene addition, the bonding HOMO–LUMO interactions are *suprafacial* relative to the $(4n+2)$ π-system; therefore, the nonlinear carbene approach leads to the *allowed*, concerted and *stereospecific* formation of the two σ bonds.

The Simmons–Smith reaction is an organic cheletropic reaction in which an organozinc carbene reacts with an alkene to form a cyclopropane. In the total synthesis of the antifungal FR-900848, Barrett and coworkers demonstrated the *stereospecific* Simmons–Smith double cyclopropanation of bis[(*E*)-olefin] **2** to afford (2*S*,3*S*,4*R*,5*R*,6*S*,7*S*)-bis[methylcyclopropyl]-1,3-dioxolane **3**. By the same token, the bis[(*Z*)-olefin] analogue **4** afforded the all-(*R*) stereoisomer **5** as a single diastereoisomer (Figure 8.6). These reactions confirm the *stereospecific* cheletropic cyclopropanation of (*E*) olefins to the *trans*-disubstituted cyclopropanes and the *stereospecific* cyclopropanation of (*Z*) olefins to give the corresponding *cis*-disubstituted cyclopropanes.

Carbene chemistry plays an important role in organic synthesis due to the unique reactivity and versatility of carbenes. Indeed, the reaction between a carbene and an alkene produces cyclopropane rings, which are found in many natural products and pharmaceuticals. This transformation is widespread owing to its high stereocontrol and efficiency. There are various synthetic methods for the generation of carbenes, for further study, the reader is advised to consult Clayden, Greeves and Warren (2012); Barrett et al. (1996); Onoda et al. (1996).

The MO conservation of symmetry theory predicts that the [2+2] cheletropic reaction between olefins and carbon monoxide (CO) to produce cyclopropanone derivatives is *forbidden* (Figure 8.7). In this reaction, the olefin π orbital and the lone pair in the *n*-orbital of CO could in principle interact to form two new sigma (C—C) orbitals. In the process, a π CO orbital is converted to a *p*-type orbital.

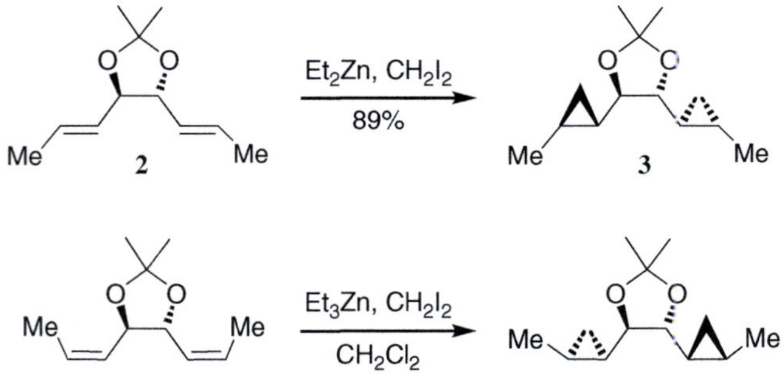

FIGURE 8.6   Stereospecificity in the Simmons–Smith reaction for the formation of cyclopropanes.

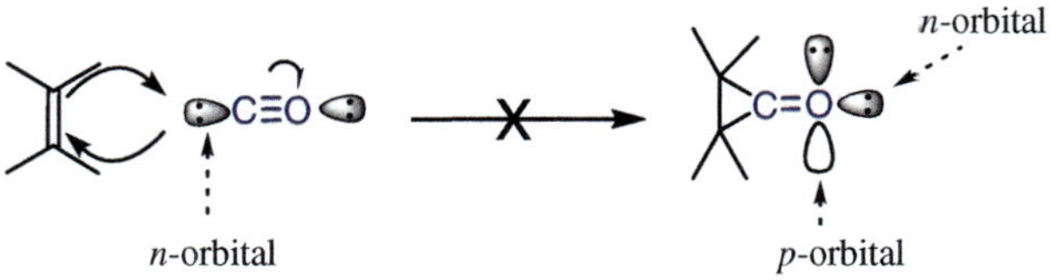

FIGURE 8.7   [2+2] *Forbidden* cheletropic reaction between olefins and carbon monoxide (CO).

For examination of the symmetry properties of each participant orbital it is convenient to use a σ plane bisecting ethylene and passing linearly through the CO molecule (Figure 8.8a). It can be appreciated that both MOs are symmetric (S) relative to the σ plane. However, Figure 8.8b shows a qualitative orbital energy level distribution of the MOs in both reactants, as well as their corresponding symmetry (A or S) with respect to the σ plane. An analysis of the frontier molecular orbitals (FMO) shows that neither the combination of the HOMO π orbital in the olefin (S) with the LUMO of carbon monoxide, CO (A), nor the combination of the LUMO π^* orbital in the olefin (A) with the HOMO of the CO (S) present the same symmetry. Therefore, the HOMO–LUMO analysis indicates that the reaction is *forbidden*.

Furthermore, Figure 8.9 shows how, in the linear approach of carbon monoxide to an olefin, both HOMO–LUMO frontier orbital combinations generate out-of-phase overlap, which is an anti-bonding *forbidden* MO interaction for cheletropic cycloaddition.

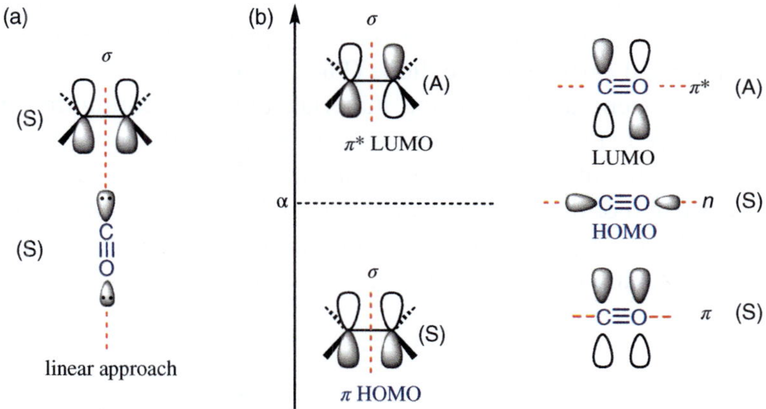

FIGURE 8.8 *Forbidden* cheletropic reaction between CO and one olefin. (a) Linear approach. (b) FMO analysis of the reaction.

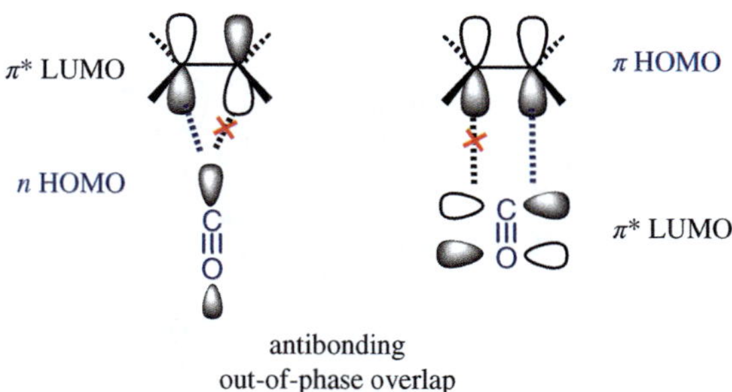

FIGURE 8.9 HOMO–LUMO frontier orbital interactions in the linear approach of carbon monoxide to olefins.

[4 + 2] CHELETROPIC REACTIONS

An illustrative example of concerted [4+2] cheletropic cycloaddition reactions is the reversible insertion of SO_2 into 1,3-butadiene to produce cyclic sulphones (Figure 8.10). In the SO_2 molecule, the HOMO corresponds to the lone pair of electrons in the plane of the molecule, while the LUMO corresponds to the *p*-orbital perpendicular to the plane formed by the O—S—O atoms.

Figure 8.11 shows that the HOMO–LUMO interactions in the *suprafacial* approximation of SO_2 to the π-system are associated with in-phase MO bonding interactions. Importantly, it can be appreciated that in the transition state, the diene terminal carbons must rotate in a *disrotatory* manner to allow the diene ψ_3 LUMO to interact constructively (in-phase MO overlap) with the *n*-orbital of SO_2, which corresponds to its HOMO. A similar conclusion is reached when one considers the *disrotatory* rotation required for ring formation, involving the interaction of the ψ_2 HOMO of the diene interacting with the SO_2 *p*-orbital LUMO.

The [4+2] cheletropic reaction is *stereospecific*. For instance, with (2E,4E)-hexa-2,4-diene as the starting material, the cheletropic addition reaction of SO_2 affords the *cis*-configured product *cis*-**6** as a consequence of the *disrotatory* rotation of the terminal carbons in the diene (Figure 8.12a, top). By contrast, with (2Z,4E)-hexa-2,4-diene as the starting material, the *trans*-configured product *trans*-**6** is produced (Figure 8.12a, bottom). As mentioned above, the reaction is reversible and this process is also *stereospecific*. Thus, under thermal conditions, SO_2 extrusion of the *cis* compound leads to the exclusive formation of the stereoisomer (2E,4E)-hexa-2,4-diene. According to the principle of microscopic reversibility, in a state of equilibrium, the forward and backward pathways of

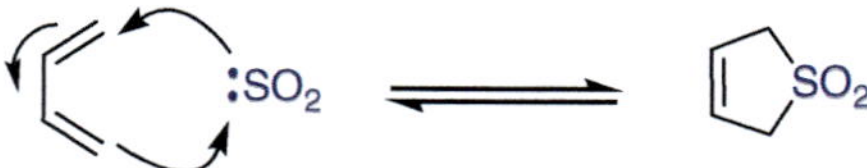

FIGURE 8.10 [4+2] cheletropic reaction between SO_2 and 1,3-butadiene.

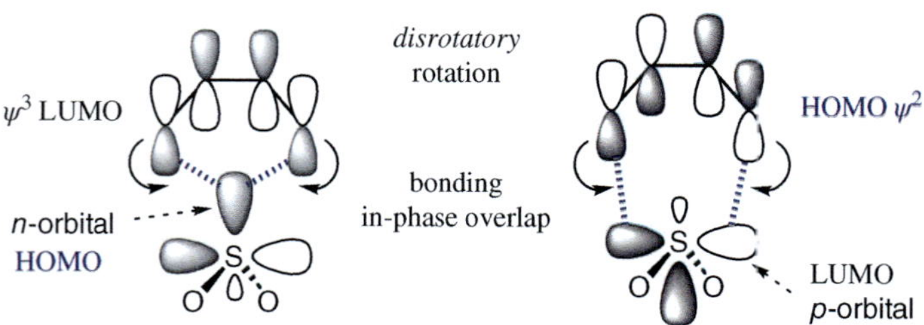

FIGURE 8.11 HOMO–LUMO frontier orbital interactions in the linear approach of SO_2 to 1,3-butadiene, that is *allowed* for *disrotatory* rotation of the π-system.

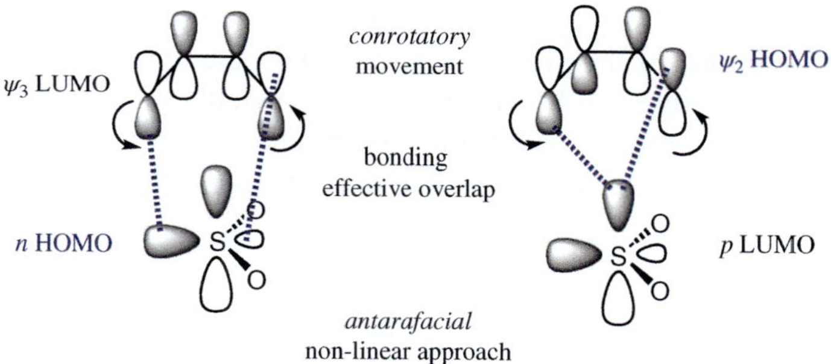

FIGURE 8.12 Stereochemical outcome of the cheletropic reaction between SO_2 and hexa-2,4-diene isomers.

FIGURE 8.13 HOMO–LUMO interactions in the nonlinear approach of SO_2, *antarafacial* relative to the diene.

each of the elementary steps must be the same. Therefore, SO_2 extrusion proceeds with a *disrotatory* movement, which produces the (*E*,*E*) isomer (Figure 8.12b, top).

As it turns out, under photochemical conditions, the configuration of the products is exactly opposite (Figure 8.12b, bottom). Therefore, SO_2 extrusion under light irradiation of sulphone *cis*-**6** provides the diastereomeric (2*Z*,4*E*)-hexa-2,4-diene.[1]

It is important to note that the nonlinear approach of the SO_2 is required to be *antarafacial* relative to the π-system in order to be *allowed* by symmetry (Figure 8.13). In this case, to attain an effective in-phase overlap between the HOMO–LUMO FMOs, the rotation of the terminal carbons must be *conrotatory*

---

[1] In very general terms, the photochemical promotion of an electron to a higher energy orbital changes the HOMO of the system, and thus its reactivity (in this case from a *disrotatory* to a *conrotatory* rotation). Nevertheless, the mechanism of the photochemical reaction is more complex and is beyond the scope of this text (see Saltiel and Metts 1967 in the 'Further reading' section at the end of this chapter).

and would, therefore, afford the opposite stereoisomer by comparison with the linear approach activated under thermal conditions.

In an illustrative example, Nicolaou, Barnette and Ma (1980) reported the synthesis of steroidal polycycles based on the intramolecular capture of *o*-quinodimethanes generated by cheletropic SO_2 extrusion from strategically designed starting materials such as **7** (Figure 8.14). The *o*-quinodimethanes intermediates reacted intramolecularly via a [4+2] cycloaddition reaction with the pendant dienophile. Using this strategy, the total synthesis of estra-1,3,5,(10)-trien-17-one **8** was achieved in good yield.

The irreversible extrusion of nitrogen from cyclic diazene (Figure 8.15a) and the easy loss of carbon monoxide from norbornadienone (Figure 8.15b) are important examples of six-electron cheletropic reactions. The driving force for these reactions is often the entropic gain during carbon monoxide detachment, and this effect is augmented by the fact that nitrogen and carbon monoxide are rather stable molecules from an enthalpic point of view. In this context, Pirrung's and McGeehan's (1983) suggestion that ethylene biosynthesis proceeds via cheletropic fragmentation of the precursor nitrene **9** is noteworthy (Figure 8.15c).

Another illustrative study is the [4+2] cheletropic reaction between 1,3-butadiene and carbon monoxide to produce 3-cyclopentenone. The butadienyl π-system and the occupied *n*-orbital on carbon in CO react in a

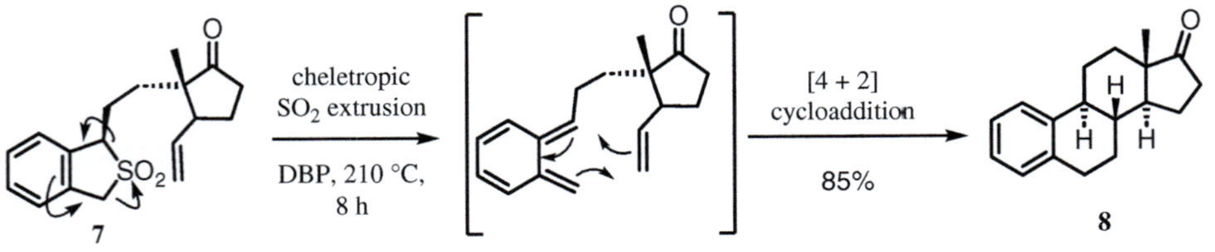

FIGURE 8.14   Synthesis of steroids *via* intramolecular capture of *o*-quinodimethanes generated by cheletropic elimination of SO_2.

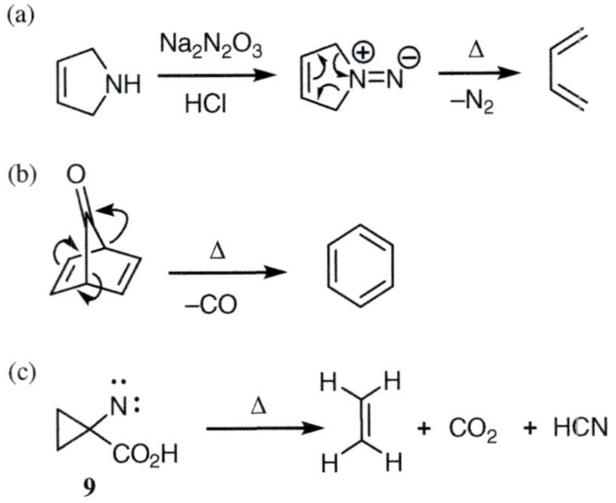

FIGURE 8.15   Salient examples of cheletropic reactions.

cheletropic pericyclic concerted reaction to generate two C—C sigma orbitals. Simultaneously, the π-type orbital on CO is converted to a p-orbital on the carbonyl oxygen (Figure 8.16).

Figure 8.17 presents the MOs involved in the classical cheletropic cycloaddition reaction between 1,3-butadiene and carbon monoxide. Figure 8.17 also includes the symmetry of the corresponding MOs with respect to a σ-plane perpendicular to the plane of the 1,3-butadiene, which divides the 1,3-butadiene π-system into two ethylene units and passes linearly across the CO molecule. The σ-plane is also located in the direction of a linear approach of CO towards the 1,3-butadiene. From this diagram it can be concluded that the reaction should be *allowed* because the occupied MOs in the reactants and in the product present the same symmetry with respect to the σ plane: [S_2A_2]. The symmetry of the MOs of the reactants can also be assigned as indicated in the diagrams shown in Figure 8.17.

On the other hand, Figure 8.18 shows the symmetry of the FMO of interest in the reactants relative to the reference σ plane. It can be appreciated that the interaction between the HOMO orbital of carbon monoxide (S) and the LUMO

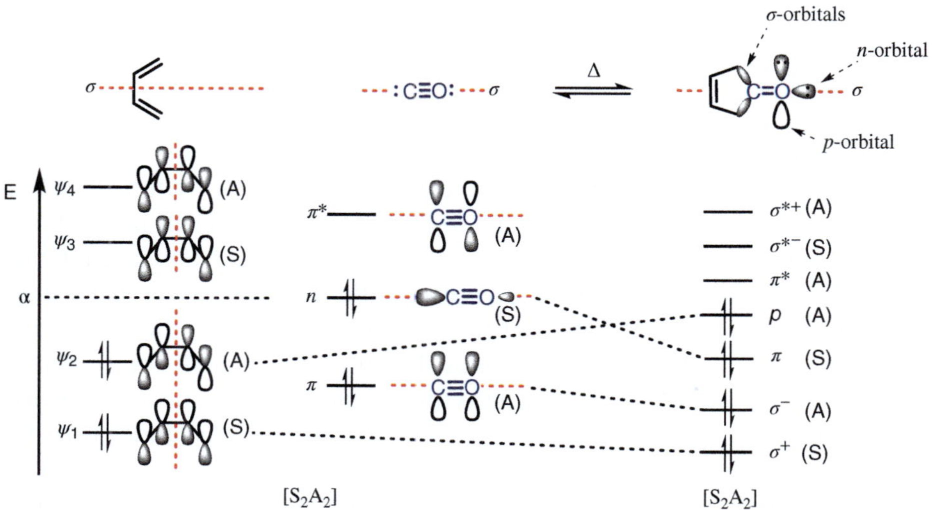

FIGURE 8.16   [4+2] cheletropic cycloaddition reaction between 1,3-butadiene and CO.

FIGURE 8.17   MO symmetry analysis of the cheletropic cycloaddition reaction between 1,3-butadiene and carbon monoxide.

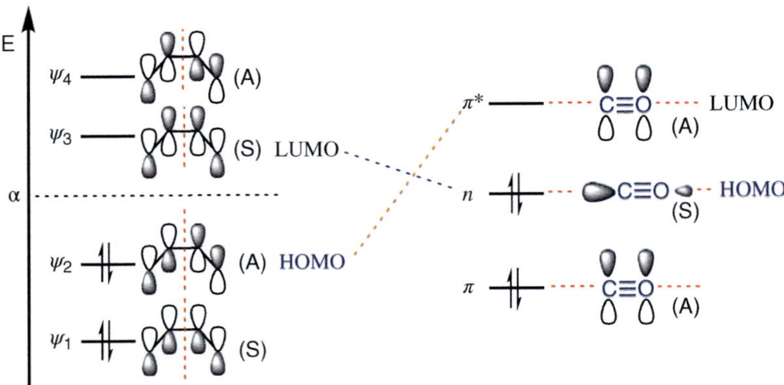

FIGURE 8.18 FMO symmetry analysis in the cheletropic cycloaddition reaction between 1,3-butadiene and carbon monoxide.

sulphone *cis*-**10**   (2Z,4Z,6E)-octa-2,4,6-triene   sulphone *trans*-**10**   (2E,4Z,6E)-octa-2,4,6-triene

FIGURE 8.19 Stereospecific [6+2] cheletropic extrusion reactions of seven-membered cyclic sulphones.

of 1,3-butadiene (S) is congruent by MO symmetry [S↔S], which is in line with an *allowed* process. Similarly, the interaction of the HOMO in 1,3-butadiene (A) with the LUMO of CO (A) presents the correct symmetry [A/A] for a constructive MO overlap.

[6 + 2] CHELETROPIC REACTIONS

The [6+2] cheletropic reaction between conjugated trienes and SO_2 is also *allowed* according to the rules of the conservation of MO symmetry. The [6+2] extrusion process is more common than the [6+2] cheletropic cycloaddition reaction. Furthermore, the thermal cheletropic extrusion reactions of seven-membered ring sulphones are stereospecific (Figure 8.19). In particular, as evidenced by the configuration of the π extrusion products, the ring opens in a *conrotatory* manner. For instance, as shown in Figure 8.19, the sulphone *cis*-**10** affords the (2Z,4Z,6E)-octatriene as the product, whereas the *trans*-**10** provides the (2E,4Z,6E)-isomer of octatriene via extrusion of SO_2 under thermal conditions.

Cycloreversion (extrusion) corresponds to the microscopic reversal of the [6+2] cycloaddition reaction. The principle of microscopic reversibility can be used to analyse and predict the stereochemical outcome of the thermal cheletropic addition of SO_2 to conjugated trienes by consideration of the MO conservation of symmetry theory (Figure 8.20a). For a constructive overlap of the frontier HOMO–LUMO orbitals, in which the terminal carbons rotate in a *conrotatory* manner, the approach of SO_2 must be *antarafacial* to the π-system (Figure 8.20b).

The concerted cheletropic cycloaddition of SO_2 to a conjugated polyene depends on the configuration of the polyene as demonstrated by Mock and coworkers. For example, the cheletropic addition of SO_2 to 1,3,5-cyclooctatriene gives a cycloadduct derived from the *suprafacial* approach of SO_2 towards the 4π diene system, which rotates in a *disrotatory* motion process to provide the cycloaddition product **11** (Figure 8.21a). By contrast, for the formation of the 1,6 adduct **12**, the SO_2 addition must be *antarafacial* via *conrotatory* rotation of the π-system, leading to a transition state with a highly constrained geometry (Figure 8.21b).

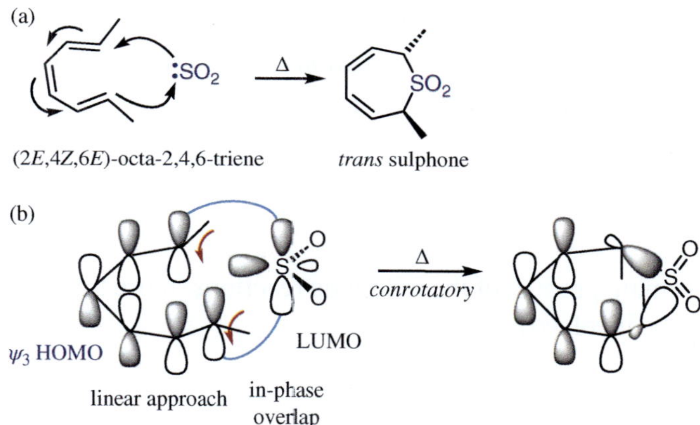

FIGURE 8.20   HOMO–LUMO interactions in the *antarafacial* approach of SO_2 to conjugated trienes.

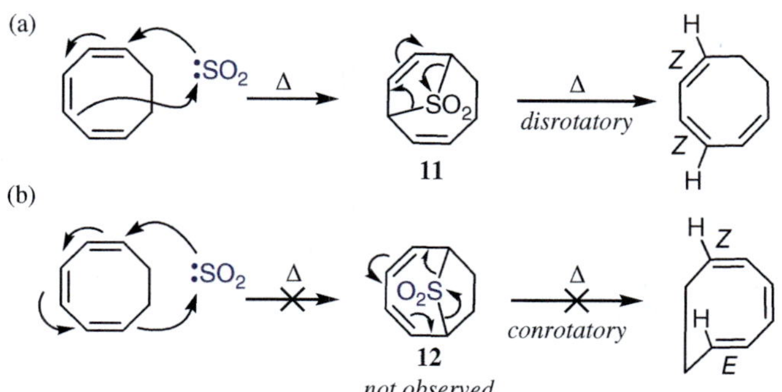

FIGURE 8.21   Cheletropic cycloaddition of SO_2 to 1,3,5-cyclooctatriene followed by extrusion.

FURTHER READING

A. G. M. Barrett, W. W. Doubleday, K. Kasdorf, G. J. Tustin, *J. Org. Chem.* **1996**, *61*, 3280.

B. M. Trost and A. C. Lavoie, *J. Am. Chem. Soc.* **1983**, *105*, 5075.

E. V. Anslyn, D. A. Dougherty, *Modern Physical Organic Chemistry*, University Science, New York, **2005**.

J. Clayden, N. Greeves, S. Warren, *Synthesis and reactions of carbenes* in *Organic Chemistry*, Oxford University Press, Oxford, UK, **2012**.

J. Saltiel and L. Metts, *J. Am. Chem. Soc.* **1967**, *89*, 2232.

K. C. Nicolaou, W. E. Barnette and P. Ma, *J. Org. Chem.* 1980, *45*, 1463.

M. C. Pirrung and G. M. McGeehan, *J. Org. Chem.* **1983**, *48*, 5143.

S. Kumar, V. Kumar and S. P. Singh, *Pericyclic Reactions: A Mechanistic and Problem-Solving Approach*, Academic Press, New York, **2015**.

S. Sankararaman, *Pericyclic Reactions- A Textbook: Reactions, Applications and Theory*; Wiley-VCH, Weinheim, **2005**.

T. Onoda, R. Shirai, Y. Koiso, S. Iwasaki, *Tetrahedron Lett.* **1996**, *37*, 4397.

W. L. Mock, *J. Am. Chem. Soc.* **1975**, *97*, 3666.

EXERCISES

8.1 Draw the main product of the following reactions (more than one pericyclic step may be involved):

(a)

(b)

(c)

(d)

(e)

(f)

(g)

8.2 Propose a mechanism for the following reactions (more than one pericyclic step may be involved):

(a)

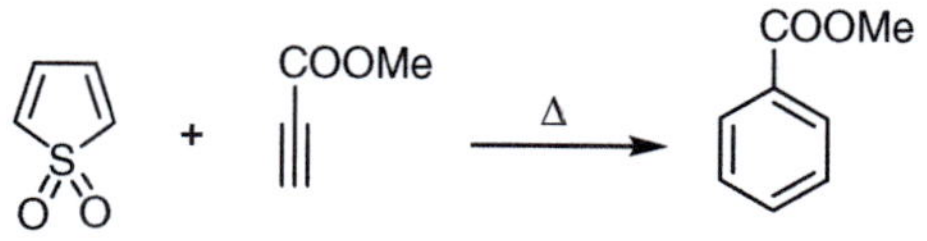

(b)

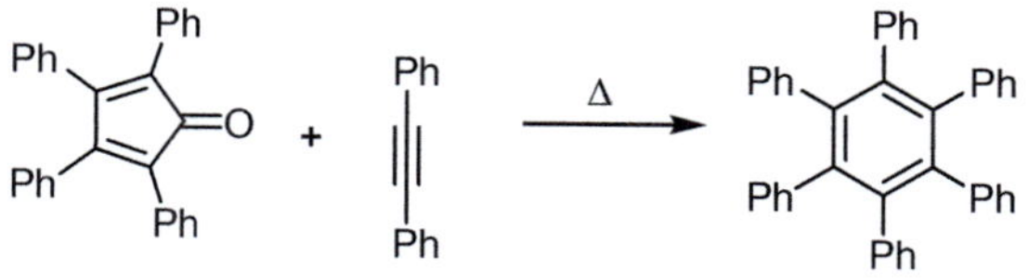

8.3 Please draw the structure of the product, the type of approach by the reagents (linear *vs.* nonlinear) and the required rotation of the MOs (*disrotatory vs. conrotatory*):

(a)

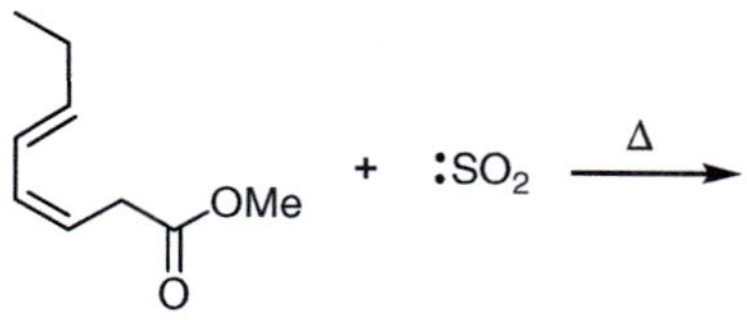

(b)

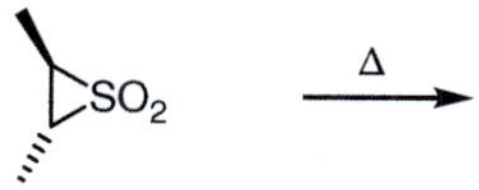

(c)

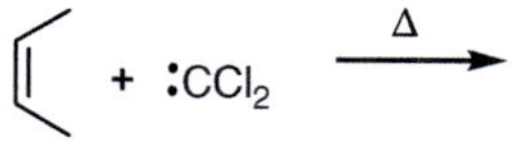

(d)

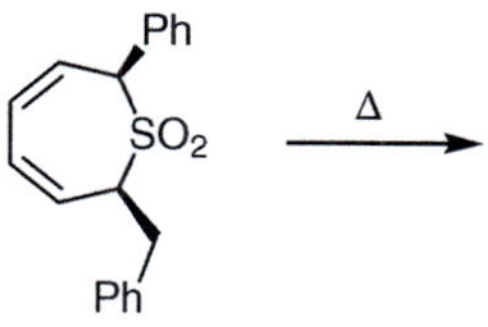

8.4 Suggest a plausible mechanism for the formation of ethylene in Figure 8.15c, including a diagram showing the frontier molecular orbitals involved.

CHAPTER 9

Electrocyclic Reactions

INTRODUCTION

Ring-forming electrocyclic reactions involve the conversion of two terminal p orbitals into a σ bond to form a ring (Figure 9.1a). The reverse reaction (ring-opening) involves the cleavage of a C—C σ bond to form two C—C π bonds and corresponds also to an electrocyclic process (Figure 9.1b).

In general, the cyclic products are anticipated to prevail in a reversible process because σ bonds are stronger than π bonds (*ca.* 90 *vs.* 60 kcal mol^{-1}, respectively); therefore, thermodynamic control shifts the equilibrium towards the cyclic compounds. An illustrative example is the electrocyclic reaction that takes place when 1,3,5-hexatriene is heated to afford the corresponding cyclic diene 1,3-cyclohexadiene (Figure 9.2a). Nevertheless, in some cases, ring strain favours the open species, as in the 1,3-butadiene ⇌ cyclobutene equilibrium, where angle strain overrides bond strength (Figure 9.2b).

Electrocyclic reactions are *stereospecific* as they are subject to the rules of the conservation of orbital symmetry. For instance, it has been found that the thermal electrocyclic ring-opening reaction of *cis*-3,4-dimethylcyclobutene **1** affords the (*E,Z*)-**2** diene (Figure 9.3a), whereas *trans*-3,4-dimethylcyclobutene provides the (*E,E*)-**2** diene isomer (Figure 9.3b).

Electrocyclic ring closure is also *stereospecific*. For example, in Figure 9.4a it is appreciated that for the formation of the observed product cyclobutene (*cis*)-**1**, the terminal methylene groups of the (*E,Z*)-**2** starting material must have rotated in a *conrotatory* manner; that is, in the same direction. On the other hand, in order for the methyl groups in product **4** to present the relative *cis* configuration, the terminal methylene groups of (*E,Z,E*)-octatriene **3** must rotate in a *disrotatory* way, that is, in opposite directions (Figure 9.4b).

It should be noted that steric factors or ring strain occasionally dictate which one of the two possible motions takes place. For example, in Figure 9.5a, one possible mode of the *disrotatory* rotation of the 1,3-cyclohexadiene ring in

(a)　Electrocyclic ring closure

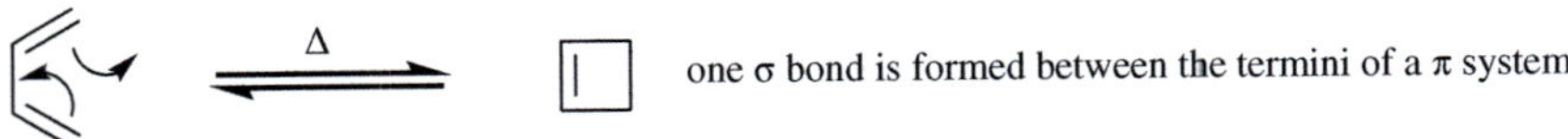

one σ bond is formed between the termini of a π system

(b)　Electrocyclic ring opening

one σ bond is broken to form a conjugated π system

FIGURE 9.1　Examples of electrocyclic reactions.

(a)　　　　　　　　　　　　　　　　　　　(b)

500 °C　　　　　　　　　　　　　　　200 °C

FIGURE 9.2　Examples of electrocyclic ring-closure and ring-opening reactions.

(a)　　　　　　　　　　　　　　　　　　(b)

cis-**1**　　　　(*E,Z*)-**2**　　　　*trans*-**1**　　　　(*E,E*)-**2**

FIGURE 9.3　Examples of *stereospecific* ring-opening in electrocyclic reactions.

(a)　　　　　　　　　　　　　　　　　　(b)

conrotatory　　　　　　　　　　　　*disrotatory*

(2*E*,4*Z*)-**2**　　　　*cis*-**1**　　　　(2*E*,4*Z*,6*E*)-**3**　　　　*cis*-**4**

FIGURE 9.4　Examples of *stereospecific* ring-closure electrocyclic reactions.

(a)　　　　　　　　　　　　　　　　　　(b)

disrotatory　　　　　　　　　　　　*disrotatory*

5　　　　(1*E*,3*Z*,5*E*)-**6**　　　　**5**　　　　(1*Z*,3*Z*,5*Z*)-**6**

FIGURE 9.5　Strain effects in some electrocyclic reactions.

5 would lead to the formation of product (1*E*,3*Z*,5*E*)-**6**, where the configuration orienting both terminal hydrogens towards the inside of the ring is too high in energy as a consequence of steric repulsion and molecular strain. By contrast, the alternative mode of *disrotatory* rotation leads to the formation of stable (1*Z*,3*Z*,5*Z*)-1,3,5-cycloheptatriene, (1*Z*,3*Z*,5*Z*)-**6**, with the two terminal hydrogens oriented outside the ring (Figure 9.5b).

Below are some examples of how the stereochemistry of electrocyclic reactions can be explained by the rules of the conservation of orbital symmetry, (1) by frontier molecular orbitals (FMO) analysis and (2) by examination of the nodal properties of the corresponding transition states. In particular, let us analyse two stereospecific electrocyclic reactions: 1,3-butadiene ⇋ cyclobutene and 1,3,5-hexatriene ⇋ 1,3-cyclohexadiene.

1,3-BUTADIENE ⇋ CYCLOBUTENE

Orbital Symmetry Correlation Diagrams

Disrotatory Electrocyclic Reactions

Figure 9.6 shows the molecular orbital (MO) correlation diagram of the orbitals involved in the *disrotatory* 1,3-butadiene ⇋ cyclobutene electrocyclic reaction. Upon consideration of the principle that the MOs in the reactant must lead to MOs of the same symmetry in the product, it can be appreciated that the π

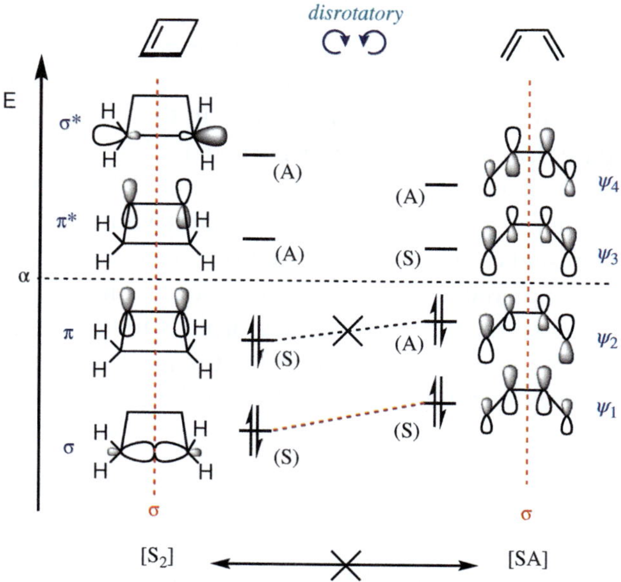

FIGURE 9.6   Molecular orbital correlation diagram in the *disrotatory* 1,3-butadiene ⇋ cyclobutene electrocyclic reaction. (S) or (A) indicates whether the orbital is symmetric or antisymmetric with respect to a σ-plane bisecting the molecule.

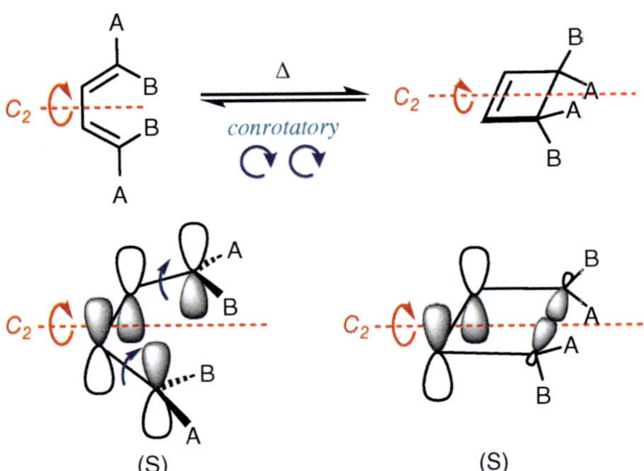

FIGURE 9.7   The symmetry of the reactants and products with respect to a C_2-axis is maintained [S] ↔ [S] during an electrocyclic reaction that proceeds with *conrotatory* motion.

orbital of the ethylene segment in cyclobutene (S = symmetrical with respect to the σ plane) evolves into the bonding ψ_2 orbital (A) of the diene (dashed grey line), which makes the thermal *disrotatory* rotation *forbidden*. Furthermore, the symmetry properties of the occupied orbitals in butadiene and cyclobutene are not the same [S$_2$] ↮ [SA], again leading to the conclusion that the *disrotatory* ring-opening reaction is *forbidden*.

Conrotatory Electrocyclic Reactions

During a *conrotatory* rotation in the 1,3-butadiene ⇌ cyclobutene electrocyclic reaction, it is an *axis of symmetry* C_2 (instead of a σ-plane) that makes evident that symmetry (or antisymmetry) is maintained in the reactant to product MO conversion (Figure 9.7). It can be appreciated that if the molecular structure remains unchanged (undistinguishable from the initial structure) after 180° rotation, it is symmetric (S), otherwise, it will be antisymmetric (A).

Figure 9.8 presents the orbital correlation diagram for the *conrotatory* cyclobutene ⇌ 1,3-butadiene electrocyclic reaction, which clearly shows that *conrotatory* rotation is *allowed*: the occupied orbitals in the reactant and product both present [SA] symmetry. Furthermore, the reaction proceeds smoothly through a low-energy transition state, always in the bonding level (below the reference α level).

FMO Analysis

Reaction analysis in terms of FMO leads to a similar prediction of the stereochemistry of the products in the cyclobutene ⇌ 1,3-butadiene electrocyclic reaction. It is appropriate to focus on the diene's highest occupied molecular orbital (HOMO) (Ψ_2) as shown in Figure 9.9. The orbitals that give rise to the σ bond in

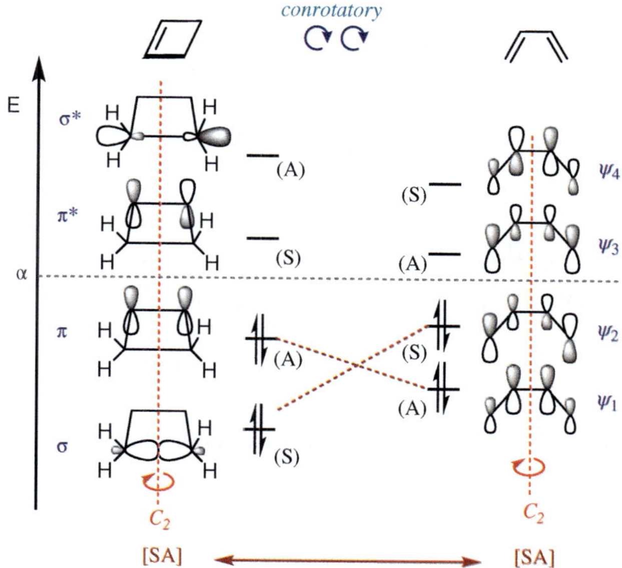

FIGURE 9.8 Molecular orbital correlation diagram in the *conrotatory* cyclobutene ⇋ 1,3-butadiene electrocyclic reaction. (S) or (A) labels indicate whether the orbital is symmetric or antisymmetric with respect to the C_2-axis of reference.

FIGURE 9.9 FMO analysis of the thermal electrocyclic ring closure of 1,3-butadiene.

the product correspond to the p orbitals on C(1) and C(4) in the 1,3-butadiene substrate, which break their π bonds with C(2) and C(3) in order to participate in the formation of the new σ bond (Figure 9.9a). For the reaction to be allowed, there must be a constructive in-phase orbital overlap between the terminal p orbitals during the formation of the σ bond. This can only be achieved if the terminal p orbitals rotate in the same direction (clockwise or counterclockwise); that is, in a *conrotatory* manner (Figure 9.9b). By contrast, upon rotation of the

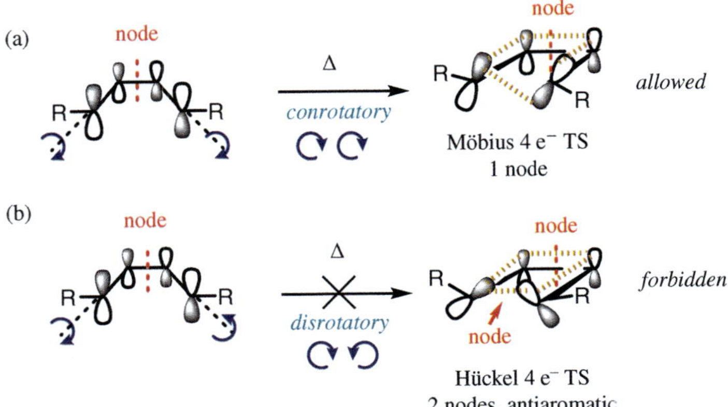

FIGURE 9.10   Nodal properties in the transition state of the thermal electrocyclic ring closure of 1,4-disubstituted-1,3-butadiene.

Ψ_2 MO in a *disrotatory* way, the orbital lobes will join out of phase, and therefore no σ-bond is formed (Figure 9.9c).

Analysis of the Nodal Properties in the Transition State

Figure 9.10a shows the transition state during the interconversion of substituted 1,3-butadiene to substituted cyclobutene via a *conrotatory* motion. This mode of orbital rotation is *allowed* since it gives rise to a Möbius-type transition state (one change of phase sign between adjacent p orbitals), which is allowed with 4π electrons. By contrast, the *disrotatory* rotation would lead to a Hückel-type transition state (2 nodes) with 4π electrons, which is forbidden for a 4-electron system (Figure 9.10b).

One can now understand the origin of the *stereospecificity* in electrocyclic reactions. Figure 9.11a shows the electrocyclic ring-closure reaction of diene (2*E*,4*Z*)-**2**: rotation of the two methyl groups in the same direction (*conrotatory* fashion), affords product **1** with *cis*-relative configuration. By the same token, and according to the principle of microscopic reversibility, the opening of *cis*-cyclobutene *cis*-**1** will afford the product (2*E*,4*Z*)-**2** in stereospecific manner (Figure 9.11b).

1,3,5-HEXATRIENE ⇌ 1,3-CYCLOHEXADIENE

Orbital Correlation Diagram

It can be appreciated that an appropriate symmetry element for the construction of the MO correlation diagram for the *disrotatory* rotation in the 1,3,5-hexatriene ⇌ 1,3-cyclohexadiene electrocyclic reaction is a σ plane (Figure 9.12). Thus, the symmetry of the Ψ_1–Ψ_6 MOs in hexatriene is assigned with respect to a σ

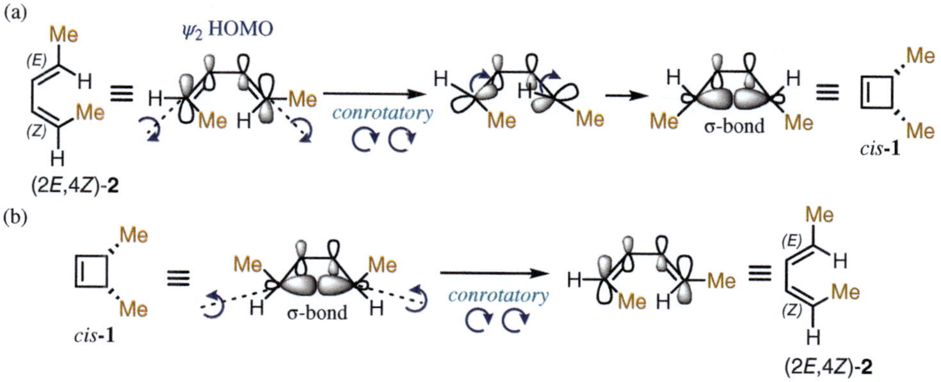

FIGURE 9.11 Examples of stereospecificity in the electrocyclic reactions involving substituted cyclobutenes.

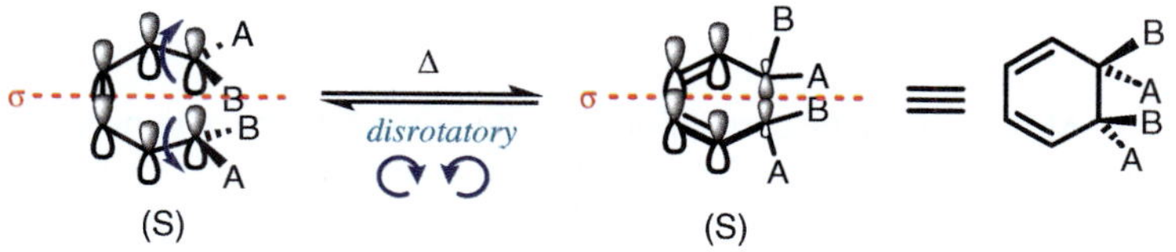

FIGURE 9.12 *Disrotatory* 1,3,5-hexatriene ⇋ 1,3-cyclohexadiene electrocyclic reaction.

plane that bisects the involved π molecular orbitals and takes into account their relative energy as dictated by the corresponding nodal properties; that is, the number of nodes as shown in Figure 9.13.

Thus, both the correlation of the symmetry of the MOs in reactants and products $[S_2A]\leftrightarrow[S_2A]$ and the analysis of the energy profile of the reactant ⇌ product transformation (no MO evolves into an antibonding orbital) indicate that the *disrotatory* motion is *allowed* in the 1,3,5-hexatriene ⇌ 1,3-cyclohexadiane electrocyclic reaction. It is important to focus attention on the effect of the number of p electrons involved in the electrocyclic reaction. For the 1,3,5-hexatriene ⇌ 1,3-cyclohexadiane process, the system consists of 6π [that is, $(4n + 2)$ electrons], and the *allowed* thermal process is the *disrotatory* one. By contrast, in the case of the cyclobutene ⇌ 1,3-butadiene electrocyclic reaction, the process involves 4π electrons [that is, 4n electrons], and it is allowed when *conrotatory*.

FMO Analysis

Figure 9.14 shows how the terminal p-type lobes at the HOMO orbital Ψ_3 of 1,3,5-hexatriene must rotate in a *disrotatory* manner for effective overlap in the electrocyclic ring-closing reaction during σ-bond formation. Consequently, the electrocyclic reaction of (2E,4Z,6E)-octatriene **3** to afford 1,3-cyclohexadiene is *stereospecific* and leads to the *cis*-**4** product under thermal conditions. By contrast, isomeric (2E,4Z,6Z)-**3** affords the *trans* isomer.

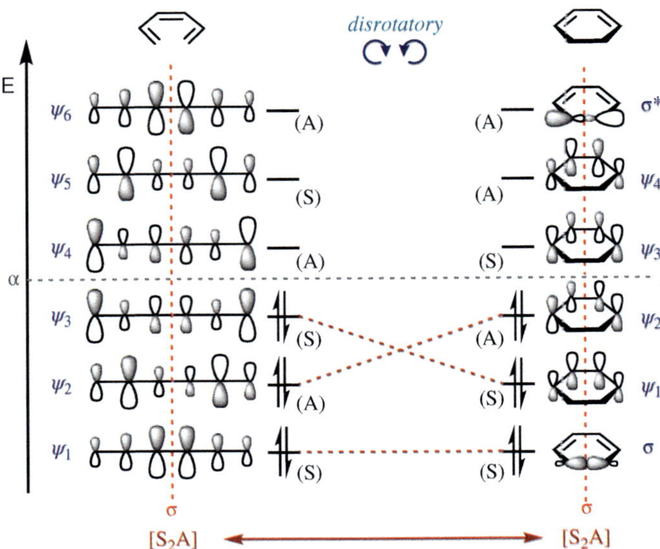

FIGURE 9.13 Molecular orbital correlation diagram in the 1,3,5-hexatriene ⇌ 1,3-cyclohexadiane electrocyclic reaction with *disrotatory* motion. (S) or (A) indicates whether the orbital is symmetric or antisymmetric with respect to the σ-plane.

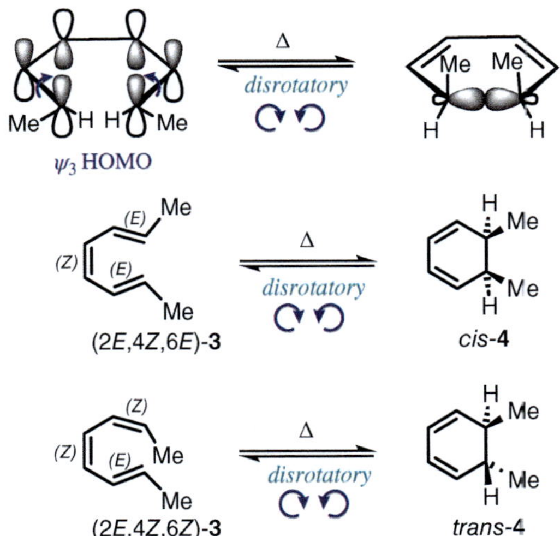

FIGURE 9.14 Examples of *stereospecific* electrocyclic reactions involving 6 π electrons.

Analysis of Nodal Properties in the Transition State

Figure 9.15 shows that the *disrotatory* motion in the conversion of 1,3,5-hexatriene to 1,3-cyclohexadiene proceeds through a 6 π-electron Hückel transition state and is therefore *allowed*. By contrast, the *conrotatory* motion proceeds by a Möbius transition state, which is *forbidden* for a [4n + 2] electron system.

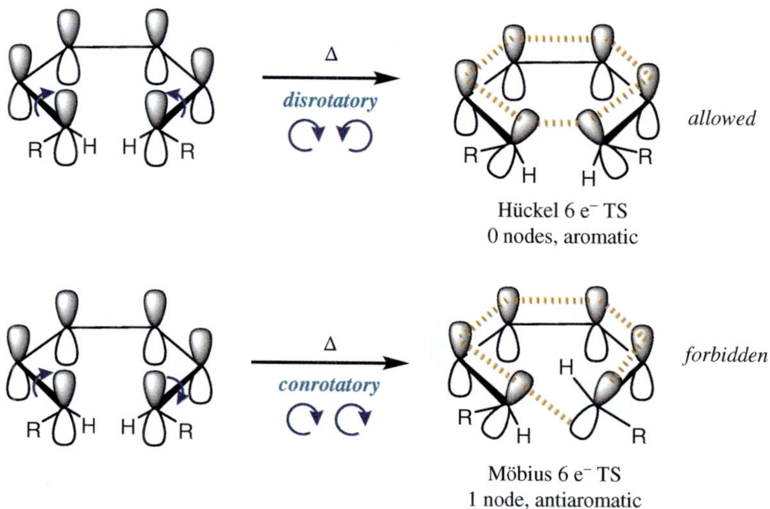

FIGURE 9.15 Nodal properties in the transition state of the thermal electrocyclic ring closure of substituted 1,3,5-hexatriene affording substituted 1,3-cyclohexadiene.

PHOTOCHEMICAL ELECTROCYCLIC REACTIONS

In photochemical electrocyclic reactions, the rules for *conrotatory* and *disrotatory* cyclisation are inverse. This is because, in terms of FMO, light irradiation promotes an electron from the HOMO to the LUMO in the excited state of the π system. Let us consider the case of 1,3-butadiene, if one compares the shape of the HOMO (Ψ_2) with the shape of the next higher energy orbital (Ψ_3) MO, it can be appreciated that Ψ_3 presents two nodes instead of one and that the terminal p orbitals in the π-system present different symmetry (Figure 9.16). In particular, in 1,3-butadiene's HOMO (Ψ_2) the C(1) and C(4) p-type orbitals are oriented in opposite directions, whereas in the Ψ_3 MO the C(1) and C(4) p-type orbitals are oriented in the same direction. Therefore, photoexcitation changes the phase properties of the orbitals involved in the electrocyclic transformation, modifying the molecular dynamics of C—C bond formation.

In this context, in the ground state the 4π-electron diene (2*E*,4*E*)-hexadiene (*E,E*)-**2** undergoes thermal *conrotatory* electrocyclic ring closure to give the *trans*-dimethylcyclobutene *trans*-**1** (Figure 9.17, left side). By contrast, under photochemical conditions, (2*E*,4*E*)-hexadiene (*E,E*)-**2** reacts *via* a *disrotatory* process to afford *cis*-**1** (Figure 9.17, right side).

A symmetry correlation diagram can help understand why under photochemical conditions *disrotatory* motion is *allowed* in the cyclobutene ⇋ 1,3-butadiene electrocyclic reaction. Following light irradiation, 1,3-butadiene's first excited state Ψ_2, which is A with respect to the reference σ-plane, correlates by symmetry with cyclobutene's excited state π^* (A with respect to the reference σ-plane) (Figure 9.18). Therefore, in a photochemically activated reaction, the *disrotatory* ring-closing (or opening) process will

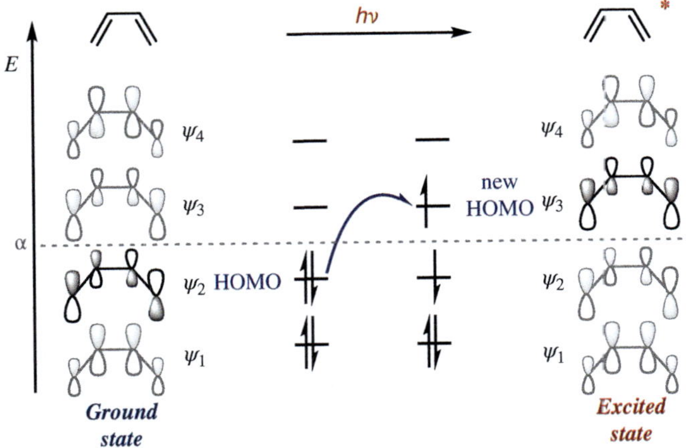

FIGURE 9.16 MO diagram of 1,3-butadiene in the ground and excited state.

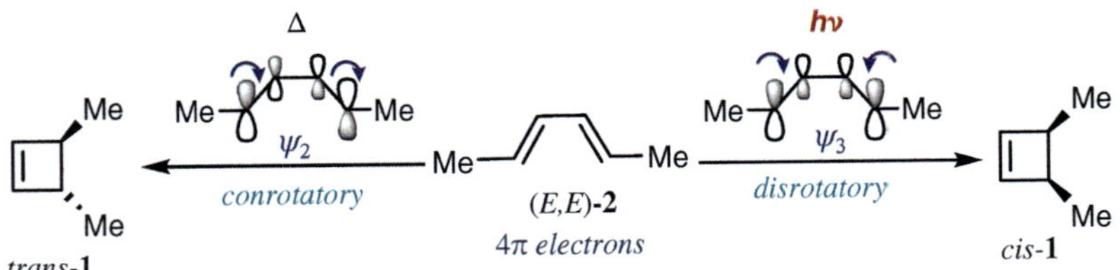

FIGURE 9.17 Thermal and photochemical electrocyclic ring closure of (2E,4E)-hexadiene.

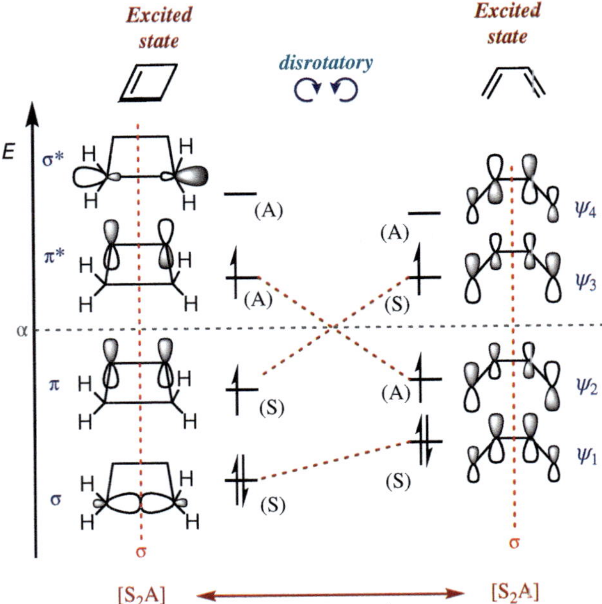

FIGURE 9.18 Orbital correlation diagram in the photochemical cyclobutene ⇌ 1,3-butadiene electrocyclic reaction with *disrotatory* motion. (S) or (A) indicates whether the orbital is symmetric or antisymmetric with respect to the σ plane.

be *allowed* by symmetry (determined with respect to the reference σ-plane). Furthermore, the symmetry of the reactant's and product's MOs is also congruent: $[S_2A] \leftrightarrow [S_2A]$ (Figure 9.18).

In the photochemically activated 1,3,5-hexatriene ⇆ 1,3-cyclohexadiene electrocyclic reaction, the HOMO of the open-chain π system corresponds to Ψ_4 (Figure 9.19). As anticipated, the reaction proceeds in *conrotatory* fashion under

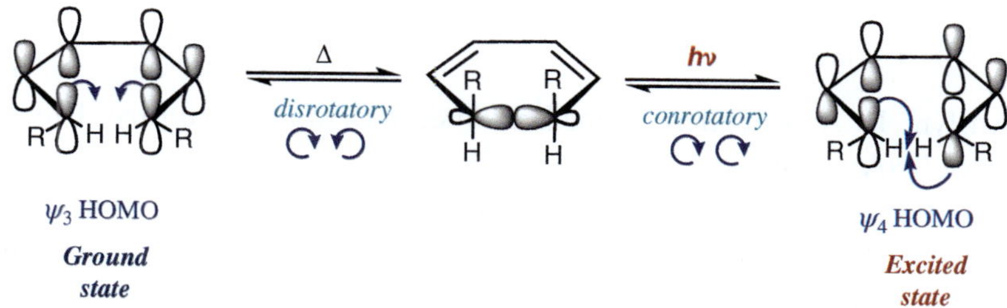

FIGURE 9.19 Thermal and photochemical electrocyclic ring closure and opening in the 1,3,5-hexatriene ⇆ 1,3-cyclohexadiene system.

TABLE 9.1 Selection Rules for Electrocyclic Reactions

π-system	Allowed reaction	
n (number of electrons)	Thermal (Δ)	Photochemical (**hν**)
4n	Conrotatory	Disrotatoy
4n + 2	Disrotatoy	Conrotatory

TABLE 9.2 Examples of the Application of the Selection Rules for Electrocyclic Reactions

Reaction	π-system number of electrons	Thermal Δ	Photochemical hν
	4 (4n)	*Conrotatory*	*Disrotatory*
	6 (4n +2)	*Disrotatory*	*Conrotatory*
	8 (4n)	*Conrotatory*	*Disrotatory*
	2 (4n +2)	*Disrotatory*	*Conrotatory*

light irradiation to achieve effective orbital overlap during the formation of the σ bond (Figure 9.19, right side). This is in contrast with the thermally activated reaction, which proceeds by *disrotatory* rotation of the Ψ_3 MO (Figure 9.19, left side).

With the information discussed in this Chapter, the selection rules for opening and closing electrocyclic reactions can be summarised (Table 9.1). Table 9.2 provides some examples of the application of these rules.

FURTHER READING

E. V. Anslyn, D. A. Dougherty, *Modern Physical Organic Chemistry*, University Science Books, Sausalito, California, **2005**.

I. Fleming, *Pericyclic Reactions*, Oxford University Press, Oxford, UK, **1998**.

J. Clayden, N. Greeves, S. Warren, *Organic Chemistry*, Oxford University Press, Oxford, UK **2012**.

R. B. Woodward, R. Hoffmann, *J. Am. Chem. Soc.* **1965**, *87*, 395.

S. Kumar, V. Kumar, S. P. Singh, *Pericyclic Reactions: A Mechanistic and Problem-Solving Approach*, Academic Press, Cambridge, UK, **2015**.

S. Sankararaman, *Pericyclic Reactions—A Textbook: Reactions, Applications and Theory*, Wiley-VCH, Weinheim, **2005**.

EXERCISES

9.1 Construct the orbital correlation diagram for the conversion of 1,3,5-hexatriene to 1,3-cyclohexadiene for a thermally activated *disrotatory* process. Using the diagram, explain why the reaction is forbidden in the *conrotatory* mode.

9.2 Construct the orbital correlation diagram for the photochemical electrocyclic ring closure of 1,3,5-hexatriene. Using the diagram, explain why the reaction is allowed in the *conrotatory* mode.

9.3 Predict the direction of rotation of the following electrocyclic reactions.

9.4 Draw the structure of the corresponding products indicating their stereochemistry.

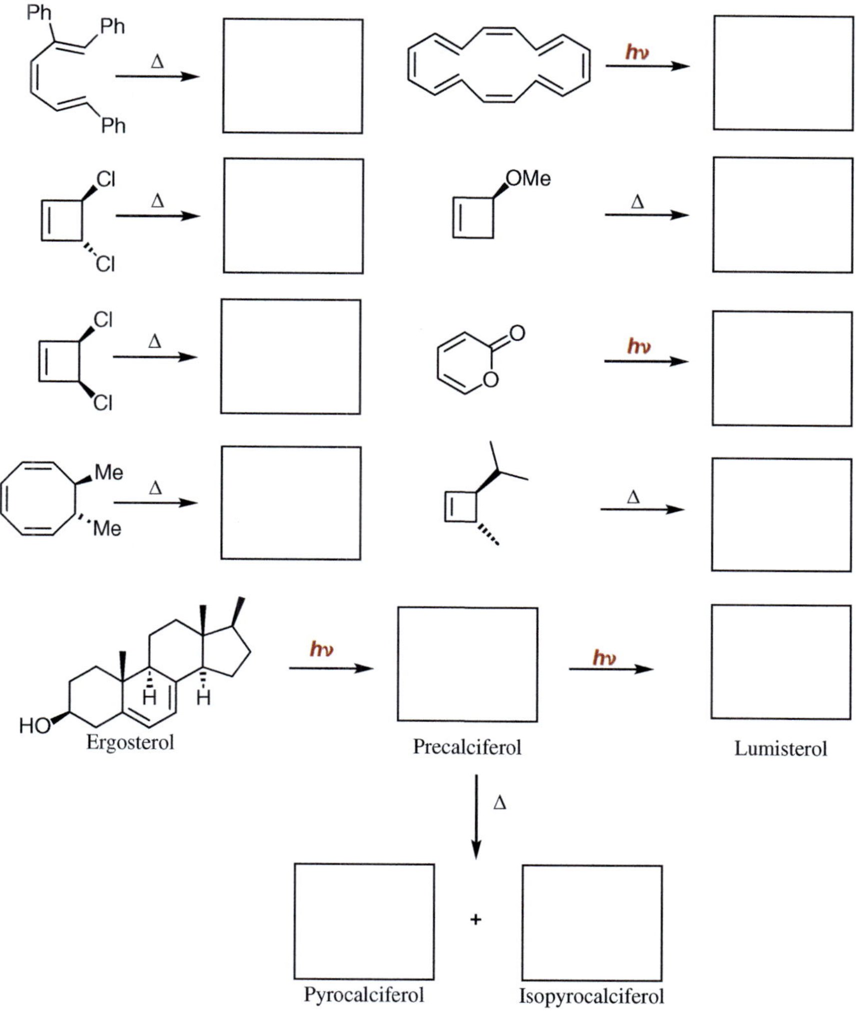

Ergosterol

Precalciferol

Lumisterol

Pyrocalciferol

Isopyrocalciferol

9.5 Draw the mechanism with 'arrow-pushing' for the following reactions. Justify the observed stereochemistry.

CHAPTER 10

Sigmatropic Reactions

INTRODUCTION

The term 'sigmatropic shift' refers to the displacement of a sigma bond along a π-system; in particular, sigmatropic reactions are concerted processes in which a sigma bond, adjacent to one or more conjugated π-systems, shifts to a new position within the π-system. For the reaction to take place, the symmetry of the conjugated π-system must allow for in-phase orbital interaction leading to effective orbital overlap. Typically, the migrating group is a hydrogen atom or an alkyl group. It is crucial to keep in mind that in a sigmatropic rearrangement, two sigma bonds are involved: one that will be broken, and another one that will be formed.

For the classification of a sigmatropic rearrangement, the π-chain(s) in the reactant molecule is (are) numbered starting from the location of the bond being broken to the position where the new bond is being formed. The reaction is then denoted as [i,j], where the sigma bond moves from position [1,1] to position [i,j]. Figure 10.1a illustrates the numbering of π-chains in the reactant molecule. It can be appreciated, for instance, that when the new sigma bond shifts from position 1 to position 3 on both molecular chains, the rearrangement is classified as a [3,3] sigmatropic shift. In [i,j] sigmatropic rearrangements where i = 1, one end of the migrating bond remains attached to its original carbon atom. This type of reaction is commonly observed with migrating hydrogen atoms in C—H bonds bonded to the π-chain, as it is the case of a [1,3] hydrogen shift (Figure 10.1b). Additional illustrative examples of sigmatropic rearrangements are presented in Figure 10.1c, including the Cope reaction, an emblematic example of [3,3] sigmatropic shifts. Additional examples include [1,3], [1,5], and [2,5] sigmatropic shifts, as depicted in Figure 10.1c.

As it was mentioned earlier, one of the most common sigmatropic rearrangements corresponds to the [3,3] type. In this context, the reaction presented in Figure 10.2a is exothermic because the product molecule does not suffer the

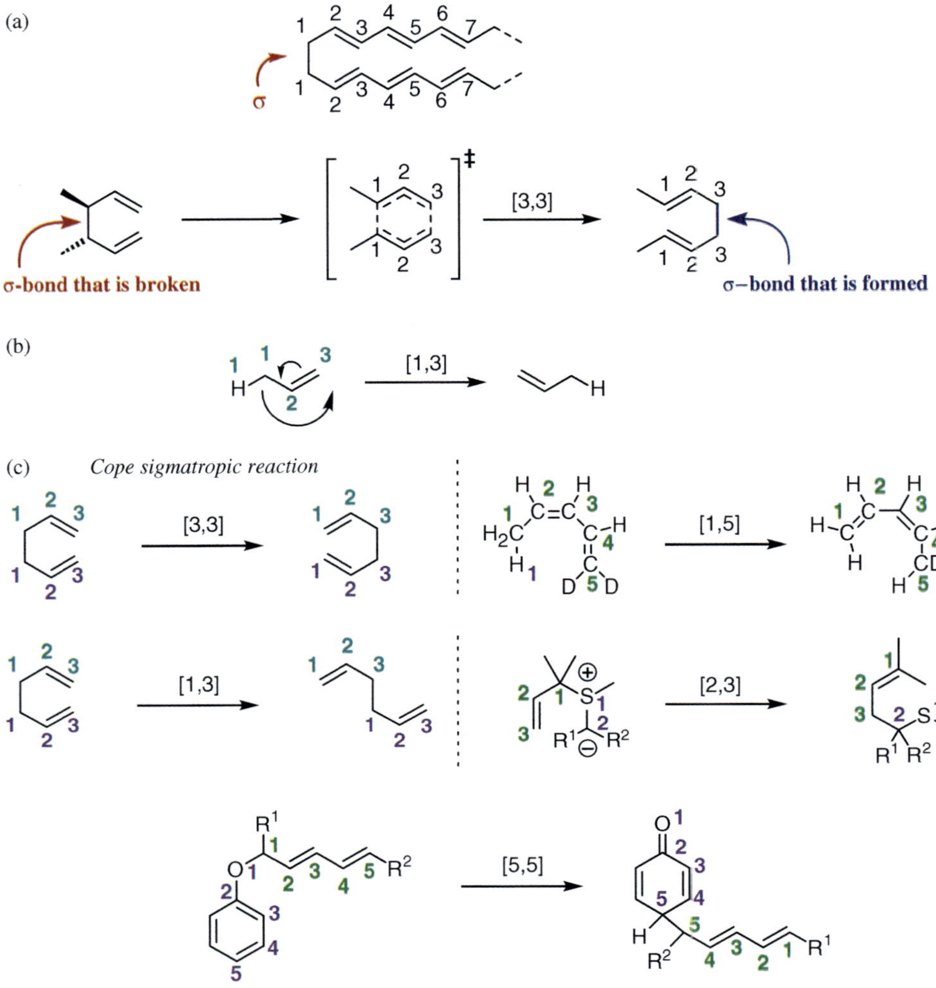

(a)

σ-bond that is broken   [3,3]   σ−bond that is formed

(b)

[1,3]

(c)   *Cope sigmatropic reaction*

[3,3]

[1,3]

[1,5]

[2,3]

[5,5]

FIGURE 10.1   Numbering at the reactant molecule and illustrative examples of how the sigmatropic rearrangement is classified.

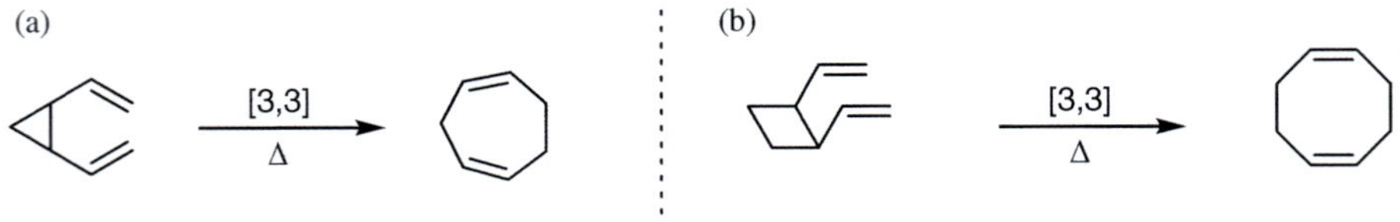

(a)   [3,3]   Δ

(b)   [3,3]   Δ

FIGURE 10.2   [3,3] Sigmatropic rearrangements with liberation of ring strain.

ring strain associated with the reactant. The same is true for the conversion of *cis*-1,2-divinylcyclobutane to 1,5-cyclooctadiene (Figure 10.2b).

It is worthy of note that *trans*-1,2-divinylcyclobutane also gives rise to 1,5-cyclooctadiene (Figure 10.3). However, in this case, the reaction requires

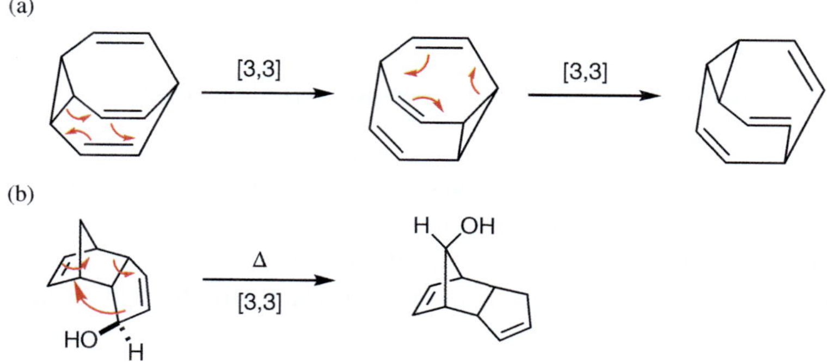

FIGURE 10.3 [3,3] Sigmatropic rearrangement of *trans*-1,2-divinylcyclobutane to 1,5-cyclooctadiene.

FIGURE 10.4 [3,3] Sigmatropic rearrangements.

heating to 210 °C, and it has been established that the mechanism proceeds *via* a diradical intermediate that affords the *cis*-1,2-divinylcyclobutane stereoisomer, which is now suitable for the concerted [3,3] sigmatropic shift.

Special mention deserves bulvalene and analogous compounds: the molecular geometry is so well suited to undertake Cope rearrangements that the activation energy for the process is a mere 11.6 kcal mol^{-1} (Figure 10.4a). It should be noted that some of these rearrangements are not easily recognised (Figure 10.4b).

There can be important stereochemical implications in sigmatropic rearrangements. In particular, in the case of hydrogen [1,n] migrations, the hydrogen atom may migrate to its new terminal position from the same face of the π-system to which the relevant C—H bond was originally attached (*suprafacial* displacement), or it may migrate to the opposite face of the π-chain (*antarafacial* displacement). This contrasting mode of displacement would result in the opposite configuration of the newly formed chiral sp^3 carbon (Figure 10.5).

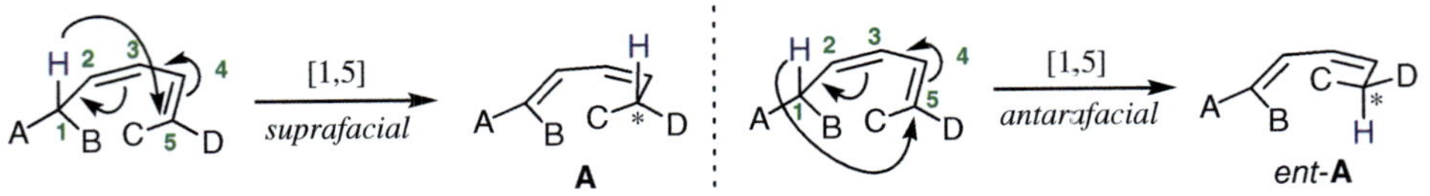

FIGURE 10.5 *Suprafacial vs. antarafacial* shift modes of H atom migration in [1,5] sigmatropic reactions.

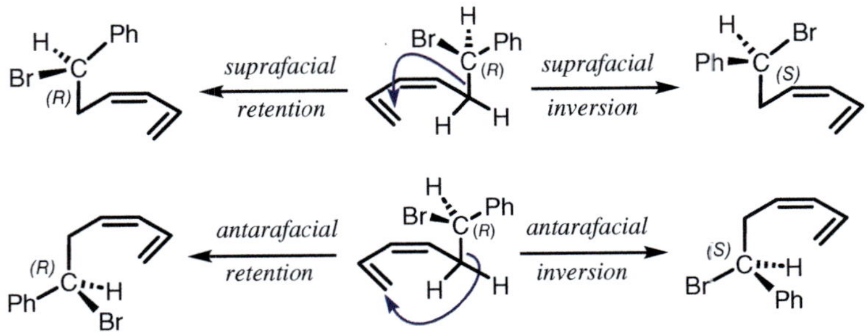

FIGURE 10.6   Modes of chiral alkyl group migration in [1,5] sigmatropic reactions.

Furthermore, in the migration of a chiral carbon centre, it is necessary to specify the stereochemical mode of migration of the carbon atom with respect to itself and with respect to the π-system. In other words, the displacement of an alkyl group may proceed in *suprafacial* or *antarafacial* mode, and in the process, it may undergo *inversion* or *retention* of the configuration of the chiral carbon. Figure 10.6 illustrates the four possible stereochemical outcomes for the sigmatropic rearrangement of a chiral alkyl group. As it can be anticipated, both symmetry and structural constrains will influence the mode of migration of the sigma bond.

[3,3] SIGMATROPIC REARRANGEMENTS

This type of reaction has been used to induce the formation of a new centre of chirality as a consequence of the sigmatropic rearrangement. For this reason, it is an important tool in the synthesis of organic compounds, and it is not surprising that a significant number of variants have been developed since Ludwig Claisen first reported this rearrangement in 1912 (known as Claisen rearrangement; see the 'further reading' section at the end of the chapter). For instance, when a vinyl allyl ether is heated, the Claisen rearrangement affords a chiral γ,δ-unsaturated carbonyl derivative (Figure 10.7a). On the other hand, the [3,3] sigmatropic rearrangement of 1,5-dienes (developed by Cope and Hardy in 1940) is known as the Cope rearrangement (Figure 10.7b). In this regard, it was

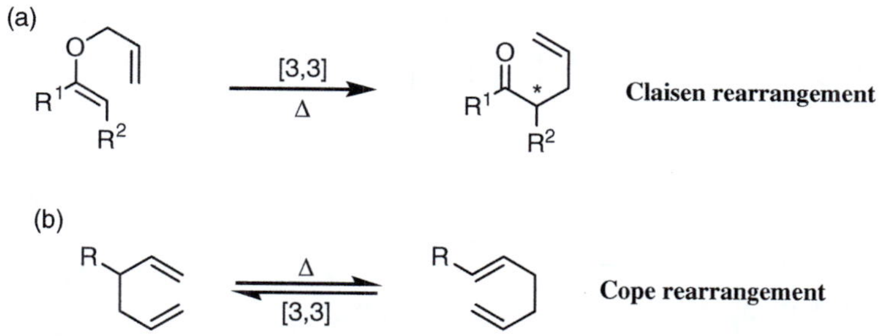

FIGURE 10.7   Iconic [3,3] sigmatropic rearrangements.

noticed that the Claisen rearrangement often leads to the carbonyl derivative in an irreversible manner, whereas in the Cope rearrangement, the reagent and the product usually interconvert via a reversible process.

The molecular orbital symmetry analysis for the Cope reaction is shown in the following.

REACTANT ⇌ PRODUCT Correlation Diagram

Since both the left and right sides of the equilibrium consist of two π-systems and a σ carbon-carbon bond joining them, symmetry elements that normally enable the elaboration of a correlation diagram involving the partici- pant orbitals in the reactant and the product are not useful here as they present the same symmetry properties. As it turns out, the most convenient approach for the analysis of molecular orbital symmetry in [3,3] sigmatropic rearrange- ments is based on examination of the highest occupied molecular orbital (HOMO)/lowest unoccupied molecular orbital (LUMO) frontier molecular orbitals (FMOs).

HOMO/LUMO Interaction

The transition state (TS) of the [3,3] sigmatropic rearrangement involved in the Cope rearrangement can be described as the result of the interaction bet- ween two allylic π-systems (see Figure 5.6 in Chapter 5). In particular, one approach to carry out the analysis of the HOMO/LUMO interaction is to assume one of the allylic systems in the TS as the *donor* moiety with four elec- trons (anion) and the second allylic system as the *acceptor* species with two electrons (cation). As it can be appreciated in Figure 10.8, with respect to a σ reference plane, the HOMO/LUMO interaction is *allowed* by symmetry and therefore stabilizing. In fact, both the HOMO and LUMO consist of the same orbital (Ψ_2), and their in-phase combination implies that both components react in a *suprafacial* manner in a *thermally allowed* process (Figure 10.8).

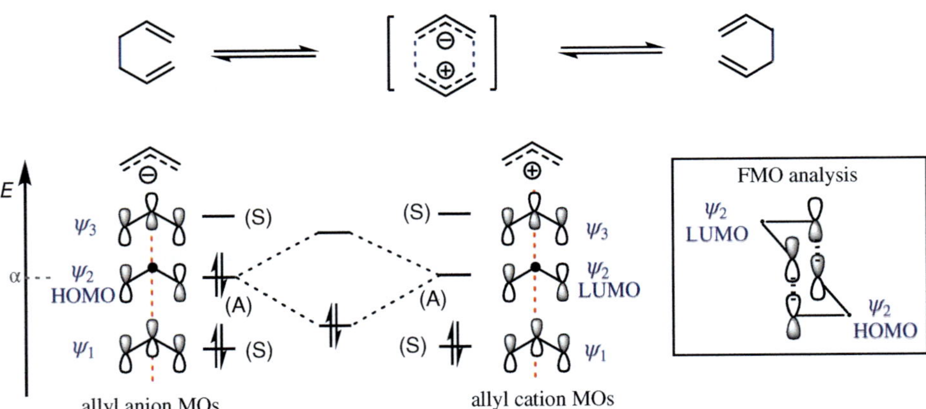

FIGURE 10.8   FMO analysis of the [3,3] sigmatropic rearrangement employing two allylic ionic fragments. The pericyclic process is *allowed* since both the HOMO and LUMO are antisymmetric (A) with respect to the σ reference plane. Furthermore, the *suprafacial* approach on both components involves in-phase interaction of the donor and acceptor MOs.

Analysis of Nodal Properties in the Transition State

The TS in [3,3] sigmatropic rearrangements may be seen, as already indicated, as consisting of two allylic systems giving rise to a six electrons Hückel-type cyclic π-system with no orbital phase changes and therefore *allowed*. Figure 10.9 presents two MO representations of the TS, being reasonable to anticipate a chair-like TS. The two allylic fragments form a closed loop of *p* orbitals interacting in a *supra-supra* fashion, without sign inversions in the phase of the interacting orbitals. The fact that the [3,3]-sigmatropic rearrangements take place *via* a chair-like TS allows us to predict the (*E*) or (*Z*) configuration of the new double bond (see below).

As it turns out, in [3,3] sigmatropic shifts, the TS can adopt either chair or boat conformations, since both are allowed by symmetry. Energetically, however, a chair TS is preferred. For example, the rearrangement of *meso* diene **1** proceeds via the low-energy chair-like TS to afford the (*2E,6Z*) isomer of the product (Figure 10.10).

[1,3] SIGMATROPIC REARRANGEMENTS OF ALKYL GROUPS

HOMO/LUMO Interaction

Figure 10.11a shows the orbitals used for FMO analysis in [1,3] rearrangements of alkyl groups. In a thermally activated *suprafacial* [1,3] sigmatropic displacement of a chiral carbon, symmetry is maintained only if the migrating carbon in

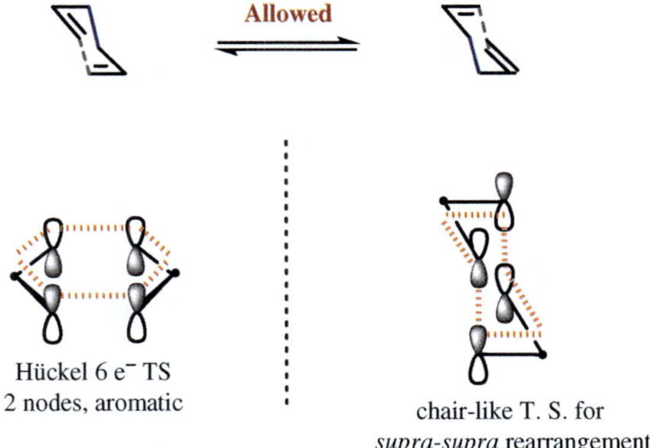

FIGURE 10.9   Analysis of the transition state for the [3,3] sigmatropic rearrangement, involving in-phase combination of the MOs of the allylic segments.

FIGURE 10.10   Stereochemical consequences [(*E*) or (*Z*) configuration of the double bonds in the substrate] on the transition state conformation for the [3,3] sigmatropic rearrangement.

the TS interacts with *opposite lobes* of the Ψ_2 MO (Figure 10.11b). That is, if the migrating carbon is initially bonded through its negative lobe, the new carbon bond is formed by means of the positive lobe. The result of the whole process is an *inversion* of the configuration of the migrating carbon atom.

An alternative way to understand the reason for the inversion of the configuration of the migrating alkyl group in [1,3] sigmatropic rearrangements is

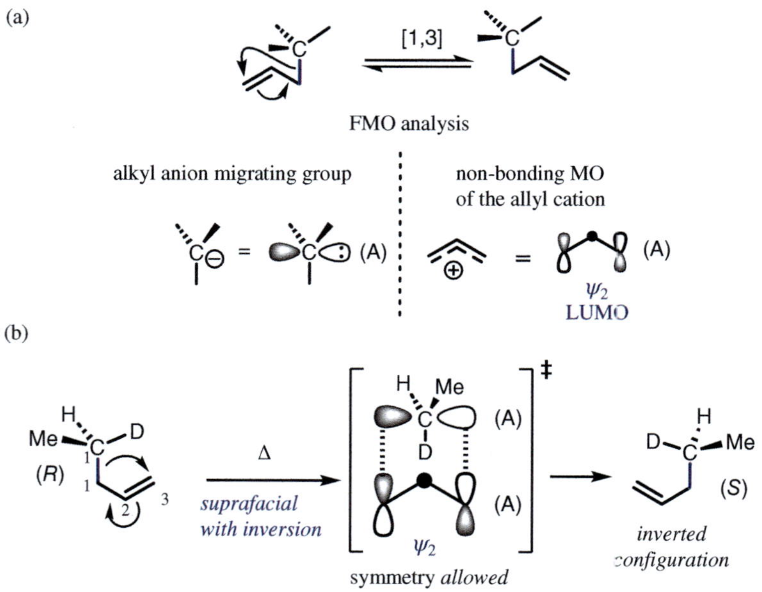

(a)

FMO analysis

alkyl anion migrating group | non-bonding MO of the allyl cation

Ψ_2 LUMO

(b)

(R)

Δ

suprafacial with inversion

symmetry *allowed*

(S)

inverted configuration

FIGURE 10.11   FMO analysis of the [1,3] sigmatropic rearrangement of a chiral group with inversion of the configuration.

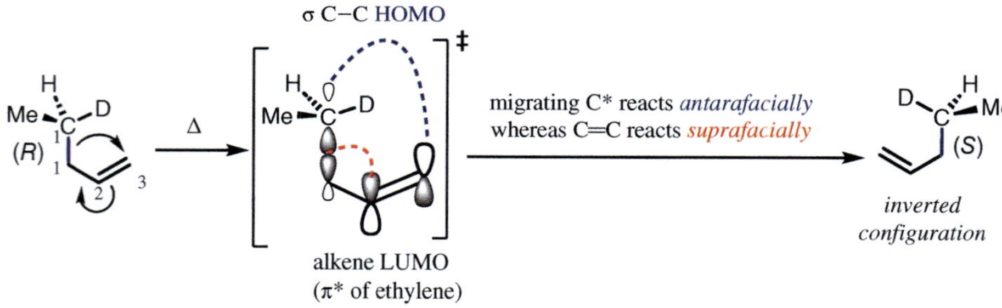

σ C–C HOMO

(R)

Δ

migrating C* reacts *antarafacially* whereas C=C reacts *suprafacially*

(S)

inverted configuration

alkene LUMO
(π^* of ethylene)

FIGURE 10.12   FMO analysis of the [1,3] sigmatropic rearrangement of chiral alkyl groups.

by considering the TS structure shown in Figure 10.12. Here, the HOMO corresponds to the σ C–C bond, whereas the LUMO is the antibonding π^* MO of ethylene. It can be seen that the alkene segment reacts in a *suprafacial* fashion during the course of the migration; nevertheless, the migrating carbon reacts on the back lobe, that is, in an *antarafacial* fashion with respect to the *p* orbital, in order to attain efficient in-phase orbital overlap during the rearrangement.

By contrast, the *suprafacial* [1,3] sigmatropic rearrangement involving the interaction of the same lobe on the migrating anionic carbon atom with the empty Ψ_2 π^* MO of the allyl segment and leading to the product with *retention* of configuration is *forbidden* by MO symmetry. Indeed, Figure 10.13a shows that the TS in such a process would involve an out-of-phase interaction, which

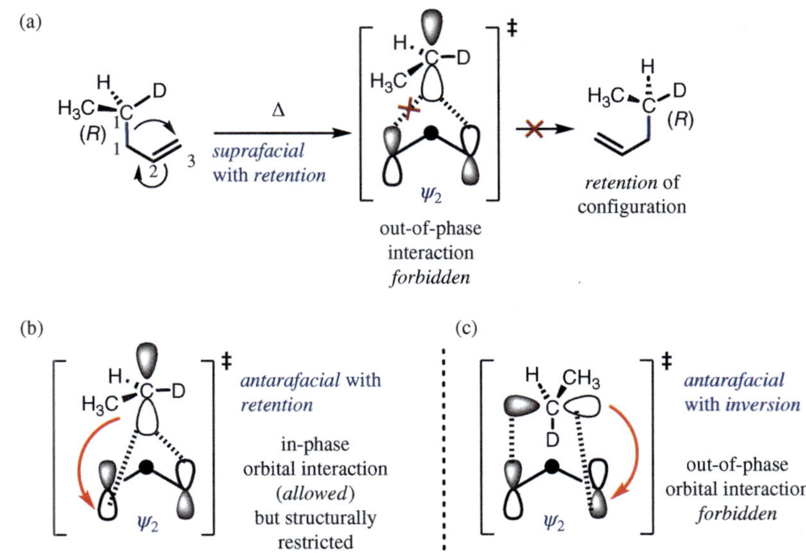

FIGURE 10.13   FMO analysis of the [1,3] sigmatropic rearrangement of chiral alkyl groups.

prevents bond formation. On the other hand, the thermally activated [1,3] *antarafacial* displacement with retention of configuration is *allowed* by symmetry, although it is structurally restricted (Figure 10.13b). Finally, the *antarafacial* process with inversion of configuration is *forbidden* by orbital symmetry (Figure 10.13c).

ANALYSIS OF NODAL PROPERTIES IN THE TRANSITION STATE

Figure 10.14 summarises the nodal properties of the TSs involved in different modes of [1,3] sigmatropic rearrangements of chiral alkyl groups. Note that the resulting prediction is in agreement with FMO analysis.

[1,5] SIGMATROPIC REARRANGEMENTS OF ALKYL GROUPS

HOMO/LUMO FMO Interaction

In the [1,5] *suprafacial* thermal process, orbital symmetry is preserved only if the same orbital lobe at the migrating carbon (p orbital, S symmetry) interacts with the acceptor Ψ_3 molecular orbital of the pentadienyl moiety (S symmetry). Thus, if the migrating carbon was originally bonded in the starting material

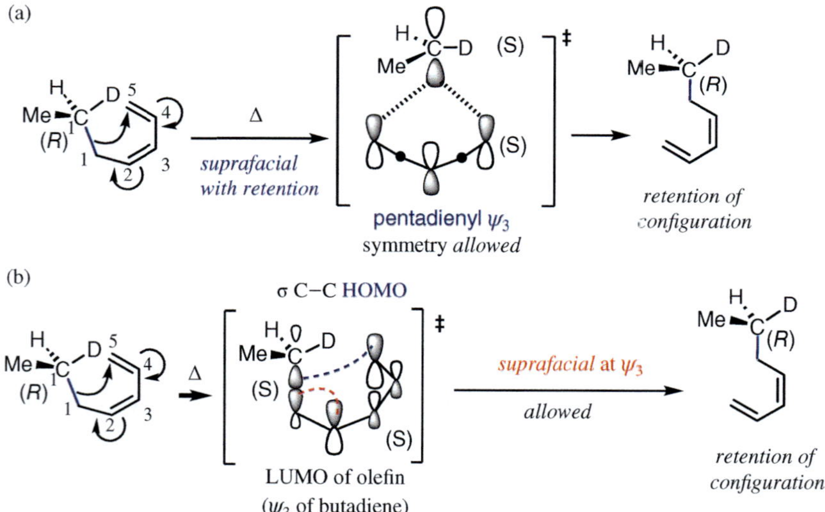

FIGURE 10.14   Nodal properties in the transition state of the [1,3] sigmatropic rearrangement of chiral alkyl groups by different modes of migration.

FIGURE 10.15   FMO analysis of the [1,5] sigmatropic rearrangement of chiral alkyl groups with retention of configuration.

by means of its positive lobe, it must use the same positive lobe to form the new C—C bond, as illustrated in Figure 10.15a. This in-phase orbital interaction results in the *retention* of the configuration at the migrating carbon. The [1,5] rearrangement can also be analysed using the σ C—C orbital as the HOMO (S symmetry) and the Ψ_3 orbital of the butadienyl π system (S symmetry) as the LUMO. In this scenario, the allowed reaction is *suprafacial* with regard to the

π-system, and interaction through the front lobe of the migrating carbon leads to the product with *retention* of configuration (Figure 10.15b).

If the migrating carbon uses the opposite lobes of its *p* orbital, the product would be obtained with *inversion* of configuration, but the process is *forbidden* by the presence of an out-of-phase orbital interaction (Figure 10.16a). By the same token, the thermal [1,5] *antarafacial* process with *retention* of configuration at the migrating alkyl group is *forbidden* by the presence of an out-of-phase orbital interaction (Figure 10.16b). On the other hand, the *antarafacial* process with inversion of configuration at the chiral carbon is structurally restricted (Figure 10.16c).

Analysis of Nodal Properties in the Transition State

In congruence with the FMO symmetry analysis discussed earlier, Figure 10.17a shows that [1,5] sigmatropic shifts that are *suprafacial* and take place with *retention* of configuration are *allowed* when analysis of the nodal properties at the

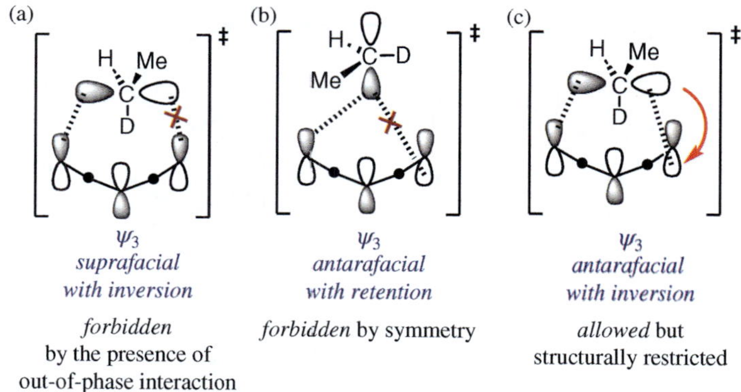

FIGURE 10.16   FMO analysis of the [1,5] sigmatropic rearrangement of chiral alkyl groups.

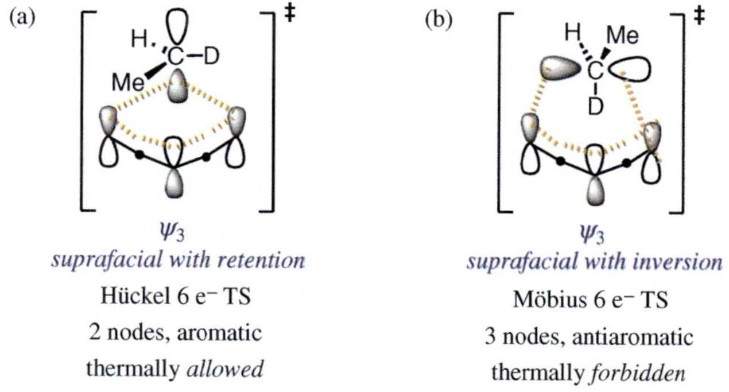

FIGURE 10.17   Nodal properties in the transition state of the [1,5] sigmatropic rearrangement of chiral alkyl groups.

TSs is performed. On the other hand, [1,5] sigmatropic displacements that are *suprafacial* on the π-system and take place with *inversion* of configuration at the chiral carbon are thermally *forbidden* according to the analysis of the nodal properties at the TSs (Figure 10.17b).

In photochemical sigmatropic processes, the reactivity rules are reversed as a consequence of the promotion of an electron to a higher energy anti-bonding molecular orbital in the π-system. For example, in the pentadienyl radical, light excitation promotes an electron from the Ψ_3 to the Ψ_4 orbital, making Ψ_4 the interacting orbital in the sigmatropic process (Figure 10.18a).

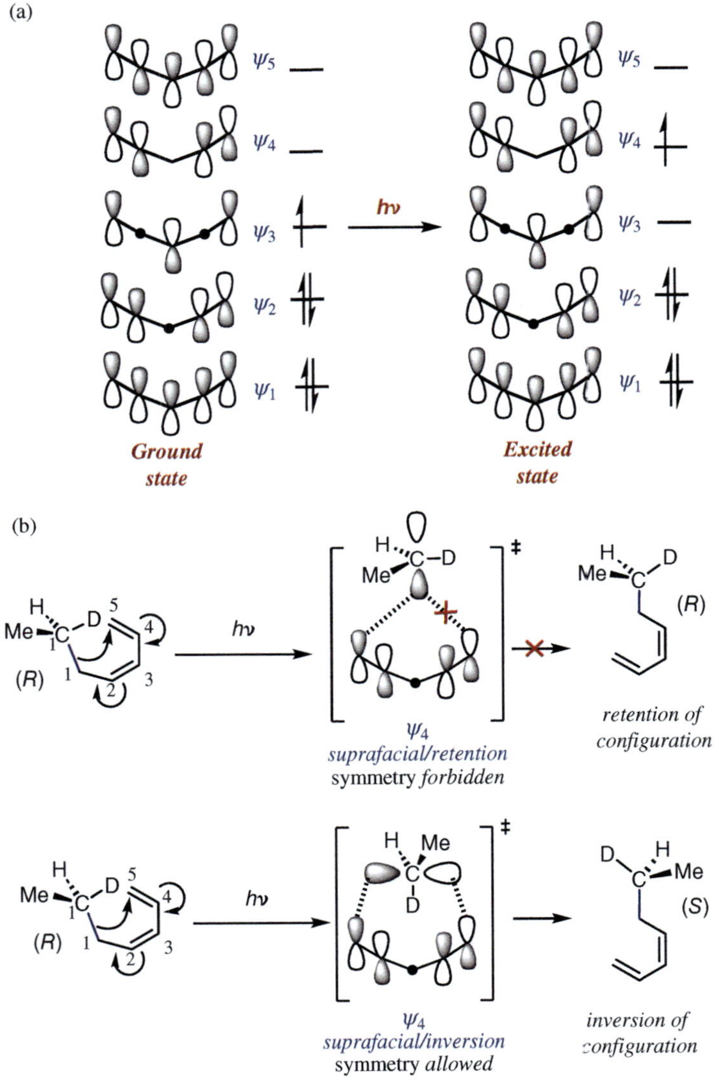

(a)

(b)

FIGURE 10.18　FMO analysis of the photochemical [1,5] sigmatropic rearrangement of chiral alkyl groups.

This excitation alters the symmetry of the FMOs. As shown in Figure 10.18b, under photochemical conditions, Ψ_4 interacts with opposite lobes of the migrating carbon fragment, allowing the [1,5] *suprafacial* rearrangement of an alkyl group with *inversion* of configuration in accordance with orbital symmetry rules.

[1,2] SIGMATROPIC REARRANGEMENTS OF ALKYL GROUPS

An illustrative example of this type of rearrangement consists of the displacement of a chiral alkyl group to an adjacent cationic carbon with *retention* of configuration (Figure 10.19). This is a special case where the migrating group and the π-system termini are all carbon atoms. Rearrangements of an alkyl group (or hydrogen atom) to an adjacent atom in a carbocation are called 1,2-rearrangements or Wagner–Meerwein rearrangements.

HOMO/LUMO Interaction

The TS can be visualised as the addition of a carbenium ion to ethylene (π) (Figure 10.20). It can be seen that both the π-orbital (HOMO) and the empty p orbital (LUMO) are symmetric (S) with respect to a plane of symmetry; therefore, it is concluded that the reaction is *allowed*.

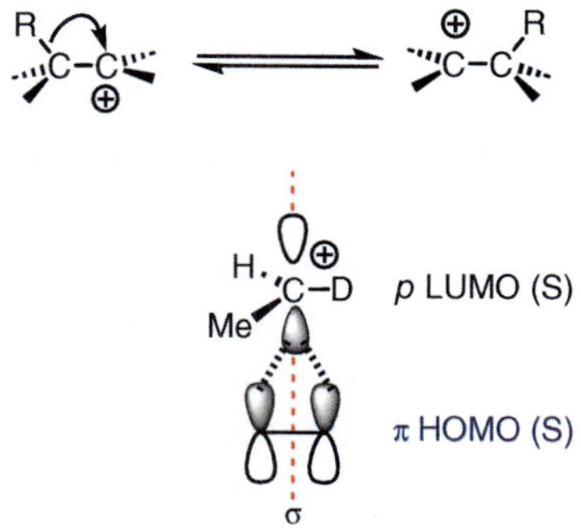

FIGURE 10.19   FMO analysis of thermally *allowed* two-electron [1,2] carbon sigmatropic rearrangement with *retention* of configuration at the chiral migrating alkyl group.

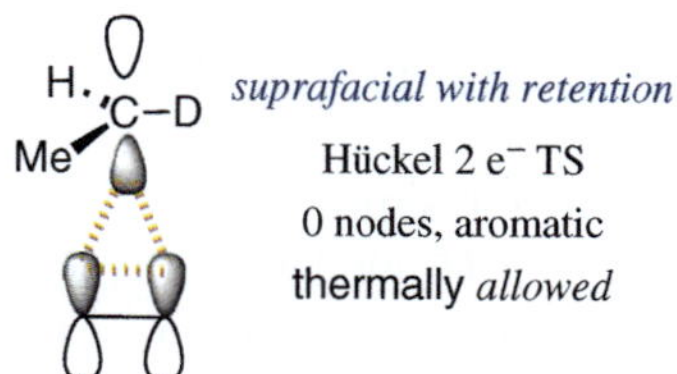

FIGURE 10.20   Analysis of the nodal properties of the transition state of the [1,2] sigmatropic shift of a chiral alkyl group to an adjacent cationic carbon.

(a) *p* HOMO (S)

π^* LUMO (A)

forbidden with **retention**

(b) HOMO (A)

π^* LUMO (A)

allowed with **inversion**

FIGURE 10.21   Four-electron thermal [1,2] carbon sigmatropic rearrangement. FMO analysis: *allowed* with *inversion* of configuration at the migrating chiral carbon, but *forbidden* with *retention* of configuration.

Nodal Properties in the Transition State

It can be seen that the TS is quite similar to a cyclopropenyl cation, that is, a Hückel-type π-cyclic system with two electrons and therefore *allowed* (Figure 10.20).

Consider now the [1,2] rearrangements in which an alkyl group moves to an adjacent anionic carbon (Figure 10.21).

HOMO/LUMO Interaction

The reaction can be visualised as the addition of a carbanion to an ethylene. In this case, the *p* orbital is occupied by two electrons, and this negative species acts as the HOMO frontier orbital. On the other hand, the LUMO frontier orbital corresponds to the ethylene π^* orbital. It is clear from analysis of the symmetry of the involved orbitals that the interaction in the rearrangement taking place with *retention* of configuration is *forbidden* since π^* is antisymmetric (A), whereas the *p* orbital is symmetric (S) with respect to the plane of symmetry σ (Figure 10.21a). By contrast, if the carbanion fragment interacts with its opposite lobes, the process affording *inversion* of configuration at the migrating carbon is *allowed* by symmetry (Figure 10.21b).

Nodal Properties of the Transition State

A four-electron thermal [1,2] carbon sigmatropic shift taking place with *retention* of configuration corresponds to a Hückel-type cyclic π-system with two nodes and four electrons, which is *forbidden* (Figure 10.21a). The displacement with inversion of configuration is of one node Möbius-type and therefore allowed for four electrons (Figure 10.21b). Note that the result is in line with the FMO analysis. As a rule, reactions [1,*n*] that are *allowed* with configuration *retention* are *forbidden* with configuration *inversion*, and *vice versa*.

[1,3] SIGMATROPIC REARRANGEMENTS OF HYDROGEN

HOMO/LUMO Frontier Orbital Interaction Analysis

A simple approach for analysing the [1,3] sigmatropic rearrangement of a terminal hydrogen atom within a π-system is by dividing the activated TS complex into two components, assuming that the migrating bond undergoes homolytic cleavage, which results in the formation of a radical pair. This allows prediction of the reaction's stereochemical outcome by considering the spherical $1s$ orbital of hydrogen and the HOMO (actually, singly occupied molecular orbital, SOMO) Ψ_2 of the allyl radical π-system.

Figure 10.22 illustrates the two potential rearrangement modes: *suprafacial* and *antarafacial*. Analysis of the involved orbitals reveals that the *suprafacial* pathway is *forbidden* by the presence of an out-of-phase orbital interaction, while the *antarafacial* process is *allowed*. However, despite being theoretically allowed by orbital symmetry, the *antarafacial* [1,3] hydrogen shift is hardly observed owing to the strain involved in the TS.

Under light irradiation (Figure 10.23), the ground state LUMO (Ψ_3) becomes the excited state HOMO. The electronic configuration of the allyl radical in the excited state is therefore $\Psi_1^2\,\Psi_3^1$. The HOMO (Ψ_3) presents the same sign phase on the terminal p-type orbital lobes, which allows for an in phase *suprafacial* [1,3] hydrogen shift.

FIGURE 10.22 Transition state analysis for the [1,3] sigmatropic hydrogen displacement.

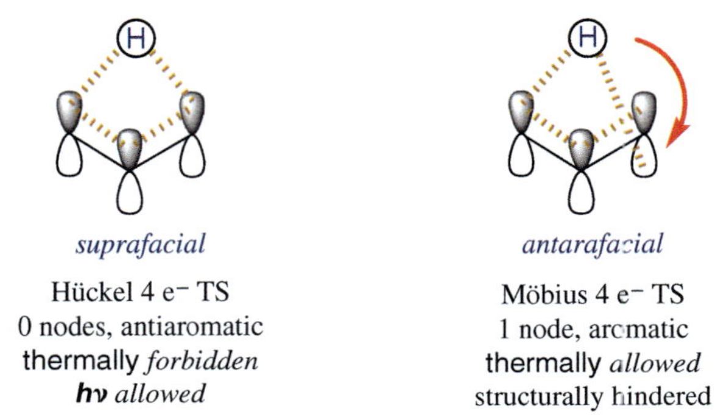

FIGURE 10.23   Transition state analysis for the photochemical [1,3] sigmatropic hydrogen rearrangement.

FIGURE 10.24   Nodal properties of the transition states for the [1,3] hydrogen shifts.

Nodal Properties of the Transition State

Figure 10.24 shows the analysis of the nodal properties in the TS for the [1,3] hydrogen shift. The conclusions are in agreement with the FMO symmetry analysis.

[1,5] SIGMATROPIC REARRANGEMENTS OF HYDROGEN

HOMO/LUMO Interaction

Under thermal activation, the [1,5] sigmatropic hydrogen shift proceeds via a *suprafacial* pathway. For FMO symmetry analysis, one can consider the pentadienyl radical fragment, whose HOMO (Ψ_3) presents terminal *p*-type orbital lobes with the same phase sign, indicating mirror symmetry and enabling *suprafacial* hydrogen displacement. Under photochemical conditions, electron promotion results in Ψ_4 becoming the HOMO, where the terminal *p*-type orbital lobes present opposite phase signs, inducing *antarafacial* migration of hydrogen (Figure 10.25).

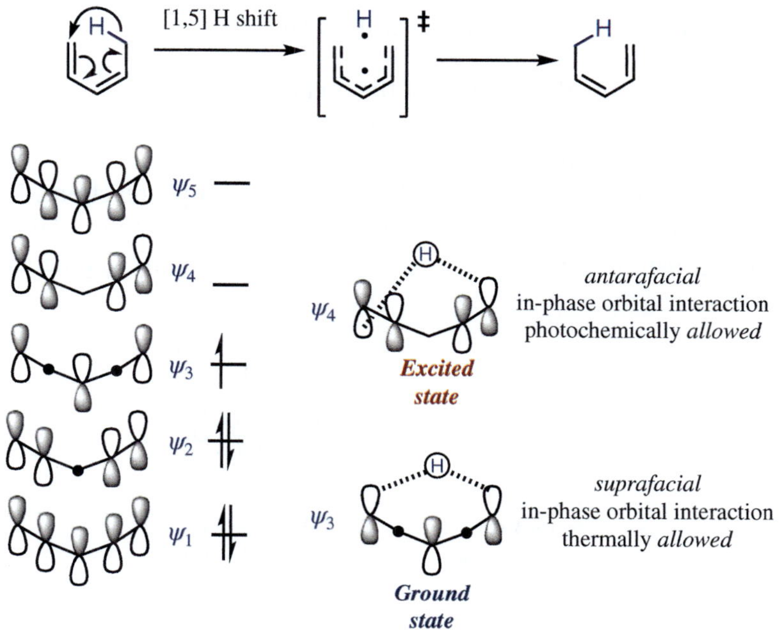

FIGURE 10.25 Transition state analysis for the [1,5] sigmatropic hydrogen rearrangement.

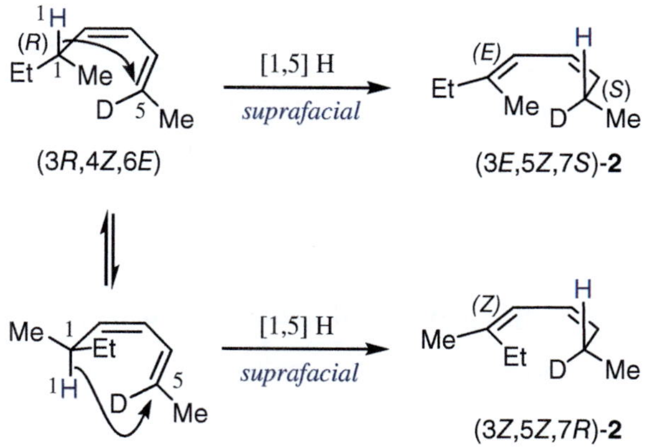

FIGURE 10.26 Illustrative suprafacial [1,5] sigmatropic hydrogen rearrangement in a chiral 2,4-octadiene.

Figure 10.26 presents illustrative examples of thermally *allowed suprafacial* [1,5] sigmatropic hydrogen displacements. The starting (2E,4Z)-octadiene exists in conformational equilibrium by rotation around the C(6)—C(7) single bond and each conformer undergoes a [1,5] *suprafacial* hydrogen shift to give a mixture of two configurational isomers (diastereoisomers).

Nodal Properties of the Transition State

Figure 10.27 presents the analysis of the nodal properties of the TSs in the [1,5] sigmatropic rearrangement of hydrogen, which is consistent with the FMO analysis discussed earlier.

Table 10.1 presents the selection rules for sigmatropic reactions. It is important to note that while some processes may be allowed by orbital symmetry, structural constraints can make those reactions challenging in practice.

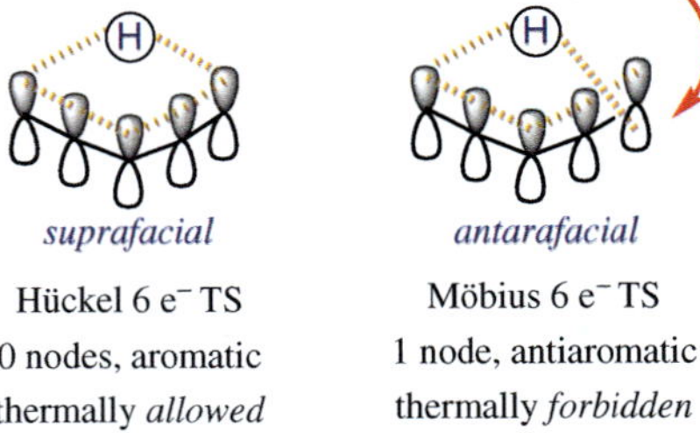

FIGURE 10.27   Nodal properties of the transition state for [1,5] sigmatropic hydrogen displacements.

TABLE 10.1   Selection Rules for the Sigmatropic Rearrangements

Selection rules for [1,n] sigmatropic displacements of hydrogen atom			
π-system	Conditions	shift	reactivity
4n Möbius	Thermal (Δ)	[1,3], [1,7]	Antarafacial (allowed)*
	Photochemical (*hv*)	[1,3], [1,7]	suprafacial (allowed)
4n+2 Hückel	Thermal (Δ)	[1,2], [1,5]	suprafacial (allowed)
	Photochemical (*hv*)	[1,2], [1,5]	antarafacial (allowed)*

Selection rules for [1,n] sigmatropic displacements of alkyl groups			
π-system (electrons)	Conditions	shift	reactivity
4n Möbius	Thermal (Δ)	[1,3], [1,7]	suprafacial/inversion antarafacial/retention
	Photochemical (*hv*)	[1,3], [1,7]	suprafacial/retention antarafacial/inversion
4n+2 Hückel	Thermal (Δ)	[1,5], [1,9]	suprafacial/retention antarafacial/inversion
	Photochemical (*hv*)	[1,5], [1,9]	suprafacial-inversion antarafacial-retention

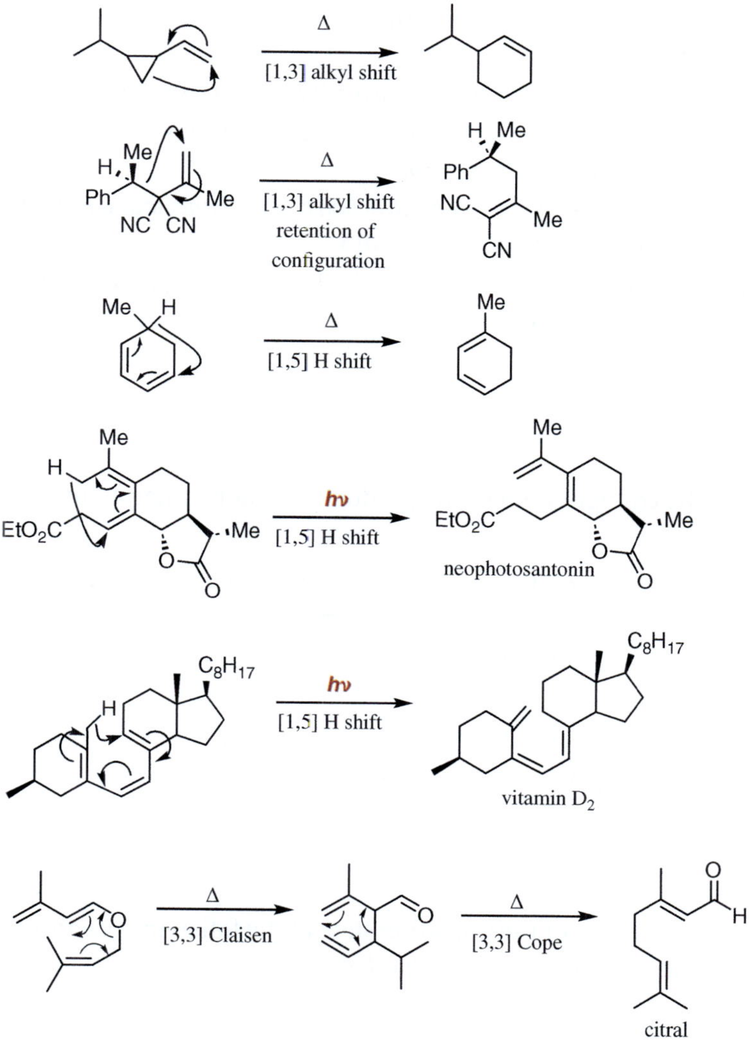

FIGURE 10.28 Sigmatropic reactions of interest.

Figure 10.28 presents several sigmatropic displacement reactions of interest. For instance, the synthesis of vitamin D_2 in the human body involves a [1,5] hydrogen sigmatropic shift. On the other hand, the industrial synthesis of citral involves two consecutive [3,3] sigmatropic rearrangements: a Claisen rearrangement followed by a Cope rearrangement.

FURTHER READING

C. Cope, E. M. Hardy, *J. Am. Chem. Soc.* **1940**, *62*, 441.

L. Claisen, *Ber. Deut. Chem. Ges.* **1912**, *45*, 3157.

D. E. Lewis. *Advanced Organic Chemistry*, Oxford University Press, Oxford, UK, **2016**, Chapter 6.

I. Fleming, *Pericyclic Reactions*, Oxford University Press, Oxford, UK, **1998**.

J. Clayden, N. Greeves, S. Warren, P. Wothers. *Organic Chemistry*, Oxford University Press, Oxford, UK, **2001**, Chapter 30.

S. Kumar, V. Kumar, S. P. Singh, *Pericyclic Reactions: A Mechanistic and Problem-Solving Approach*, Academic Press, New York, **2015**.

S. Sankararaman, *Pericyclic Reactions—A Textbook: Reactions, Applications and Theory*, Wiley-VCH, Weinheim, **2005**.

EXERCISES

10.1 Assign the configuration of the products of σ bond migration:

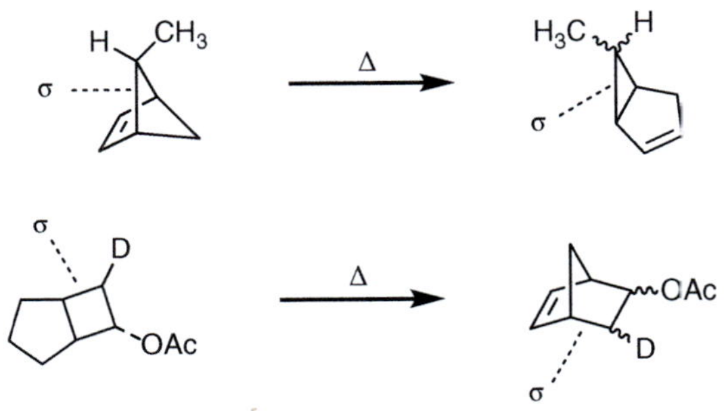

10.2 Using the analysis of the transition state of the following reactions indicate whether they are *allowed* or *forbidden*.

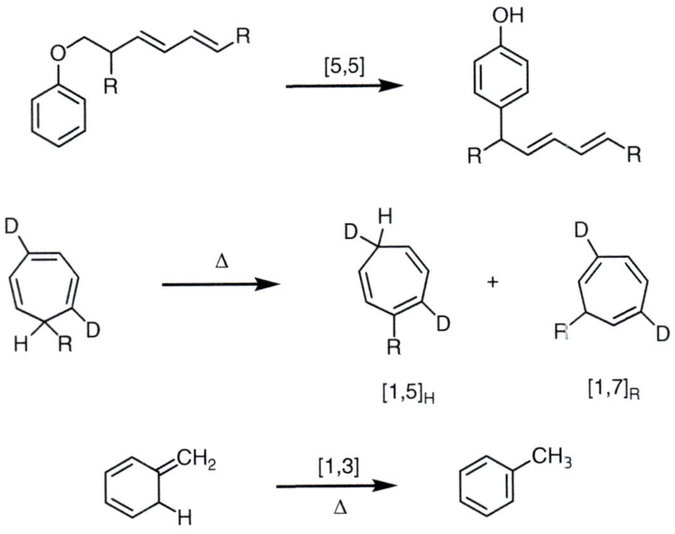

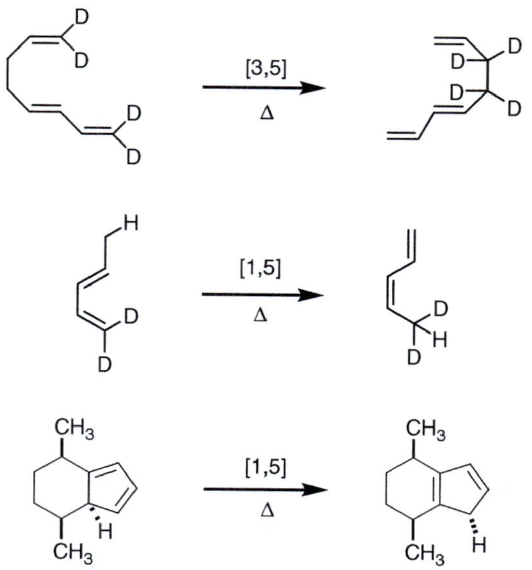

10.3 Carry out the symmetry analysis of the transition states for the [1,7] sigmatropic shifts of hydrogen in a *suprafacial* or *antarafacial* manner in a π-system.

10.4 Carry out the FMO analysis of the [1,7] sigmatropic shift of hydrogen in a π-system.

CHAPTER 11

1,3-Dipolar Cycloadditions

INTRODUCTION

1,3-Dipolar cycloadditions (1,3-DPCAs) are thermally allowed [$4\pi + 2\pi$] cycloaddition reactions involving species with formal charges. These reactions proceed *via* a six-electron Hückel aromatic transition state involving a five-membered ring (Figure 11.1). The 4π electron component corresponds to the charged 1,3-dipolar component, which presents a stable octet Lewis structure represented by zwitterionic forms. In these structures, the central atom carries a positive charge, while the negative charge is distributed across the terminal atoms. Furthermore, 1,3-dipolar reactants consist of three atoms, with at least one of them being a heteroatom. On the other hand, the uncharged 2π electron components, commonly referred to as dipolarophiles, include alkenes, alkynes, carbonyls, and nitriles, among others (Figure 11.1).

CLASSIFICATION OF 1,3-DIPOLAR REACTANTS

According to R. Huisgen, 1,3-dipolar moieties can be recognised as isoelectronic to allylic or propargylic anions. 1,3-Dipolar species present a π-electron system consisting of two filled orbitals and one empty orbital, with both ends of the dipole exhibiting nucleophilic and electrophilic properties (Figure 11.2). 1,3-Dipolar species of the allylic anion type adopt a bent structure owing to the presence of four electrons distributed across three parallel *p* orbitals, which are perpendicular to the molecular plane. By contrast, 1,3-dipolar species of the propargylic anion type are linear because of the presence of one double bond that is orthogonal to the delocalised π-system. In allylic 1,3-dipolar species, the central atom 'b' is usually a Group V element such as nitrogen or phosphorus or a Group VI element such as oxygen or sulphur. On the other hand, in the

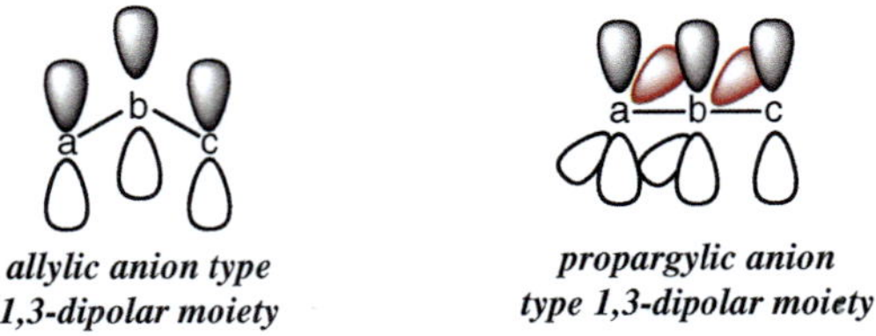

FIGURE 11.1 Top: generic 1,3-dipolar cycloaddition reactions. Bottom illustrative example of a 1,3-dipolar cycloaddition reaction: synthesis of a carboxymethyl-substituted 1-pyrazoline.

allylic anion type
1,3-dipolar moiety   *propargylic anion*
type 1,3-dipolar moiety

FIGURE 11.2 Arrangement of *p*-type orbitals in allylic and propargylic 1,3-dipolar moieties.

propargylic-type system, the central atom 'b' is always a Group V element. The terminal atoms 'a' and 'c' of the 1,3-dipolar moieties are second-row elements, carbon, nitrogen or oxygen. Table 11.1 collects some of the 1,3-dipolar systems commonly engaged in 1,3-DPCAs.

In view of the large number of known 1,3-DPCA reactions, a general analysis of their reactivity in terms of molecular orbital theory (MOT) is given first, followed by a discussion of several selected examples.

FRONTIER MOLECULAR ORBITAL ANALYSIS

In 1,3-dipolar moieties, four π-electrons are distributed over three atoms, forming an allylic-type orbital system (Figure 11.3). 1,3-Dipolar moieties can be regarded as structural analogues of the diene component in Diels–Alder reactions. Notably, the highest occupied molecular orbital (HOMO) and lowest unoccupied molecular orbital (LUMO) of a 1,3-dipolar species present similar symmetry to that encountered in a diene segment. As shown in Figure 11.3, the *p*-type orbital phases at the termini of the MOs align in a similar way in both the 1,3-dipolar moiety and the diene segment, relative to a bisecting symmetry plane σ.

TABLE 11.1 Classification of 1,3-dipolar Segments According to R. Huisgen

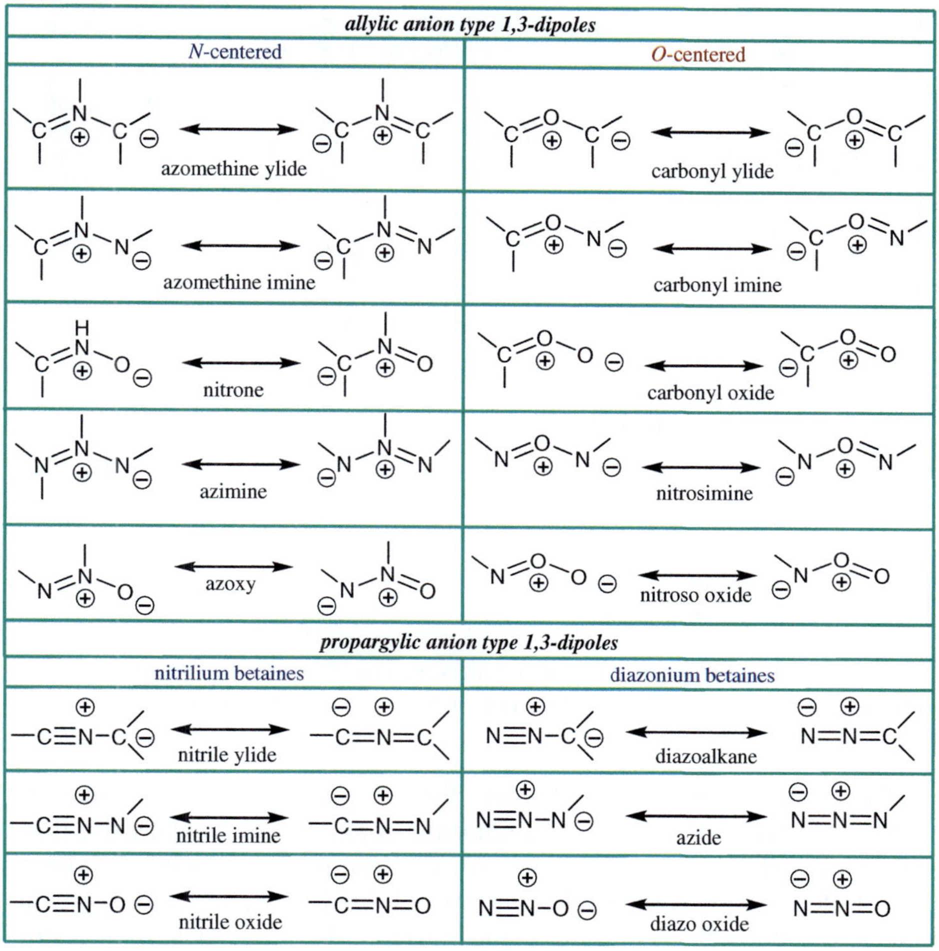
allylic anion type 1,3-dipoles
N-centered
O-centered
azomethine ylide
carbonyl ylide
azomethine imine
carbonyl imine
nitrone
carbonyl oxide
azimine
nitrosimine
azoxy
nitroso oxide
propargylic anion type 1,3-dipoles
nitrilium betaines
diazonium betaines
nitrile ylide
diazoalkane
nitrile imine
azide
nitrile oxide
diazo oxide

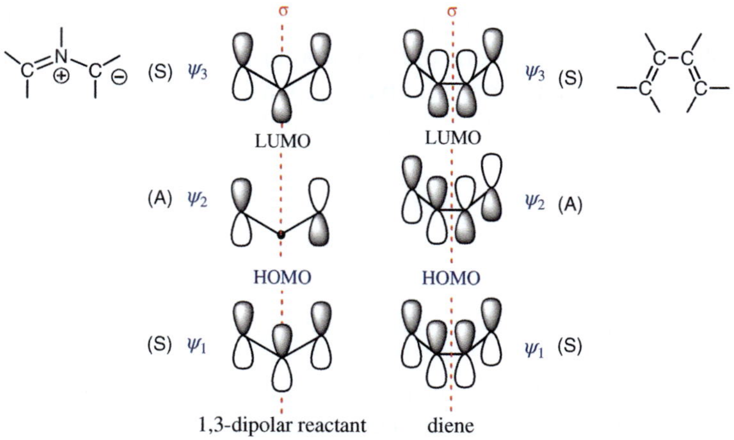
σ
σ
(S) ψ3
ψ3 (S)
LUMO
LUMO
(A) ψ2
ψ2 (A)
HOMO
HOMO
(S) ψ1
ψ1 (S)
1,3-dipolar reactant
diene

FIGURE 11.3 Comparison of the symmetry in the π MOs of a 1,3-dipolar moiety and a diene segment.

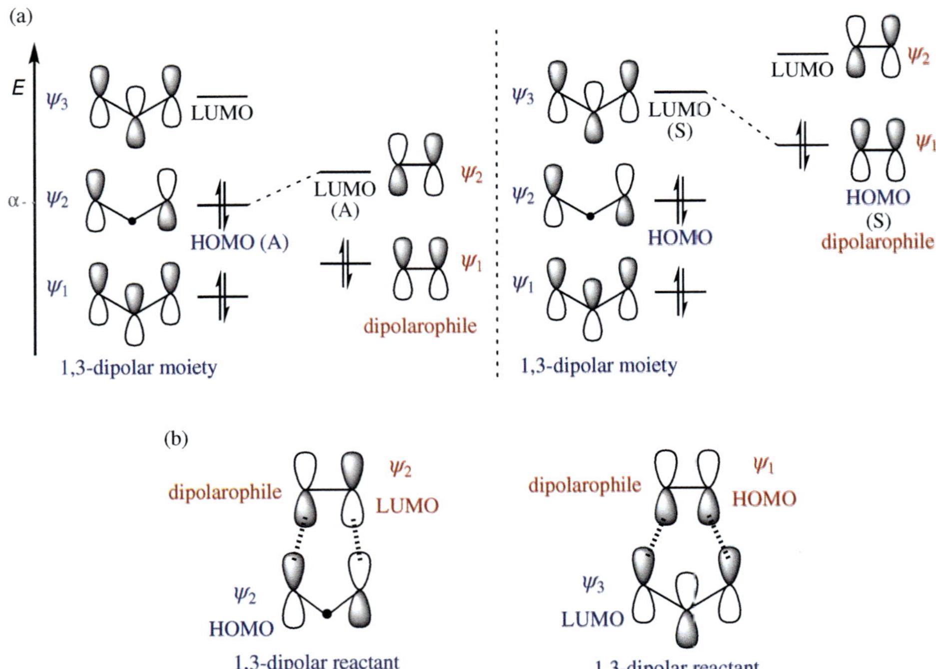

FIGURE 11.4   Frontier molecular orbital (FMO) analysis of generic 1,3-DPCAs. (a) HOMO/LUMO symmetry match. (b) In-phase *p*-type orbital interactions.

The principal difference between dienes and 1,3-dipolar moieties lies in their reactivity. Dienes act as donors/nucleophiles and typically use their HOMO in cycloaddition reactions with electron-poor dienophiles (normal electron-demand Diels Alder). In contrast, 1,3-dipolar moieties, as their name implies, exhibit both electrophilic and nucleophilic properties. As a result, they can engage either their HOMO or LUMO, depending on whether the dipolarophile is electron-rich or electron-poor (Figure 11.4a). The 1,3-DPCA reactions involve 4π electrons from the 1,3-dipolar species and 2π electrons from the dipolarophiles. The constructive, in phase orbital interactions of HOMO and LUMO in the 1,3-dipolar moiety and in the dipolarophile are shown in Figure 11.4b.

ANALYSIS OF NODAL PROPERTIES IN THE TRANSITION STATE

The transition state in 1,3-DPCAs consists of a cyclic 6 electrons system with zero nodes, and it is therefore aromatic. Thus, the process is thermally *allowed* (Figure 11.5).

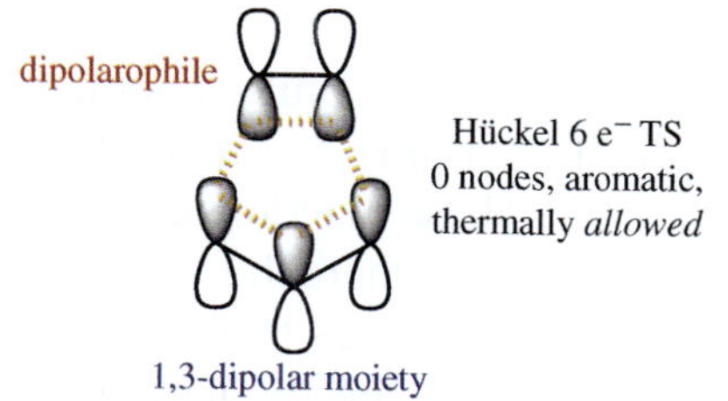

FIGURE 11.5 FMO analysis of generic 1,3-DPCAs.

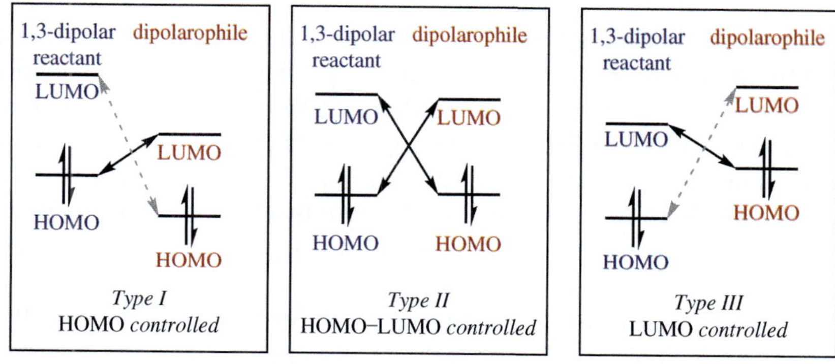

FIGURE 11.6 Types of 1,3-DPCA reactions.

TYPES of 1,3-DPCA REACTIONS AND REGIOSELECTIVITY

The regioselectivity in 1,3-DPCA reactions, like in other pericyclic reactions, is governed by the coefficients of the FMOs leading to the most efficient orbital overlap in the transition state. 1,3-DPCA reactions are typically classified into three categories: Type I, Type II and Type III.

Type I: In this case, the high-energy HOMO of the dipolar reactant overlaps with the LUMO of the dipolarophile, a situation often referred to as a *HOMO-controlled* (Figure 11.6, left). Many dipolar substrates react in this manner, including azomethylene ylides, carbonyl ylides, nitrile ylides, azomethylene imines, carbonyl imines, and diazoalkanes. This type of interaction closely resembles a normal electron-demanding Diels–Alder reaction, where the reaction is triggered when the diene's HOMO interacts with the dienophile's LUMO.

Type II: In this case, the energy gap between the two FMOs is small, allowing for two possible modes of reaction: (1) the 1,3-dipolar moiety's HOMO interacting with the dipolarophile's LUMO, or (2) the dipolarophile's HOMO interacting with the 1,3-dipolar moiety's LUMO (Figure 11.6, centre). This scenario is referred to as a *HOMO–LUMO controlled*. 1,3-Dipolar species such as nitrile imines, nitrones, carbonyl oxides, nitrile oxides, and azides typically react in this manner.

Type III: In this mode, the low-lying LUMO of the 1,3-dipolar moiety overlaps with the HOMO of the dipolarophile, a situation often referred to as a *LUMO-controlled* (Figure 11.6, right). Compounds such as nitrous oxide and ozone exhibit this behaviour. These reactions are analogous to an inverse electron-demand Diels–Alder reaction, where the reaction is triggered when the diene's LUMO interacts with the dienophile's HOMO.

When the 1,3-dipolar moiety and the dipolarophile reactant are asymmetric, the reaction can yield a mixture of stereoisomers. The stereoselectivity of the 1,3-DPCA reaction is governed by the magnitude of the interaction at the transition state (TS) between the FMOs with largest coefficients. As discussed in previous Chapters, the magnitude of the orbital coefficients is influenced by the nature of the substituents. In particular, electron donating or electron withdrawing groups alter the shape of the FMOs of the dipolarophile, as illustrated in Figure 11.7. Dipolarophiles with electron-withdrawing groups (EWG) or conjugating groups (CG) exhibit a higher orbital coefficient on the β carbon in both their HOMO and LUMO. By contrast, dipolarophiles with electron-donating groups (EDG) present the largest coefficients at the α carbon.

The nature of the substituents also influences the energy levels of the FMOs. An EWG on the 1,3-dipolar species or at the dipolarophile lowers the energy of both the HOMO and LUMO. By contrast, an EDG raises the energy of both the HOMO and the LUMO (Figure 11.7). A conjugating (CG) substituent raises the HOMO energy but lowers the LUMO energy. As a result, substituents can either increase or decrease the reaction rate, depending on how they affect the energy gap between the FMOs.

In a striking illustrative example, the reaction of azomethyl ylide with methyl acrylate is 50^3 times faster relative to that with ethylene (Figure 11.8). This reaction is controlled by the HOMO of the azomethyl ylide, which interacts with the LUMO of the dipolarophile. EWG such as the carboethoxy group on

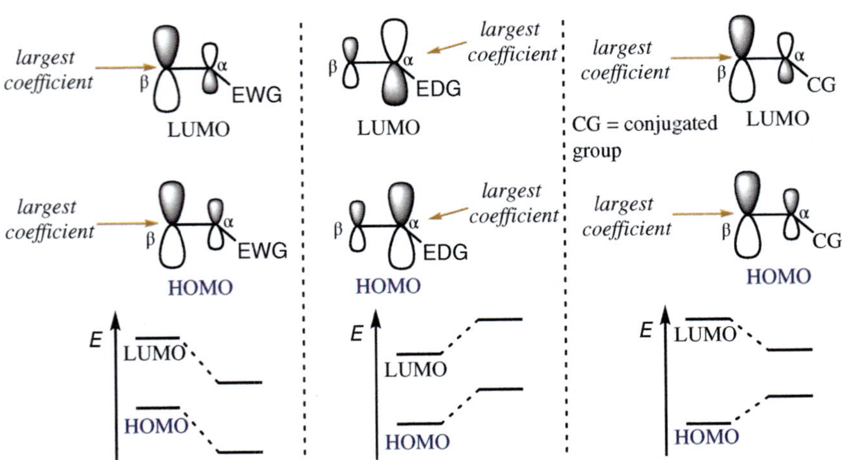

FIGURE 11.7   Effect of EWG, EDG or CG, respectively, on the dipolarophile's FMOs.

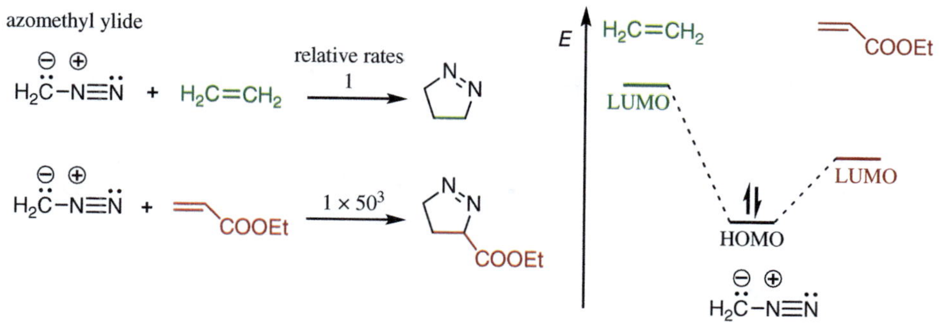

FIGURE 11.8   Effect of substituents on the energy of FMOs and consequently on the rates of reaction.

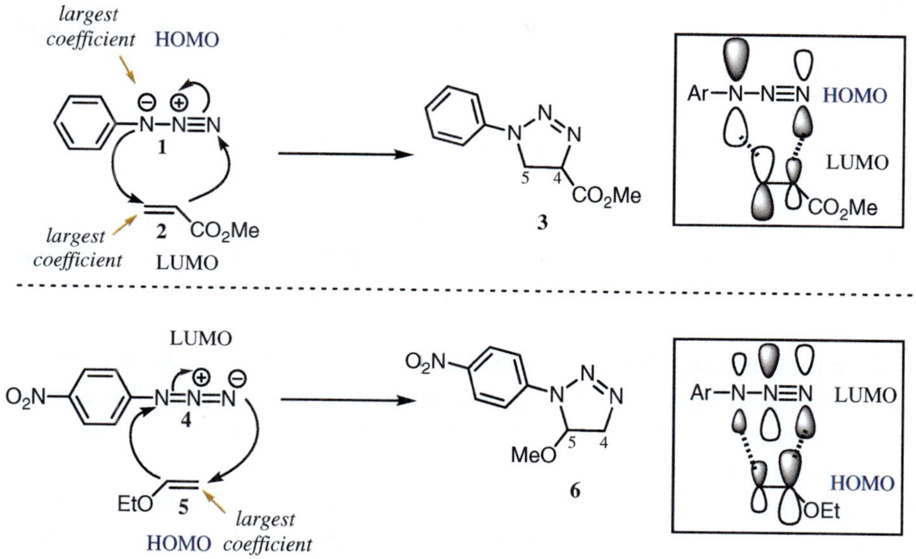

FIGURE 11.9   Regioselectivity in 1,3-DPCA reactions.

ethyl acrylate, accelerate the reaction by lowering the LUMO energy relative to that of ethylene.

An equally interesting example is the reaction between phenyl azide (**1**) and methyl acrylate (**2**). In this case, the azide utilises its HOMO, while the dipolarophile employs its LUMO (Figure 11.9). The dominant interaction takes place between the orbitals with the largest coefficients, leading to product **3**, where the methyl ester is positioned at carbon 4. Conversely, in the reaction of *p*-nitrophenyl azide (**4**) with ethoxyethylene (**5**), the azide uses its LUMO, and the dipolarophile uses its HOMO. As a result, the primary orbital interaction directs the substitution to carbon 5 affording product **6**.

1,3-DPCAs generally take place in *suprafacial* and *stereospecific* manner, leading to retention of configuration with respect to the 1,3-dipolar species

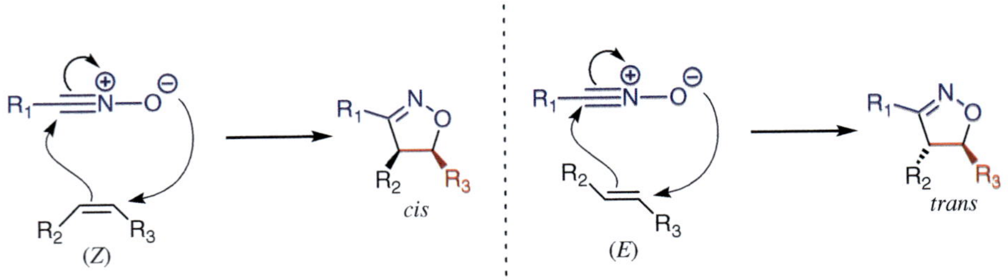

FIGURE 11.10   Stereospecificity in 1,3-DPCA reactions.

and the dipolarophile. The configuration of the alkene partners is preserved in the heterocyclic adduct, as expected for a concerted cycloaddition process. For instance, in the cycloaddition of nitrile oxides with alkenes, a *cis*-heterocycle forms from a (*Z*)-alkene, while a *trans*-heterocycle arises from an (*E*)-alkene (Figure 11.10).

1,3-DPCA REACTIONS WITH DIAZOALKANES

1,3-DPCA reactions of diazoalkanes with alkenes and alkynes afford 1-pyrazolines and 1*H*-pyrazoles, respectively. Tautomerisation of the initial products to yield 2-pyrazolines and 3*H*-pyrazoles is a thermodynamically favourable process (Figure 11.11).

When the reaction takes place with electron-deficient or conjugated alkenes, the process is controlled by the 1,3-dipolar moiety's HOMO (Type I). The carbon atom of the diazoalkane attacks the terminal carbon of the alkene, resulting in the exclusive formation of the 3-substituted pyrazolines (Figure 11.12a). Steric effects can either reinforce or compete with electronic effects. When diazomethane is added to methyl acrylate, 3-carboxylated pyrazoline is formed in excellent yield. However, increasing the steric hindrance at the β-position of the dipolarophile, as shown in Figure 11.12b, changes the ratio between the two potential regioisomers. In the extreme case, increasing the substituent size from hydrogen to *tert*-butyl completely overcomes the electronic preference and leads exclusively to the formation of the 4-substituted regioisomer.

When electron-rich alkenes participate in the DPCA reaction, the $HOMO_{dipolar\,reactant}/LUMO_{dipolarophile}$ and $LUMO_{dipolar\,reactant}/HOMO_{dipolarophile}$ interactions are comparable in magnitude (i.e. small orbital energy gap, type II, HOMO–LUMO control). Since the LUMO coefficients of the dipolar moiety are quite similar (Figure 11.13a), the regioselectivity is primarily determined by the HOMO of the 1,3-dipolar moiety (Figure 11.13b). The larger coefficients at the termini of the HOMO in the 1,3-dipolar moiety allow for better and more efficient orbital overlap, which directs the regioselective formation of 4-substituted pyrazolines.

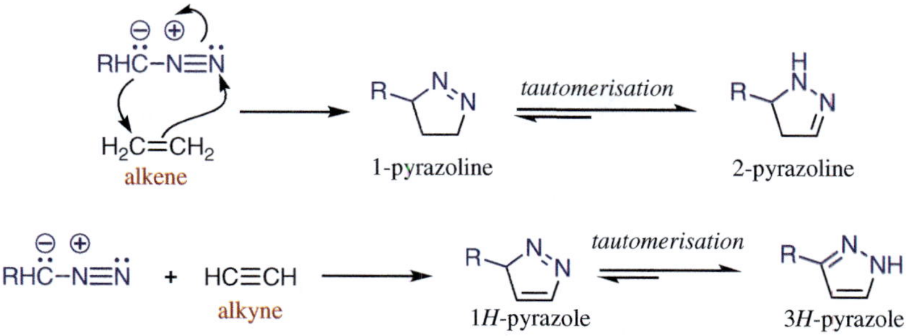

FIGURE 11.11 1,3-DPCA reactions with diazoalkanes.

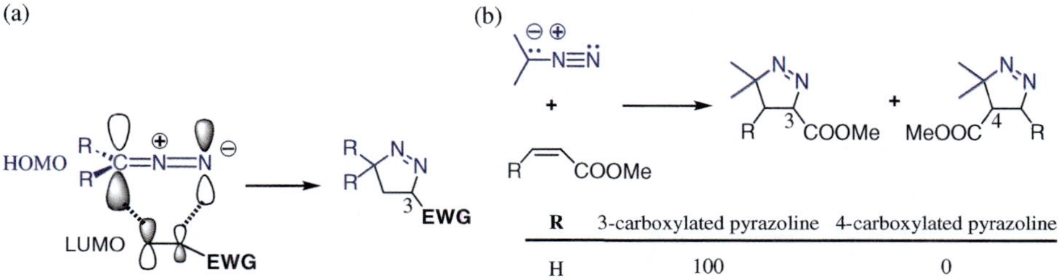

FIGURE 11.12 Steric effect in Type I (HOMO controlled) 1,3-DPCA reactions involving diazoalkanes.

R	3-carboxylated pyrazoline	4-carboxylated pyrazoline
H	100	0
Me	91	9
i-Pr	47	53
t-Bu	0	100

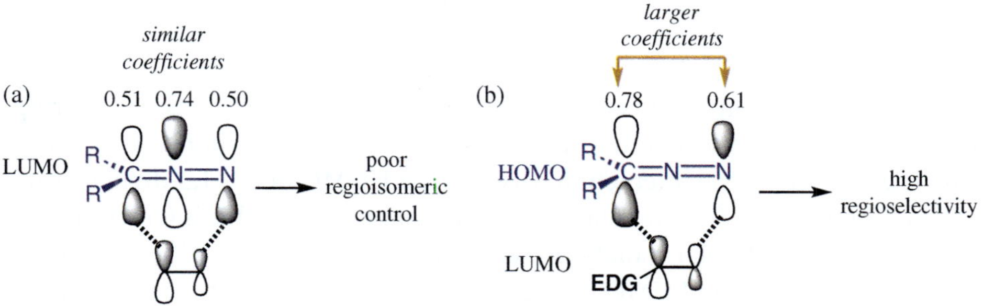

FIGURE 11.13 Type II 1,3-DPCA reactions involving diazoalkanes.

1,3-DPCA REACTIONS WITH NITRONES

Nitrones participate in cycloaddition reactions with alkenes and alkynes, resulting in the formation of isoxazolidines and isoxazolines, respectively (Figure 11.14a,b). When unsymmetrical alkenes are used, the reaction can yield two regioisomeric cycloadducts – specifically, the 4- and 5-substituted isomers, as shown in Figure 11.14c.

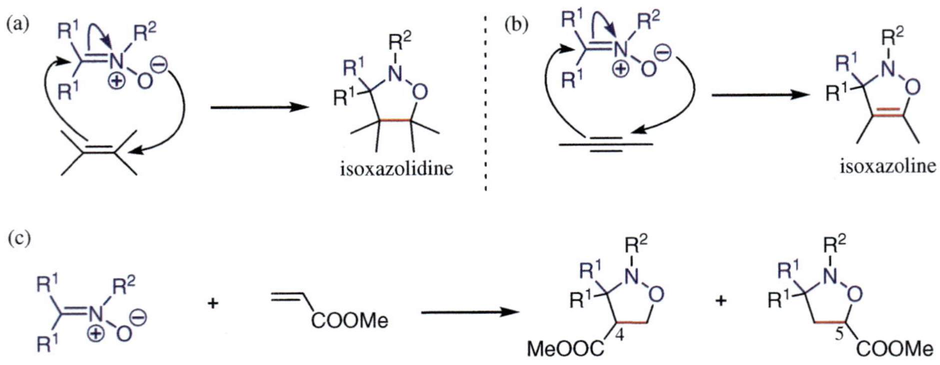

FIGURE 11.14   1,3-DPCA reactions with nitrones as 1,3-dipolar reactant.

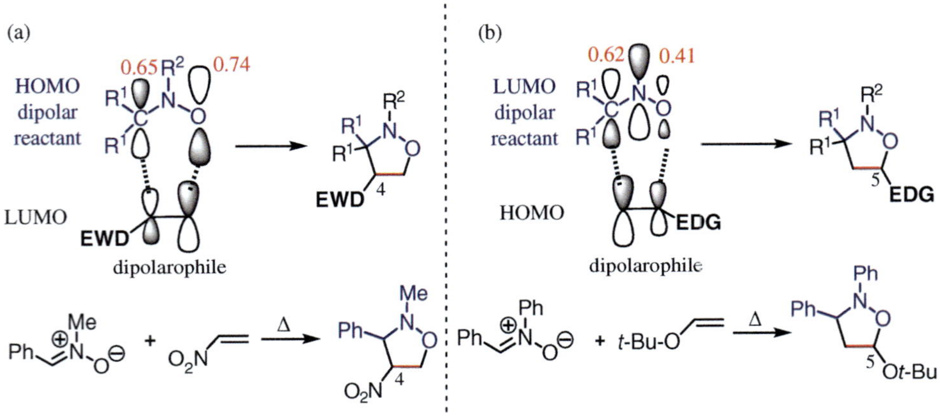

FIGURE 11.15   FMO interactions in 1,3-DPCA reactions involving nitrones as the 1,3-dipolar reactant.

The cycloaddition reaction between nitrones and terminal alkenes bearing an EWG favours the formation of the 4-substituted isoxazolidine. This selectivity is governed by the interaction between the HOMO of the 1,3-dipolar reactant and the LUMO of the dipolarophile (Figure 11.15a). In this interaction, the HOMO of the 1,3-dipolar moiety presents the largest orbital coefficient on the oxygen atom, while the LUMO of the dipolarophile presents its highest coefficient on the β-carbon (terminal carbon), leading to the preferential formation of the 4-isomeric product. Conversely, when a nitrone undergoes cycloaddition with an electron-rich terminal alkene, the reaction is controlled by the interaction between the LUMO of the 1,3-dipolar moiety and the HOMO of the dipolarophile (Figure 11.15b). In this case, the LUMO of the 1,3-dipolar moiety presents the largest orbital coefficient at the carbon atom, while the HOMO of the dipolarophile exhibits the largest orbital coefficient at the terminal carbon. This orbital-controlled interaction dictates the regiochemical outcome of the reaction.

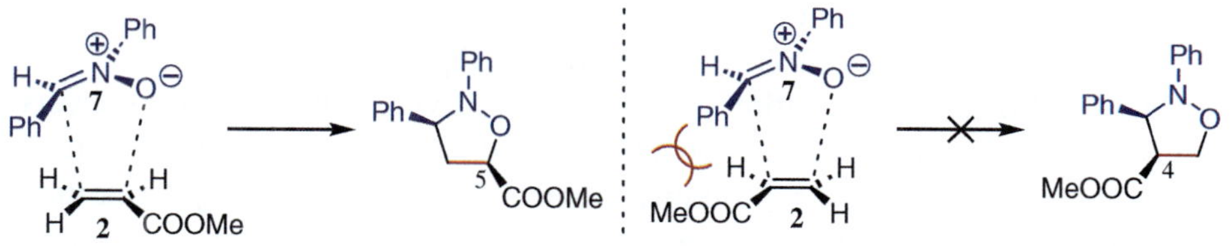

FIGURE 11.16 Regio- and stereoselectivity in 1,3-DPCA reactions involving nitrones as the 1,3-dipolar reactant.

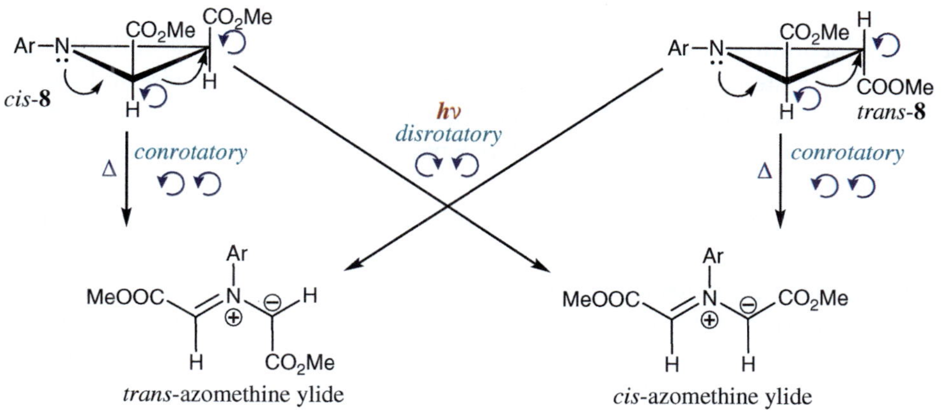

FIGURE 11.17 Thermal and photochemical ring opening of aziridines for the generation of azomethine ylides.

Steric factors must also be considered as they can significantly influence the regioselectivity of the reaction. For instance, in the cycloaddition of phenyl-*N*-phenylnitrone **7** with methyl acrylate **2**, steric hindrance prevents the formation of the electronically favoured 4-substituted isomer. Instead, steric hindrance promotes the formation of the 5-substituted isomer (Figure 11.16).

1,3-DPCA REACTIONS WITH AZOMETHINE YLIDES AS THE 1,3-DIPOLAR REACTANT

Thermolysis or photolysis of appropriately substituted aziridines provides an efficient method for generating azomethine ylides. According to orbital symmetry rules, the thermal ring opening of aziridines takes place via a *conrotatory* process, characteristic of a 4n electron cyclopropyl anion system (see Chapter 9). By contrast, photochemical ring opening proceeds via a *disrotatory* mechanism. These contrasting modes of ring opening give rise to distinct stereochemical outcomes: thermolysis of aziridine *cis*-dicarboxylic acid ester **8** leads to the formation of *trans*-azomethine ylides, whereas photolysis yields *cis*-azomethine ylides (Figure 11.17). The process is complementary with the *trans*-**8** aziridine substrate.

The 1,3-dipolar cycloaddition of azomethine ylides with alkenes and alkynes affords pyrrolidines and pyrrolines, respectively (Figure 11.18).

The 1,3-dipolar cycloaddition of an azomethine ylide with an olefin is a very useful reaction because it creates two C–C bonds in a single operation, with a high degree of regio- and stereochemical control. For instance, the cycloaddition of aziridine **9** with *tert*-butyl acrylate **10** (low-energy LUMO) exhibited complete regioselectivity, producing only the isomer with the 2,4-substitution pattern (Figure 11.19 top). By contrast, the 2,3-regioisomer was obtained in the cycloaddition of ethyl vinyl ether **11** (high-energy HOMO) with aziridine **9** (Figure 11.19 bottom).

Cycloaddition reactions of azomethine ylides with alkenes and alkynes proceed in a stereospecific manner. In an illustrative example, *cis-* and *trans-*dicarboxylic azomethine ylides, generated *via* the ring opening of their respective aziridines (**8**), are trapped in a stereospecific way by dipolarophiles such as alkynes (Figure 11.20).

Aziridines containing pendant olefinic or acetylenic groups can undergo *in situ* intramolecular 1,3-DPCA reaction with azomethine ylide intermediates. This approach provides a valuable method for the synthesis of fused bicyclic systems, offering efficient access to complex molecular architectures in a single step. An intermediate for the synthesis of acromelic acid A was obtained using this strategy (Figure 11.21).

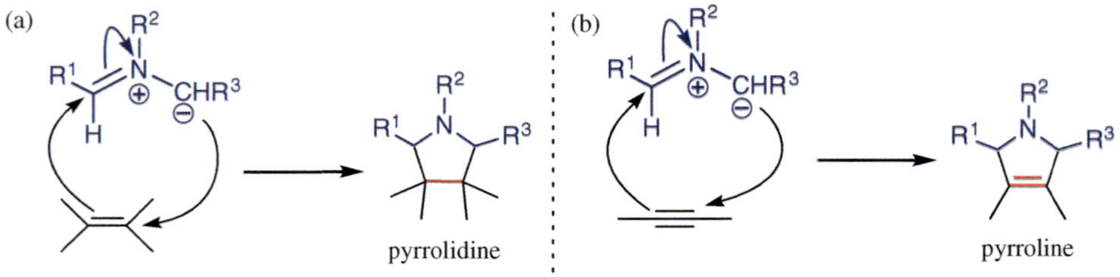

FIGURE 11.18　The 1,3-DPCA reactions of azomethine ylides.

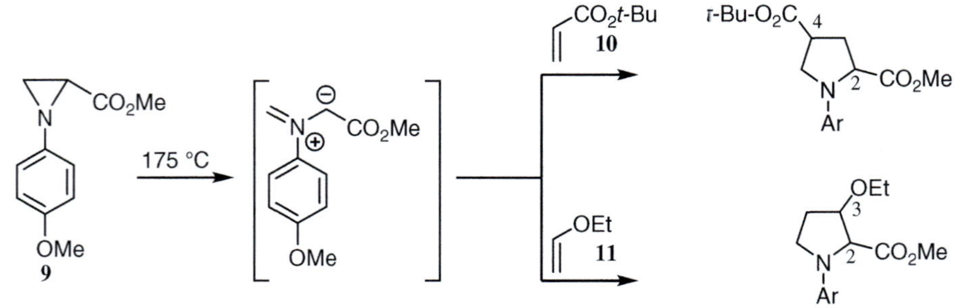

FIGURE 11.19　Regioselectivity in 1,3-DPCA reactions of azomethine ylides as 1,3-dipolar reactants with ethylene derivatives as dipolarophiles.

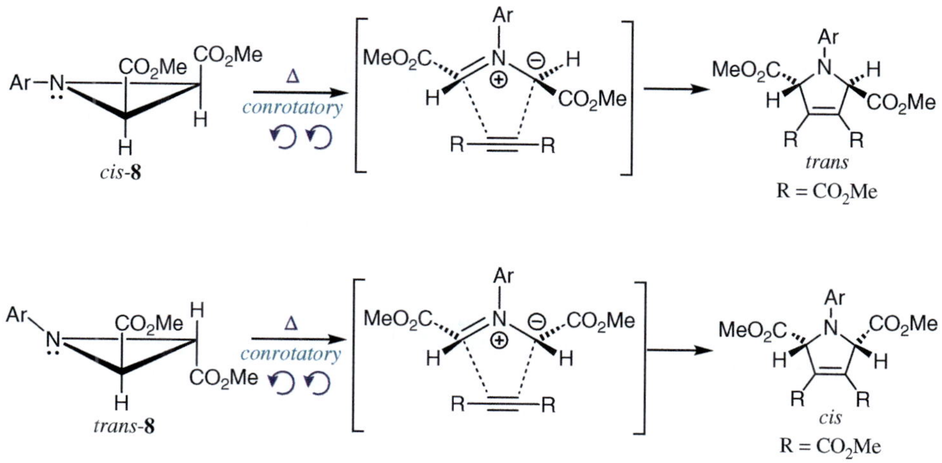

FIGURE 11.20 Stereospecificity of 1,3-DPCA reactions of azomethine ylides as 1,3-dipolar reactants.

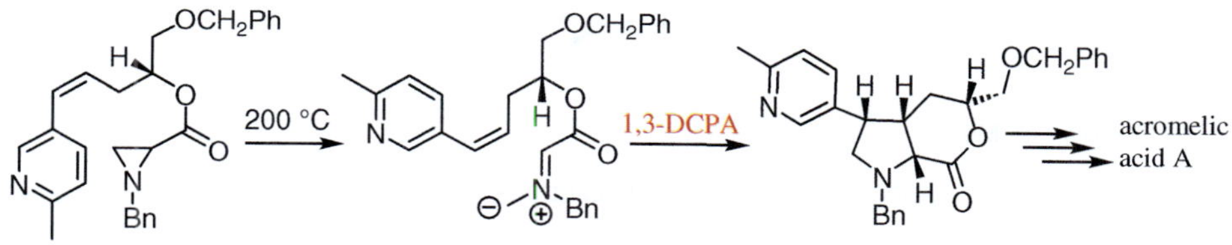

FIGURE 11.21 Intramolecular 1,3-DPCA reactions of azomethine ylides as 1,3-dipolar reactants.

1,3-DPCA REACTIONS WITH NITRILE OXIDES AS 1,3-DIPOLAR REACTANTS

Nitrile oxides are 1,3-dipolar moieties, which react readily with alkenes or alkynes to afford isoxazoline or isoxazole products (Figure 11.22).

Cycloaddition reactions between nitrile oxides and olefins are *stereospecific*. For instance, the reaction of a nitrile oxide with a *(E)*-alkene produces the *trans*-product, while the reaction with a *(Z)*-alkene yields the *cis*-product. This stereospecificity ensures that the configuration of the starting alkene is retained in the final product. Two examples are shown in Figure 11.23.

The cycloaddition reaction of nitrile oxides with monosubstituted electron-rich or conjugated alkenes conducts exclusively to the formation of the 5-substituted isoxazolines. These reactions are controlled by the LUMO of the 1,3-dipolar moiety encompassing the terminal carbon atom of the nitrile oxide, which presents the highest orbital coefficient (Figure 11.24).

1,3-DPCA reactions of nitrile oxides have been widely employed in the synthesis of bioactive compounds. For example, in the synthesis of ptilocaulin, a key intermediate was formed *via* an intramolecular 1,3-DPCA reaction

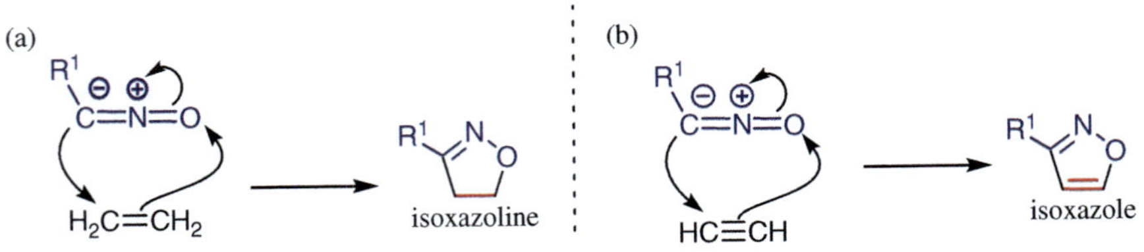

FIGURE 11.22 1,3-DPCA reactions with nitrile oxides as 1,3-dipolar reactants.

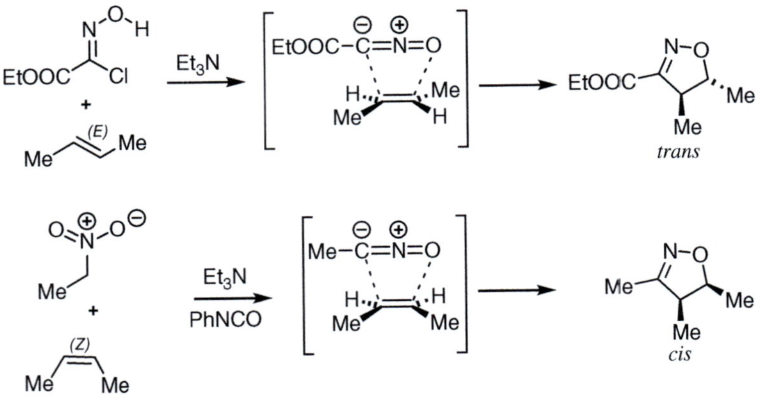

FIGURE 11.23 Stereospecificity in 1,3-DPCA reactions with nitrile oxides as 1,3-dipolar reactants.

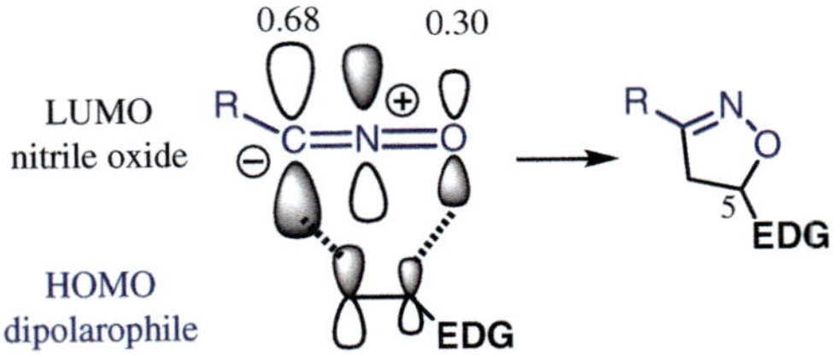

FIGURE 11.24 FMO interaction in the 1,3-DPCA reactions of nitrile oxides as 1,3-dipolar reactants with electron-rich or conjugated alkenes.

(Figure 11.25a). Similarly, an intramolecular 1,3-DPCA reaction with nitrile oxides as 1,3-dipolar reactants played a crucial role in the synthesis of the essential vitamin biotin (Figure 11.25b).

1,3-DPCA REACTIONS WITH AZIDES, OSMIUM TETROXIDE AND OZONE

The 1,3-dipolar cycloaddition reactions with azides, osmium tetroxide and ozone are of great importance in synthetic chemistry and will be briefly reviewed here.

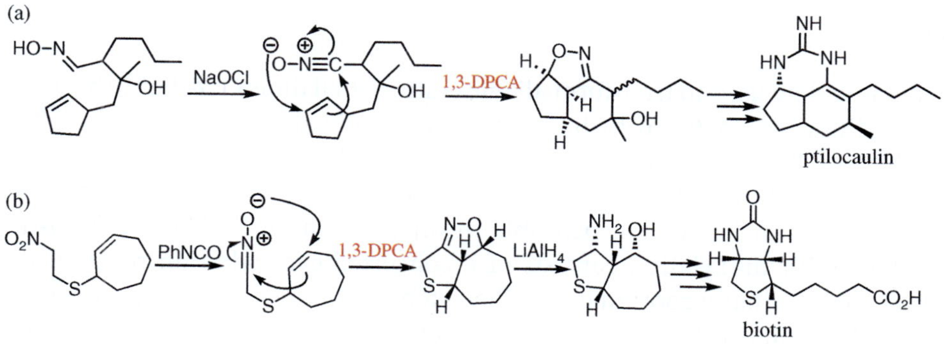

FIGURE 11.25 Synthetic application of the 1,3-DPCA reactions with nitrile oxides as 1,3-dipolar reactants.

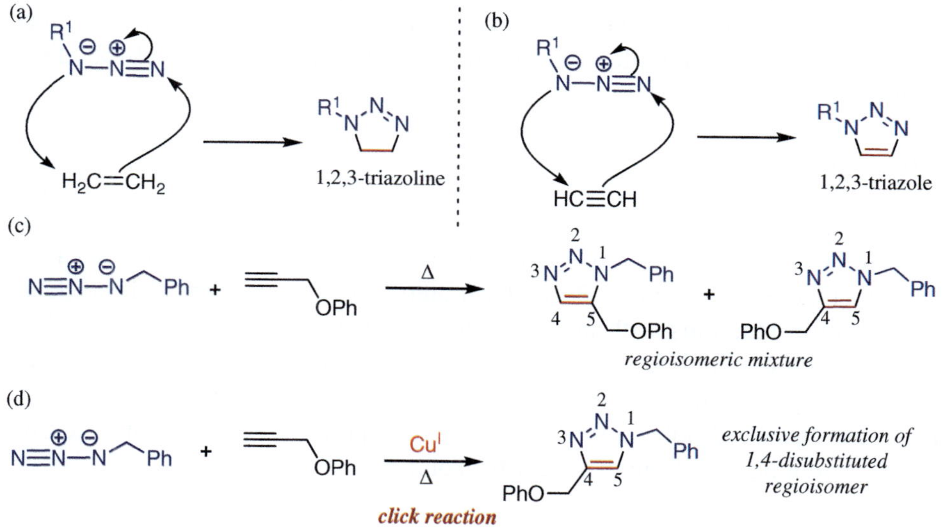

FIGURE 11.26 (a–c) Illustrative examples of 1,3-DPCA reactions of azides with alkenes or alkynes. (d) Copper(I)-catalysed azide-alkyne cycloaddition.

Azides (R-N$_3$)

Azides undergo cycloaddition with alkenes and alkynes to form 1,2,3-triazolines and 1,2,3-triazoles, respectively (Figure 11.26a,b). Owing to the comparable FMO energies of the reactants, the reactions can be driven by the HOMO of either the dipole or the dipolarophile (see Figure 11.26). This similarity in orbital energies leads to poor regioselectivity, often resulting in a mixture of regioisomeric 1,2,3-triazoles when the alkyne is dissymmetric (Figure 11.26c). However, this limitation can be overcome by metal catalysis. Copper(I)-catalysed azide–alkyne cycloaddition (CuAAC, often referred to as the *'click reaction'*) yields only 1,4-disubstituted-1,2,3-triazoles at room temperature in excellent yields (Figure 11.26d).

Osmium Tetroxide (OsO_4)

OsO_4 is widely used to transform alkenes into vicinal diols (1,2-diols) through *syn*-addition, so that two hydroxyl groups are added to the same face of a double bond. Many natural products, such as carbohydrates, contain vicinal diol functionalities. The ability of OsO_4 to introduce these groups with high stereocontrol makes it indispensable in organic synthesis. The mechanism of alkene dihydroxylation consists of a concerted cycloaddition reaction in which OsO_4 initially coordinates to the alkene, forming a cyclic intermediate called the osmate ester (Figure 11.27a). In this step, the π bond of the alkene interacts with the osmium centre and two of the oxygen atoms in OsO_4 that are transferred to the carbon atoms of the alkene in *syn* fashion. The cyclic osmate ester is then hydrolysed, typically in the presence of water. OsO_4 is a key component in the Sharpless asymmetric dihydroxylation (AD), a most useful reaction that enables the enantioselective formation of chiral diols (Figure 11.27b).

Ozone (O_3)

Ozone is a bent molecule in which the central oxygen atom carries a partial positive charge, while the two terminal oxygen atoms bear partial negative charges (Figure 11.28). As a 1,3-dipolar moiety, it undergoes typical 1,3-dipolar cycloaddition reactions with alkenes. The initial product of this reaction is a 1,2,3-trioxolane, which is highly unstable and rapidly decomposes *via* a retro 1,3-dipolar cycloaddition to give a carbonyl oxide and an aldehyde. The carbonyl oxides are also 1,3-dipolar compounds and undergo 1,3-DPCA to afford the carbonyl compound, resulting in a 1,2,4-trioxolane.

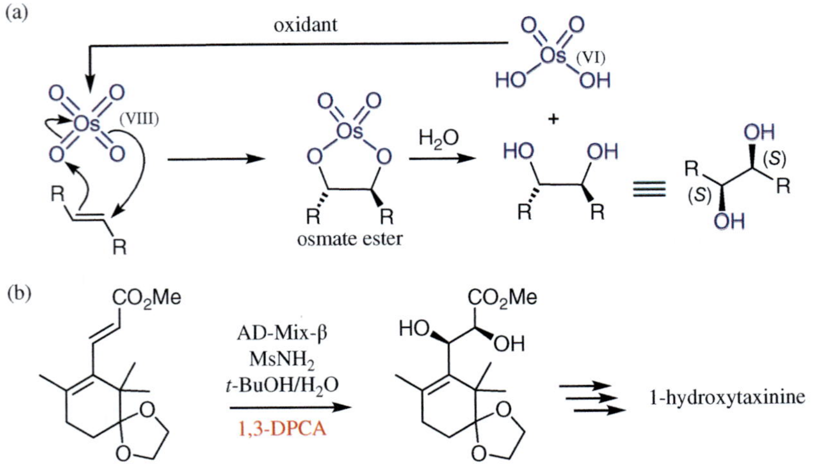

FIGURE 11.27 Application of OsO_4 for the dihydroxylation of alkenes. AD-mix β is a commercially available reagent that contains potassium osmate [$K_2OsO_2(OH)_2$] as a source of osmium, and $(DHQ)_2$-PHAL (dihydroquinidine derivative) as chiral ligand.

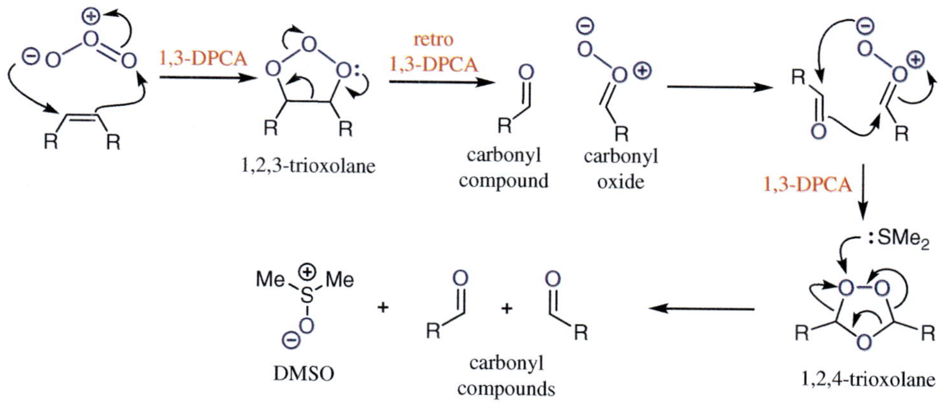

FIGURE 11.28 Plausible mechanism of the 1,3-DPCA of ozone to alkenes.

FURTHER READING

E. V. Anslyn, D. A. Dougherty, *Modern Physical Organic Chemistry*, University Science Books, Sausalito, California, **2005**.

H. Iida, C. Kibayashi, *Tetrahedron Lett.* **1981**, *22*, 1913.

J. Clayden, N. Greeves, S. Warren, *Organic Chemistry*, Oxford University Press, Oxford UK, **2012**.

K. N. Houk, Joyner. Sims, R. E. Duke, R. W. Strozier, J. K. George, *J. Am. Chem. Soc.* **1973**, *95*, 7287.

M. Meldal, F. Diness, *Trends Chem.* **2020**, *2*, 569.

P. De Shong, D. A. Kell, D. R. Sidler, *J. Org. Chem.* **1985**, *50*, 2309.

R. Huisgen, *J. Org. Chem.* **1976**, *41*, 403.

S. Kumar, V. Kumar, S. P. Singh, *Pericyclic Reactions: A Mechanistic and Problem-Solving Approach*, Academic Press, New York, **2015**.

S. Sankararaman, *Pericyclic Reactions- A Textbook: Reactions, Applications and Theory*, Wiley-VCH, Weinheim, **2005**.

S. Takano, Y. Iwabuchi, K. Ogasawara, *J. Am. Chem. Soc.* **1987**, *109*, 5523.

V. V. Rostovtsev, L. G. Green, V. V. Fokin, K. B. Sharpless, *Angew. Chem. Int. Ed.* **2002**, *41*, 2596.

Y. Imamura, S. Yoshioka, M. Nagatomo, M. Inoue, *Angew. Chem. Int. Ed.* **2019**, *58*, 12159.

EXERCISES

11.1 Write the product of the following reactions (please provide the corresponding configuration when appropriate).

(a) 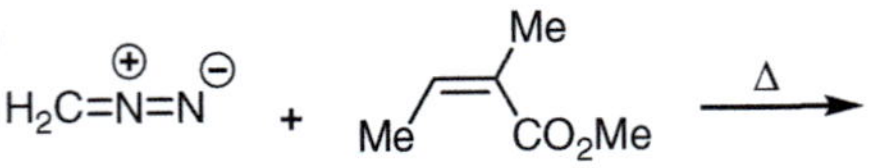

(b) 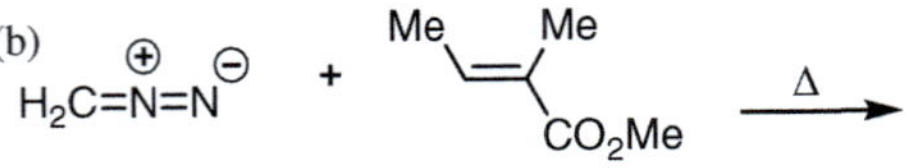

(c) 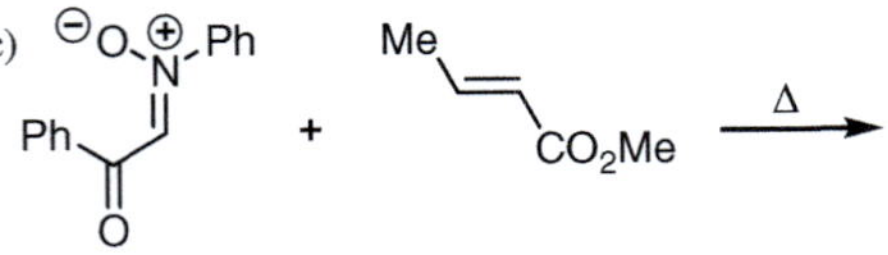

(d) 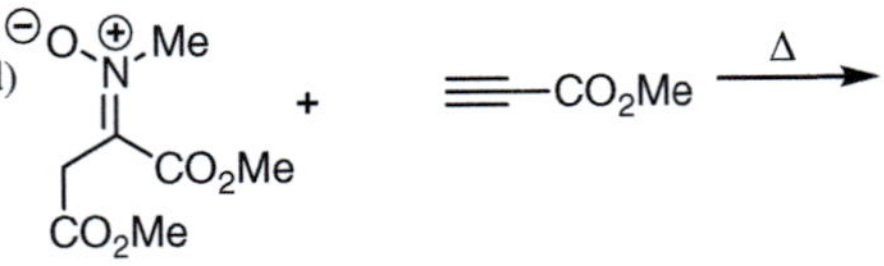

(e) 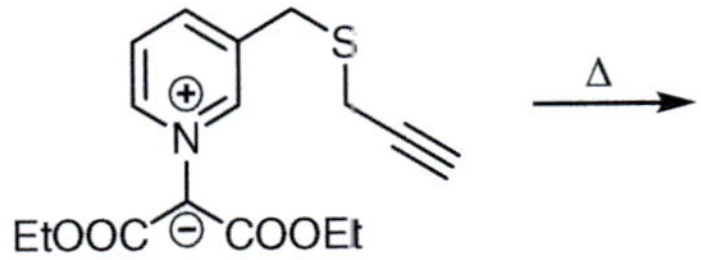

(f) 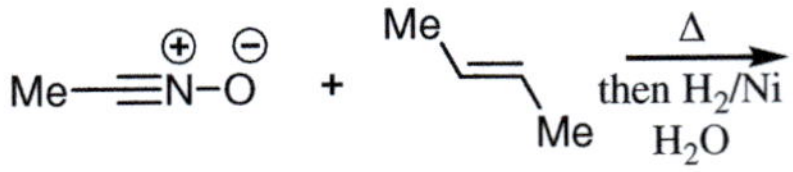

(g) 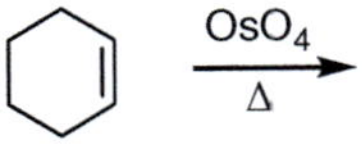

(h) 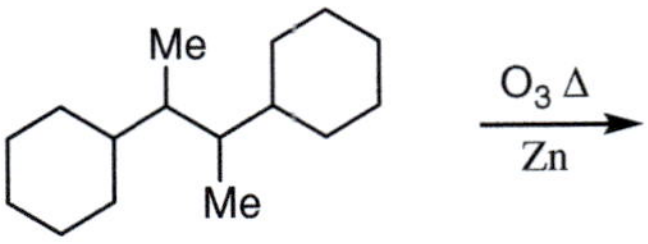

11.2 Draw the mechanism for the formation of the observed products. Please justify the observed stereochemistry when appropriate.

(a) 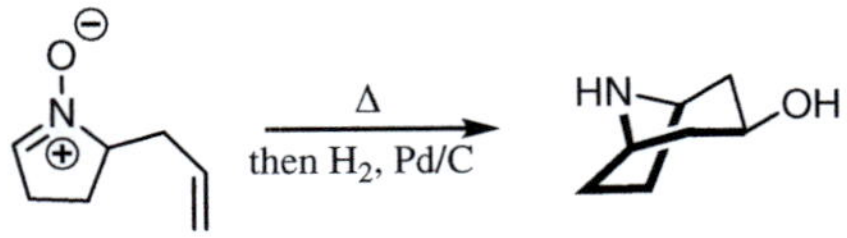

(b) 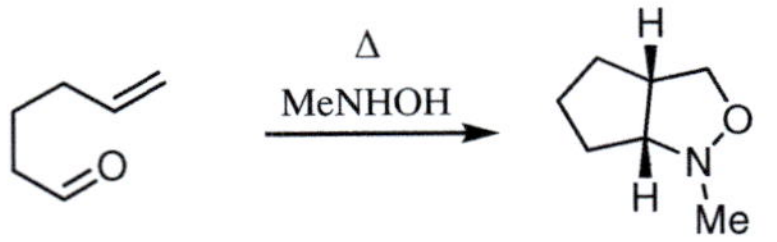

(c) 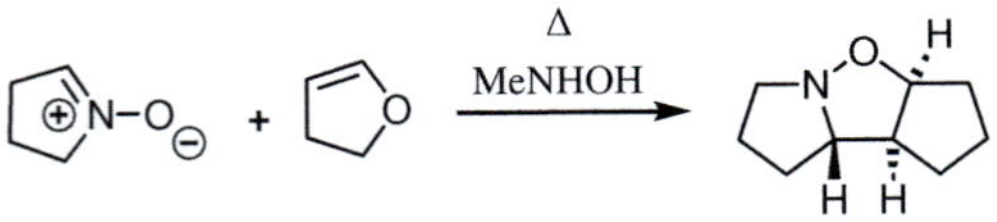

(d) 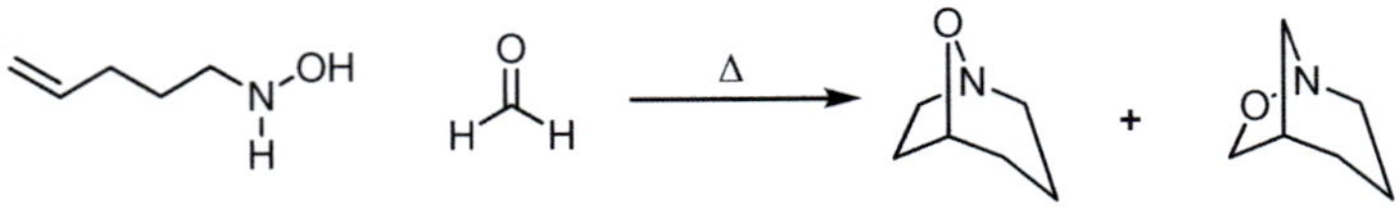

(e) 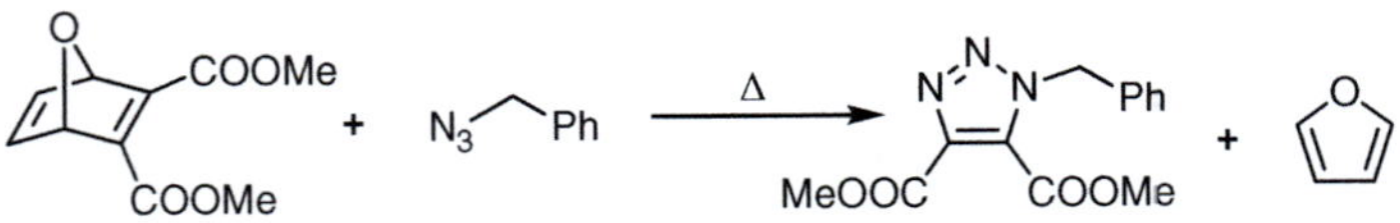

(f) 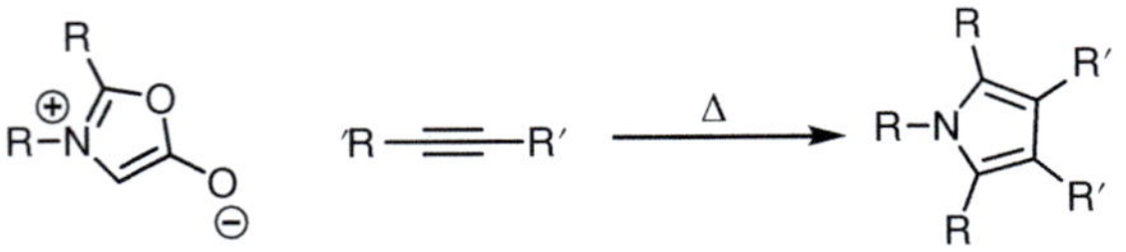

11.3 Draw the anticipated products with the correct stereochemistry and justify the stereospecificity of the reaction.

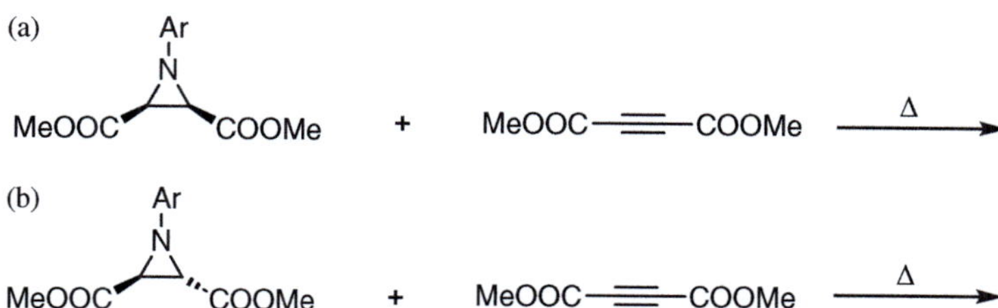

CHAPTER 12

Stereoelectronic Interactions

INTRODUCTION

The concept of *stereoelectronic interactions* originated when the so-called 'anomeric effect' was discovered some 60 years ago; indeed, in order to understand such a conformational effect, several theoretical models were advanced, one of them in terms of a stereoelectronic effect. To explain what this means, let us recall the structure and conformation of cyclohexane. Cyclohexane is not a planar molecule; instead, it actually occupies a three-dimensional space where it adopts the so-called 'chair conformation' (Figure 12.1). In the chair conformation of cyclohexane, six of the C—H bonds adopt axial orientations, highlighted with solid bars in Figure 12.1. At the same time, cyclohexane presents six 'equatorial' C—H bonds, represented with simple lines.

Cyclohexane is not a rigid structure, instead, rotation around C—C bonds causes an inversion of the chair conformation, a process that is quite rapid at ambient temperature. In this conformational process, axial hydrogens become equatorial, and equatorial hydrogens become axial. The conformational process that results in ring inversion does not have specific consequences in unsubstituted cyclohexane because the two chair conformations are identical (Figure 12.2a). However, when one replaces one of the hydrogen atoms in cyclohexane with, for example, a chlorine atom, then two distinct species are obtained. The left side of Figure 12.2b presents *axial* chlorocyclohexane, whose chair inversion affords the *equatorial* conformer (right side in Figure 12.2b). These two conformers present distinct energy, as well as different properties and reactivity.

As it turns out, the axial conformation of substituted cyclohexanes is usually less stable than the equatorial conformation, this being a consequence of steric interactions present in the axial conformer that destabilise such axial conformer so that the equatorial conformer is more stable (Figure 12.3).

Table 12.1 collects thermodynamic data for axial $\leftrightarrows$ equatorial equilibria in monosubstituted cyclohexanes. This table is arranged so that several representative

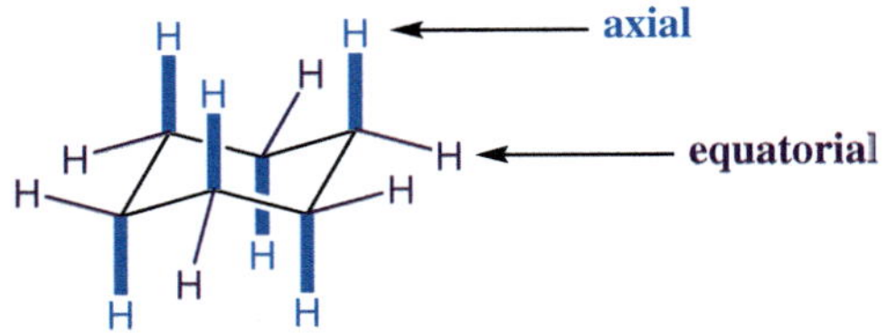

FIGURE 12.1 Distinct axial and equatorial C—H bonds in the chair conformation of cyclohexane.

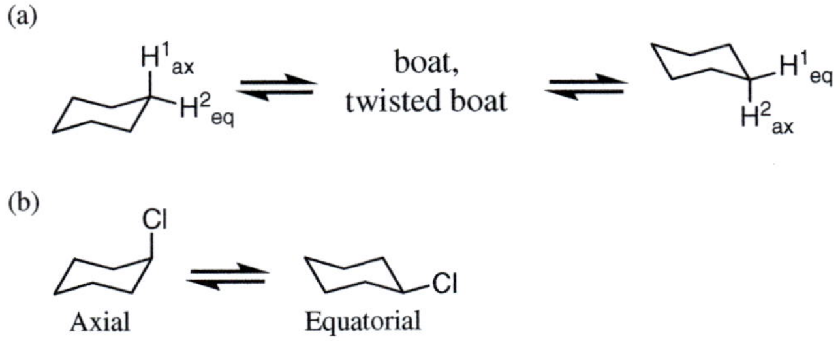

FIGURE 12.2 (a) Chair-to-chair inversion in unsubstituted cyclohexane. (b) Axial ⇆ equatorial equilibrium of chlorocyclohexane.

FIGURE 12.3 Steric repulsion in the axial conformer of a substituted cyclohexane.

TABLE 12.1 Conformational Preferences in Monosubstituted Cyclohexanes

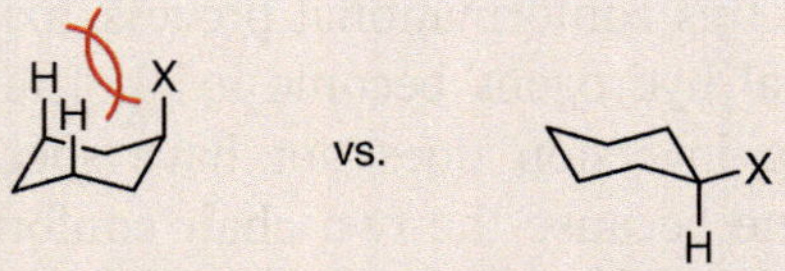

X	$-\Delta G°$ (kcal mol^{-1})	X	$-\Delta G°$ (kcal mol^{-1})
F	0.25	SH	1.00
Cl	0.40	CH$_3$	1.74
Br	0.50	CH$_2$CH$_3$	1.80
I	0.40	i-Pr	2.10
OH	0.70	t-Bu	4.90
OCH$_3$	0.80	C$_6$H$_5$	2.90
OCH$_2$CH$_3$	0.90	C≡N	0.20

substituents 'X' are located in columns 1 and 3, followed by the corresponding conformational free energy differences, $\Delta G°$ values in kcal mol^{-1} (columns 2 and 4). These $\Delta G°$ values are all negative, which means that the axial to equatorial equilibria are exothermic since the equatorial conformers are more stable. Thus, an equatorial preference is observed with halogens such as fluorine, chlorine, bromine, and iodine, as well as with oxygen-containing substituents such as the methoxy group or aliphatic groups such as methyl, ethyl, isopropyl, and *t*-butyl.

Let us focus on methoxycyclohexane: as it was mentioned earlier, the equatorial isomer is more stable and the corresponding $\Delta G°$ value is -0.8 kcal mol^{-1}, which means that the equatorial conformer is more stable than the axial conformer. The axial $\leftrightarrows$ equatorial conformational equilibrium is therefore exothermic.

What happens if we replace a methylene group adjacent to the carbon that supports the methoxy group, to afford 2-methoxy-oxane, instead of methoxycyclohexane? The conformational behaviour is rather different, now the axial conformer is more stable and $\Delta G°$ is positive, $+0.6$ kcal mol^{-1}! (Figure 12.4).

2-Methoxy-oxane is structurally related to carbohydrates with their iconic anomeric carbon and indeed, the anomeric effect was discovered in carbohydrates (Figure 12.5).

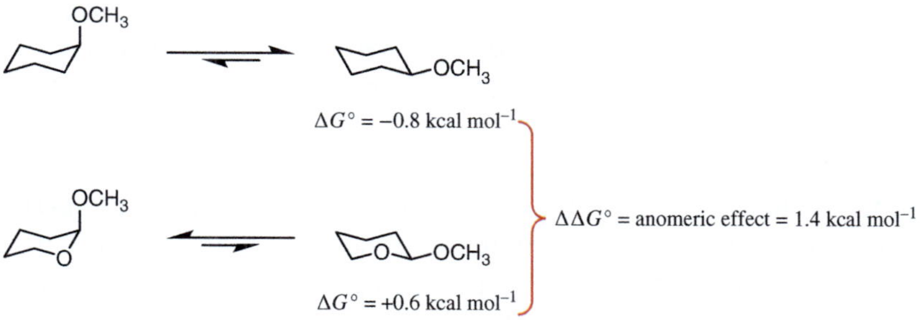

$$\Delta G° = -0.8 \text{ kcal mol}^{-1}$$

$$\Delta G° = +0.6 \text{ kcal mol}^{-1}$$

$$\Delta\Delta G° = \text{anomeric effect} = 1.4 \text{ kcal mol}^{-1}$$

FIGURE 12.4 Contrasting conformational behaviour between methoxycyclohexane that prefers the equatorial conformation, $\Delta G° = -0.8$ kcal mol^{-1}, and 2-methoxy-oxane that favours the axial conformation, $\Delta G° = +0.6$ kcal mol^{-1}. The difference $\Delta\Delta G° = 1.4$ kcal mol^{-1} corresponds to the magnitude of the anomeric effect.

36%

64% $\Delta G° = -0.34 \text{ kcal mol}^{-1}$

$$\Delta\Delta G° = -0.34 - (-1.25)$$
$$= +0.91 \text{ kcal mol}^{-1}$$

11%

89% $\Delta G° = -1.25 \text{ kcal mol}^{-1}$

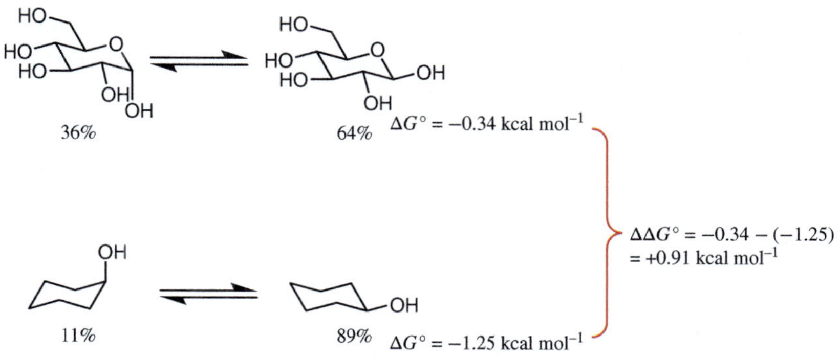

FIGURE 12.5 Magnitude of the anomeric effect in D-glucose.

In the 2-methoxy-oxane molecule, the anomeric effect is then manifested by the fact that the methoxy group at the anomeric carbon prefers to adopt the axial orientation rather than the equatorial position. In summary, in cyclohexane the methoxy group wants to be equatorial by 0.8 kcal mol^{-1} but in the heterocycle, the methoxy group prefers to be axial. Thus, the contrasting behaviour, -0.8 kcal mol^{-1} in cyclohexane *versus* $+0.6$ kcal mol^{-1} in the heterocycle, gives evidence of a rather strong conformational effect, the so-called 'anomeric effect', worth 1.4 kcal mol^{-1}. Not surprisingly, this observation caught the attention of many chemists when it was originally discovered in sugar chemistry, already over sixty years ago.

INTERPRETATION OF THE ANOMERIC EFFECT

So, the concept of the *anomeric effect* was advanced, and it was defined as 'the tendency of electronegative substituents at the anomeric position, C1 in Figure 12.6, to adopt the axial rather than the equatorial orientation'. Of course, a crucial issue was then to explain the origin of this conformational effect. This question has attracted the attention of many chemists, both experimentalists and theoreticians, and the topic is still nowadays rather challenging. As we will see, the interpretation of the anomeric effect is based on *stereoelectronic interactions*.

Nevertheless, it must be pointed out that the interpretation of the anomeric effect was initially advanced in terms of *electrostatic interactions*. In particular, if substituent X in axial monosubstituted oxane (left side in Figure 12.7a) is electronegative, the dipole originated at the C—X bond will be essentially *antiperiplanar* to the ring dipole (oxygen being more electronegative than carbon). This antiparallel orientation of the two dipoles is stabilising in terms of electrostatics. By contrast, in the equatorial conformer, the dipoles are nearly parallel, which results in a destabilising arrangement in terms of electrostatic interactions. Thus, from the electrostatic point of view, the conformation on the left side in Figure 12.7a is more favourable since the molecule prefers to have the dipoles antiparallel rather than parallel (right side in Figure 12.7a).

Nevertheless, an interpretation of the anomeric effect has been advanced in terms of *stereoelectronic interactions*. In this regard, a system that sheds light on this issue was *trans*-2,3-dichloro-1,4-dioxane, whose axial ⇌ equatorial conformational equilibrium is shown in Figure 12.7b. This equilibrium involves the

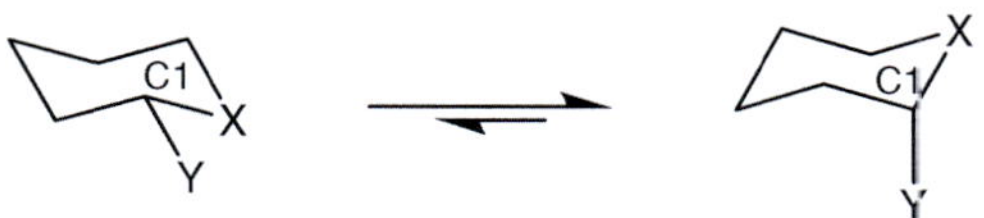

FIGURE 12.6　The anomeric effect is defined as the tendency of electronegative substituents at the anomeric carbon to adopt the axial rather than the equatorial orientation.

diequatorial conformer and the diaxial conformer. As it turns out, the diaxial conformer is more stable as a consequence of the anomeric effect. Importantly, this compound could be crystallised and from the X-ray crystallographic data that was gathered, it was found that the endocyclic carbon-oxygen bonds in the diaxial conformer are shorter than the carbon-oxygen bonds in the diequatorial conformer. Moreover, it was found that the axial carbon-chlorine bonds are longer in the diaxial conformer relative to the carbon-chlorine bonds in the diequatorial conformer. These structural data suggested the participation of a so-called canonical structure (see the right side in Figure 12.7b), where the stereoelectronic interaction corresponds to the hyperconjugation between the *antiperiplanar* donor n highest occupied molecular orbital (HOMO) orbital on oxygen and the acceptor σ^* carbon-chlorine lowest unoccupied molecular orbital (LUMO) orbital (that is, $n_O \rightarrow \sigma^*_{C-Cl}$), which gives rise to a *double bond/ no bond canonical structure*. The involvement of this canonical structure helps explain the observed shortening of the endocyclic C—O bond and the lengthening of the exocyclic C—Cl bond in the axial conformer (Figure 12.8).

(a) Dipole-dipole electrostatic interaction:

(b) Stereoelectronic effect:

FIGURE 12.7   Interpretations of the anomeric effect: (a) dipole-dipole electrostatic interaction, (b) stereoelectronic interaction.

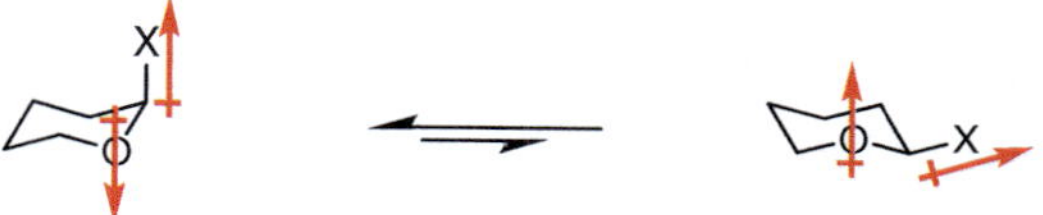

FIGURE 12.8   Stereoelectronic rationalisation of the anomeric effect: a stabilising interaction between an axial lone electron pair on the ring heteroatom 'X' and the *antiperiplanar* antibonding orbital of the bond connecting the axial substituent 'Y' at the anomeric carbon ($n_X \rightarrow \sigma^*_{C-Y}$) is proposed. This interaction induces the lengthening of the axial C—Y bond by electron transfer to its σ^* orbital, as well as the contraction of the C—X bond as a result of its increased double bond character.

This is the key stereoelectronic interaction of interest in the anomeric effect. It is *stereoelectronic* because the *antiperiplanar* orientation of the two participating orbitals is essential for its effectiveness; indeed, the orientation of the *n* (lone pair) orbital on oxygen must be *antiperiplanar* to the acceptor's C—Cl σ^* orbital. As depicted in a general manner in Figure 12.8, in the axial conformer the stereoelectronic interaction involves two orbitals and two electrons. There is an occupied lone electron pair *n* orbital on 'X' (endocyclic oxygen in 2-methoxy-oxane) that is antiperiplanar to the σ^* unoccupied C—Cl antibonding orbital. The antiperiplanar orientation of the participating orbitals is appropriate for an effective two-orbital-two electron interaction that leads to the double bond-no bond-stabilising canonical structure (right side in Figure 12.8).

Such stabilising interaction is not possible in the equatorial conformer because the orbitals are essentially orthogonal; thus, the stabilising interaction is only possible when the electronegative substituent is axial, not when this substituent is equatorial.

The anomeric effect in particular, and stereoelectronic interactions in general, are rather relevant in organic chemistry because two-orbital/two-electron interactions, such as those responsible for the anomeric effect, are also operative in many chemical reactions. For instance, one emblematic reaction in organic chemistry is the bimolecular nucleophilic S_N2 substitution reaction, where the nucleophile (occupied HOMO) approaches the leaving group (unoccupied LUMO) from the back (see Figure 5.12 in Chapter 5).

Another interesting example turns out to be the molecule of cyclohexane itself: a publication by F. Weinhold (2000), deals with an observation that brought stereoelectronic interactions into the limelight. As it is shown in Figure 12.9 the orbitals associated with occupied axial C—H carbon-hydrogen bonds interact with antiperiplanar unoccupied σ^* orbitals associated with the vicinal, axial C—H bonds. This stereoelectronic interaction leads to the double bond-no bond canonical structure shown in Figure 12.9. Indeed, as a consequence of this resonance form, the axial C—H bonds are *weaker*. In effect, X-ray crystallographic studies have demonstrated that the axial C—H bonds in cyclohexane are slightly *longer* than the equatorial C—H bonds in cyclohexane.

An additional system of interest is presented in Figure 12.10, which depicts a heterocyclic six-membered 1,3-dioxane ring that is substituted at C(4,6) by two methyl groups in the *cis* configuration. These methyl groups fix the chair conformation of the six-membered ring because an inverted chair conformation leads to a *syn*-diaxial orientation, which is prohibitive in terms of steric repulsion.

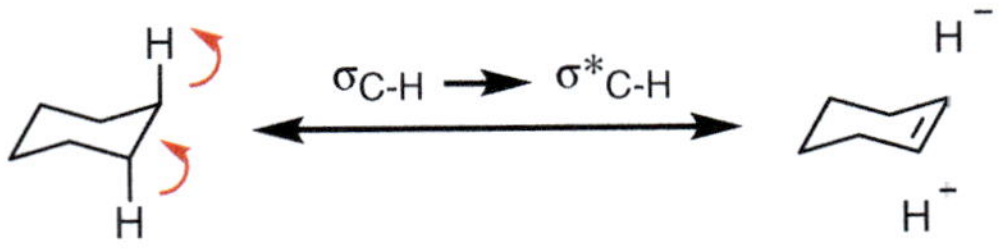

FIGURE 12.9   Double bond-no bond canonical structure in cyclohexane as the consequence of $\sigma(C{-}H) \rightarrow \sigma^*(C{-}H)$ stereoelectronic interaction in antiperiplanar diaxial vicinal carbon-hydrogen bonds.

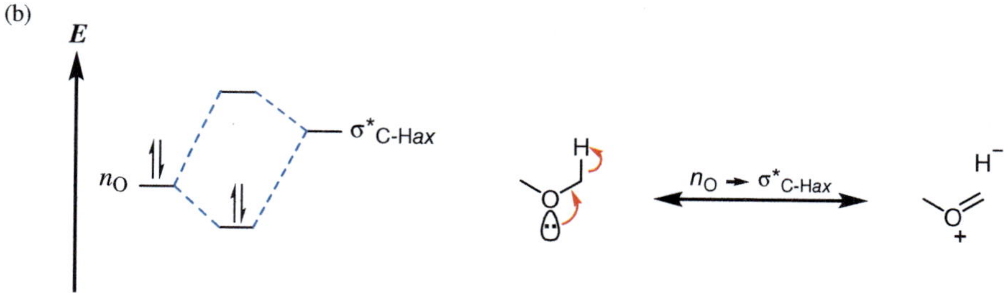

FIGURE 12.10 (a) Conformationally-fixed *cis*-4,6-dimethyl-1,3-dioxane and axial and equatorial one-bond $^1J_{C-H}$ coupling constants at C(2). (b) Two orbital-two electron stabilising interaction between the occupied HOMO n_O orbital and the antiperiplanar unoccupied σ^*_{C-Hax} orbital.

As it turned out, in the NMR study of this conformationally-fixed 1,3-dioxane reported by Perlin and Casu (1969), the measurement of the one-bond C—H coupling constants revealed a significant difference in $^1J_{C-H}$ coupling constants for the C(2)—H bonds. Indeed, the coupling constant for the axial C—H bond is significantly smaller than the $^1J_{C-H}$ coupling constant for the equatorial C—H bond. This is the result of a *stereoelectronic interaction*, involving a two orbital-two electron interaction; that is, a lone electron pair, occupied n_O orbital interacting with an empty σ^*_{C-Hax} orbital. This interaction affords two new orbitals: one occupied, lower in energy orbital and one empty, higher in energy orbital – this two-orbitas/two electrons interaction being stabilising. The right side of Figure 12.10b presents the corresponding canonical, double bond-no bond structure.

STEREOELECTRONIC INTERACTIONS IN S-C-P SEGMENTS

A heterocyclic system that was developed in our laboratories several years ago, revealed an anomeric effect that involves heteroatoms different than oxygen, in particular, sulphur in a 1,3-dithiane ring and phosphorus in a diphenylphosphi-noyl group at its anomeric-type position; that is, at C(2) in the 1,3-dithiane ring (Figure 12.11). This heterocyclic molecule had been prepared as a potential Wittig-type reagent in a project oriented to render a synthetically useful application; nevertheless, the unexpected observation was that its proton NMR spectrum exhibited a very large difference in chemical shifts for the axial and equatorial hydrogens at C(4,6), $\delta_{ax/eq} = 1.2\,\text{ppm}$. This surprising finding was explained

1-axial   1-equatorial

Δδ(Hax/Heq) = 1.2 ppm

FIGURE 12.11   Predominance of the axial isomer in the conformational equilibrium of 2-diphenylphosphinoyl-1,3-dithiane **1**. Proton NMR spectroscopic data revealed a significant deshielding effect on the *syn*-diaxial H(4,6) protons provoked by the *axial orientation* of the diphenylphosphinoyl group.

FIGURE 12.12   Structure and conformation of 2-diphenylphosphinoyl-1,3-dithiane **1** confirming the axial orientation of the phosphorus substituent. Theoretical calculations suggest that this unusual conformational behaviour is a consequence of a stereoelectronic interaction (see text).

in terms of the diphenylphosphinoyl group adopting the axial rather than the equatorial orientation, with the phosphoryl group deshielding the 1,3-diaxial hydrogens. Of course, this interpretation of the NMR observation was hard to believe in view of the rather large steric size of the phosphorus substituent.

As it was already mentioned, the conclusion that the diphenyphosphinoyl group adopts preferentially the axial orientation in **1** was very unusual given the rather large size of this phosphorus substituent. However, suitable crystals of heterocyclic compound **1** could be collected, and the X-ray diffraction data exhibited the structure and conformation presented in Figure 12.12. It can be appreciated that the phosphorus substituent, which is indeed larger in size than the dithiane heterocycle, adopts an axial orientation! This discovery was followed by several years of additional studies aiming to obtain pertinent information that could disclose the origin of this unprecedented S—C—P anomeric effect. As it happened, 35 years after the initial discovery, high-level theoretical calculations helped explain the reason for the predominance of **1**-axial in terms of a *stereoelectronic interaction*.

Examination of the X-ray diffraction structural data of **1**-ax (Figure 12.12), and comparison with the conformationally-fixed equatorial analogue (not shown here) did not exhibit the lengthening of the exocyclic axial C(2)—P bond in **1**-ax that was anticipated with base on $n(S) \rightarrow \sigma^*(C—P)_{ax}$ stereoelectronic interactions. The dilemma generated by this unexpected observation gave rise to a series of conformational and structural studies with a variety of analogues of 2-diphenylphosphinoyl-1,3-dithiane **1**. However, the information gathered from those investigations did not afford a definitive explanation.

Fortunately, with the development of powerful computational software and with access to more efficient and faster computational equipment, we were able to reproduce the S—C—P anomeric effect observed experimentally in the **1**-ax $\rightleftharpoons$ **1**-eq conformational equilibrium (Figure 12.11). In particular, it was possible to establish that Weinhold's Natural Bond Orbital (NBO) calculations do confirm that stereoelectronic interactions are responsible for the conformational equilibrium favouring **1**-axial over **1**-equatorial.

Calculations were performed with *Gaussian* computational programs and the target structures were optimised at a high level of theory. Subsequently, the involvement of hyperconjugative stereoelectronic effects was examined with the natural bond orbital (NBO) method developed by Weinhold.

In agreement with X-ray diffraction data, the calculated structural parameters summarised in Table 12.2 show that the length of the C(2)—P(O) bond in **1**-axial, 1.867 Å, is identical to the C(2)—P(O) bond length in **1**-equatorial, 1.867 Å. Similarly, the lengths of the endocyclic C(2)—S bonds in **1**-axial and **1**-equatorial are identical, 1.836 Å. Furthermore, the C(2)—S bond lengths are also identical in **1**-axial and in the conformationally-fixed equatorial analogue (not shown here), 1.809 Å. Thus, both the experimental and calculated structural data are not in agreement with the shortening of the S—C(2) endocyclic bonds and a lengthening of the exocyclic C(2)—P(O) exocyclic bond in **1**-axial relative to **1**-equatorial, that was anticipated for the $n(S) \rightarrow \sigma^*(C—P)_{ax}$ stereoelectronic interaction.

Most relevant, calculations accurately reproduce the tendency of the phosphorus substituent to adopt the axial rather than the equatorial conformation. Indeed, **1**-axial is estimated to be lower in energy than the equatorial conformer **1**-eq, $\Delta G° = +1.30\,kcal\,mol^{-1}$ (Figure 12.13). This value is rather close to the one that had been observed experimentally, $\Delta G° 294\,K = +1.0\,kcal\,mol^{-1}$.

Weinhold's NBO method is rather useful for the identification of stereoelectronic interactions. In the examination of the S—C—P anomeric effect, NBO provided theoretical insight into the magnitude of the $n(S) \rightarrow \sigma^*(C—P)_{ax}$ hyperconjugative interactions that weaken and lengthen the axial C—P bond.

TABLE 12.2   Calculated Structural Parameters of 2-Diphenylphosphinoyl-1,3-dithiane in the Axial Conformation, **1**-ax, and in the Equatorial Conformation, **1**-eq

Bond or angle	1-ax	1-eq
C(2)—P	1.867	1.867
C(2)—S	1.835–1.836	1.837–1.838
P=O	1.505	1.497
C(4)—S	1.841–1.842	1.837–1.838
C(4)—C(5)	1.528–1.529	1.529
S—C(2)—S	114.7	114.0
C(2)—P—O	113.0	114.8
C(2)—S—C(4)	100.8–100.9	98.0
S—C(4)—C(5)	114.1–114.2	114.4–114.5
C(4)—C(5)—C(6)	113.4	113.8

Note: Bond distances in Å and bond angles in degrees.

FIGURE 12.13   Calculated axial ⇌ equatorial conformational equilibrium of 2-diphenylphosphinoyl-1,3-dithiane (**1**-axial ⇌ **1**-equatorial) reproducing the predominance of the axial conformer.

In particular, energies of delocalisation (E_{del}) associated with hyperconjugative interactions can be estimated by the NBO method.

Table 12.3 summarises the energies of delocalisation E_{del} for the most salient hyperconjugative interactions that take place in dithianes **1**-axial and **1**-equatorial. Additionally, Table 12.3 collects the energy difference between the participant donor and acceptor orbitals ($\Delta E_{donor/acceptor}, \Delta E_{d/a}$). As it was anticipated, $\Delta E_{d/a}$ shows an inverse relationship between the energy gap and

TABLE 12.3 Relevant Hyperconjugative Interactions (E_{del}) in Axial and Equatorial 2-Diphenylphosphinoyl-1,3-dithiane (**1**-ax and **1**-eq)

	Donor orbital	Acceptor orbital	E_{del}(kcal mol^{-1})	$\Delta E_{d/a}$ (Hartrees)
1-ax	$n_{S(ax)}$	$\sigma_{C-P(ax)}$*	3.86	0.41
	$n_{S(eq)}$	$\sigma_{C-P(ax)}$*	1.65	0.81
	$n_{S(ax)}$	$\sigma_{C(2)-S}$*	5.31	0.37
	$n_{S(eq)}$	$\sigma_{C(2)-S}$*	1.97	0.77
	$\sigma_{C(4,6)-S}$*	σ_{C-P}*	---	---
1-eq	n_S	$\sigma_{C-P(eq)}$*	---	---
	$n_{S(ax)}$	$\sigma_{C(2)-S}$*	6.88	0.37
	$n_{S(eq)}$	$\sigma_{C(2)-S}$*	1.29	0.77
	$\sigma_{C(4,6)-S}$*	$\sigma_{C-P(eq)}$*	1.87	0.76

the magnitude of the two-electron/two-orbital hyperconjugative interaction, E_{del}; that is, the smaller the energy difference between the donor and acceptor orbitals the stronger the stereoelectronic interaction.

The most relevant observations gathered from Table 12.3 are:

(1) Significant $n(S) \to \sigma^*(C-P)_{ax}$ stereoelectronic interactions are operative in **1**-axial (large E_{del}) but are less important in **1**-equatorial (small E_{del}). This observation is in line with anticipation based on the presumed stereoelectronic interaction in the axial conformation, where the donor and acceptor orbitals are antiperiplanar to each other. By contrast, an effective $n(S) \to \sigma^*(C-P)$ interaction between the donor $n(S)$ and acceptor $\sigma^*(C-P)$ orbitals is not possible in equatorial **1**-eq since the pertinent orbitals are essentially orthogonal.

(2) Importantly, the NBO theoretical study revealed that antiperiplanar $[C(4,6)-S] \to \sigma^*(C-P)_{app}$ stereoelectronic interactions are operative in **1**-equatorial, but not in **1**-axial. These hyperconjugative interactions weaken the exocyclic equatorial $C(2)-P$ bond, which becomes longer. This finding helps explain the unexpected observation made in the early 1980s, that the $C(2)-P$ bonds are similar in length in **1**-axial and **1**-equatorial.

In summary, $n(S) \to \sigma^*(C-P)_{app}$ stereoelectronic interactions weaken the axial $C(2)-P$ bond that becomes longer, whereas $\sigma[C(4,6)-S] \to \sigma^*(C-P)_{app}$ hyperconjugation results in weaker and longer $C(2)-P$ equatorial bonds! (Figure 12.14).

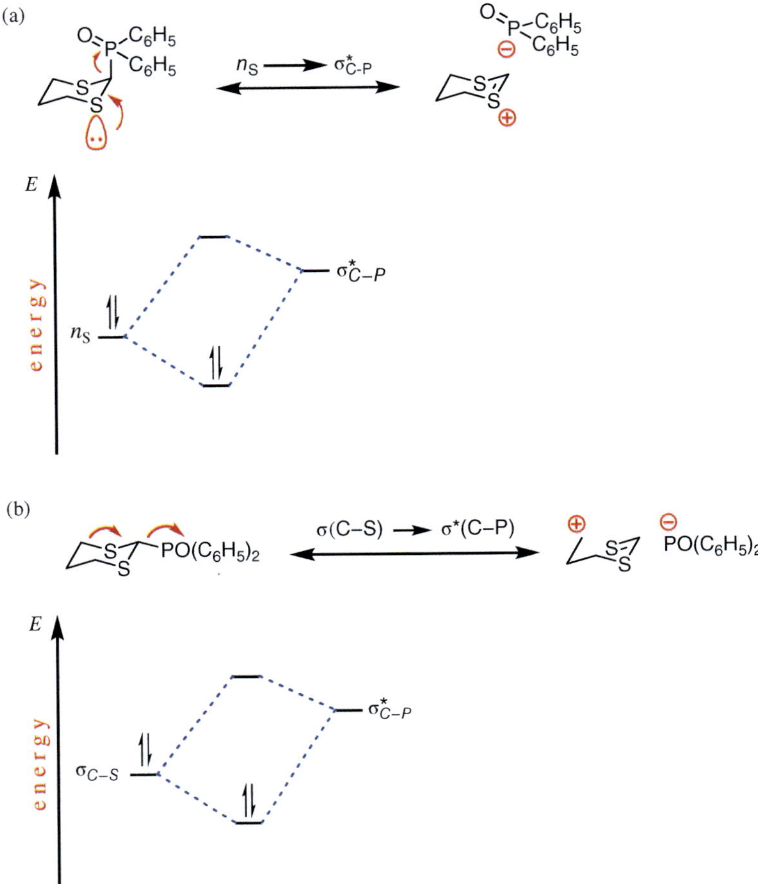

FIGURE 12.14 (a) Dominant $n(S) \rightarrow \sigma^*(C{-}P)_{app}$ stereoelectronic interaction in **1**-axial. (b) Dominant $\sigma[C(4,6){-}S] \rightarrow \sigma^*(C{-}P)_{app}$ stereoelectronic interaction in **1**-equatorial.

FURTHER READING

E. D. Glendening, A. E. Reed, J. E. Carpenter, F. Weinhold, *NBO*, University of Wisconsin, Madison, WI, **1988**.

E. Juaristi and G. Cuevas, *Acc. Chem. Res.*, **2007**, *40*, 961.

E. Juaristi and G. Cuevas, *The Anomeric Effect*, CRC Press, Boca Raton, Florida, **1995**.

E. Juaristi, *Introduction to Stereochemistry and Conformational Analysis*, Wiley, New York, **1991**.

E. Juaristi, L. Valle, B. A. Valenzuela, M. A. Aguilar, *J. Am. Chem. Soc.*, **1986**, *108*, 2000.

E. Juaristi, R. Notario, *J. Org. Chem.*, **2015**, *80*, 2879.

F. Weinhold, *Nature*, **2000**, *411*, 539.

I. V. Alabugin, *Stereoelectronic Effects: The Bridge between Structure and Reactivity*, Wiley, Chichester, UK, **2016**.

S. Perlin and B. Casu, *Tetrahedron Lett.*, **1969**, 2921.

EXERCISES

12.1 Please explain the predominance of the axial conformer of 2-chloro oxane in terms of potential electrostatic and stereoelectronic interactions.

12.2 Explain the contrasting conformational behaviour in the following equilibria:

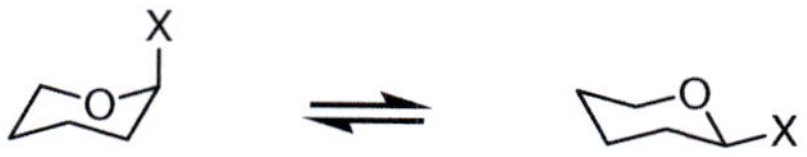

$$X = Cl \qquad \Delta G° = +1.8 \text{ kcal mol}^{-1}$$
$$X = NHMe \qquad \Delta G° = -0.9 \text{ kcal mol}^{-1}$$

12.3 Explain the preference of the *gauche* conformation over the *anti*-conformation in dimethoxymethane:

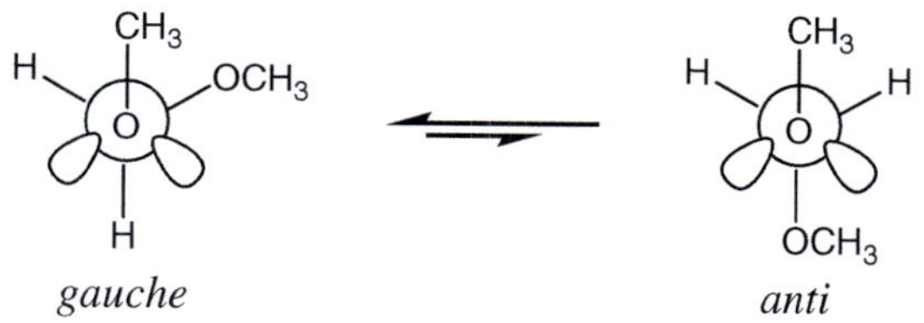

gauche   *anti*

12.4 Explain the decreasing magnitude of $\Delta G°$ values in the conformational equilibrium of 2-methoxy-oxane:

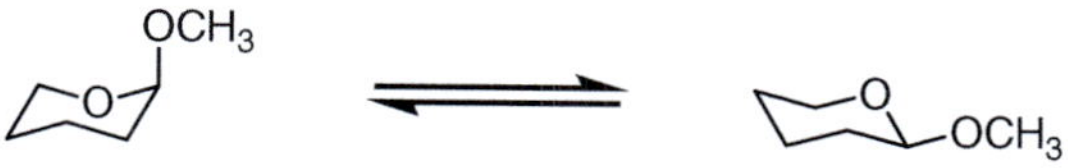

Solvent	Dielectric constant (ε)	% axial conformer
CCl_4	2.2	83
$CHCl_3$	4.7	71
CH_3OH	32.6	69
H_2O	78.5	52

Index

Printed and bound by CPI Group (UK) Ltd, Croydon, CR0 4YY

25/06/2025

14694111-0001

Increase your understanding of molecular properties and reactions with this accessible textbook

The study of organic chemistry hinges on an understanding and capacity to predict molecular properties and reactions. Molecular Orbital Theory is a model grounded in quantum mechanics deployed by chemists to describe electron organization within a chemical structure. It unlocks some of the most prevalent reactions in organic chemistry.

Basic Concepts of Orbital Theory in Organic Chemistry provides a concise, accessible overview of this theory and its applications. Beginning with fundamental concepts such as the shape and relative energy of atomic orbitals, it proceeds to describe the way these orbitals combine to form molecular orbitals, with important ramifications for molecular properties. The result is a work which helps students and readers move beyond localized bonding models and achieve a greater understanding of organic chemical interactions.

In *Basic Concepts of Orbital Theory in Organic Chemistry* readers will also find:

- Comprehensive explorations of stereoelectronic interactions and sigmatropic, cheletropic, and electrocyclic reactions,
- Detailed discussions of hybrid orbitals, bond formation in atomic orbitals, the Hückel Molecular Orbital Method, and the conservation of molecular orbital symmetry
- Sample exercises for organic chemistry students to help reinforce and retain essential concepts

Basic Concepts of Orbital Theory in Organic Chemistry is ideal for advanced undergraduate and graduate students in chemistry, particularly organic chemistry.

Eusebio Juaristi, PhD, is Professor at Centro de Investigacion de Estudios Avanzados del Instituto Politecnico Nacional, Mexico City, Mexico. He has produced influential research in numerous areas of physical organic chemistry, particularly conformational analysis and stereochemistry, as well as computational chemistry, asymmetric organocatalysis, and sustainable chemistry.

C. Gabriela Ávila-Ortiz, PhD, is a Research Assistant at Centro de Investigacion de Estudios Avanzados del Instituto Politecnico Nacional, Mexico City, Mexico. She works in Professor Juaristi's research group studying the asymmetric synthesis of organic compounds, organocatalysis, and green chemistry.

Alberto Vega-Peñaloza, PhD, is Serra Hunter Lecturer in the Section of Organic Chemistry at the University of Barcelona, Spain. He has worked as Senior Scientist I at Selvita S.A., Poland, as a postdoctoral fellow at the Faculty of Chemistry of the National Autonomous University of Mexico (UNAM), as a postdoctoral researcher at ICIQ in Spain, and at the University of Padova, Italy, where he was awarded the Seal of Excellence UniPD grant to work on the development of photocatalytic systems for sustainable synthetic methods.

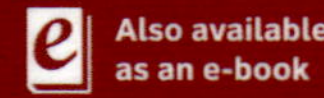

PEATE'S BODY SYSTEMS

12

THE
EYES

IAN PEATE

WILEY